Springer Desktop Editions in Chemistry

T0223281

L. Brandsma, S.F. Vasilevsky,
H.D. Verkruijsse
Application of Transition Metal Catalysts
in Organic Synthesis
ISBN 3-540-65550-6

H. Driguez, J. Thiem (Eds.)
Glycoscience, Synthesis of Oligosaccharides
and Glycoconjugates
ISBN 3-540-65557-3

H. Driguez, J. Thiem (Eds.)
Glycoscience, Synthesis of Substrate Analogs
and Mimetics
ISBN 3-540-65546-8

K. Faber (Ed.)
Biotransformations
ISBN 3-540-66949-3

W.-D. Fessner (Ed.)
Biocatalysis, From Discovery to Application
ISBN 3-540-66970-1

S. Grabley, R. Thiericke (Eds.)
Drug Discovery from Nature
ISBN 3-540-66947-7

H.A.O. Hill, P.J. Sadler, A.J. Thomson (Eds.)
Metal Sites in Proteins and Models,
Iron Centres
ISBN 3-540- 65552-2

H.A.O. Hill, P.J. Sadler, A.J. Thomson (Eds.)
Metal Sites in Proteins and Models,
Phosphatases, Lewis Acids and Vanadium
ISBN 3-540-65553-0

H.A.O. Hill, P.J. Sadler, A.J. Thomson (Eds.)
Metal Sites in Proteins and Models,
Redox Centres
ISBN 3-540-65556-5

F.J. Leeper, J.C.Vederas (Eds.)
Biosynthesis, Polyketides and Vitamins
ISBN 3-540-66969-8

A. Manz, H. Becker (Eds.)
Microsystem Technology
in Chemistry and Life Sciences
ISBN 3-540-65555-7

P. Metz (Ed.)
Stereoselective Heterocyclic Synthesis
ISBN 3-540-65554-9

H. Pasch, B. Trathnigg
HPLC of Polymers
ISBN 3-540-65551-4

J. Rohr (Ed.)
Bioorganic Chemistry, Deoxysugars,
Polyketides and Related Classes: Synthesis,
Biosynthesis, Enzymes
ISBN 3-540-66971-X

T. Scheper (Ed.)
New Enzymes for Organic Synthesis,
Screening, Supply and Engineering
ISBN 3-540-65549-2

F.P. Schmidtchen (Ed.)
Bioorganic Chemistry,
Models and Applications
ISBN 3-540-66978-7

Springer

Berlin
Heidelberg
New York
Barcelona
Hong Kong
London
Milan
Paris
Singapore
Tokyo

W.-D. Fessner (Ed.)

Biocatalysis

From Discovery to Application

Springer

Professor Dr. Wolf-Dieter Fessner
Institut für Organische Chemie
Technische Universität Darmstadt
Petersenstr. 22
64287 Darmstadt, Germany
E-mail: fessner@tu-darmstadt.de

Description of the Series

The Springer Desktop Editions in Chemistry is a paberback series that offers selected thematic volumes from Springer chemistry series to graduate students and individual scientists in industry and academia at very affordable prices. Each volume presents an area of high current interest to a broad non-specialist audience, starting at the graduate student level.

Formerly published as hardcover edition in the review series
Topics in Current Chemistry (Vol. 200) ISBN 3-540-64942-5

Cataloging-in-Publication Data applied for

ISBN 3-540-66970-1
Springer-Verlag Berlin Heidelberg New York

Die Deutsche Bibliothek - CIP-Einheitsaufnahme
Biocatalysis: from discovery to application / W.-D. Fessner (ed.) - Berlin; Heidelberg;
New York; Barcelona; Hong Kong; London; Paris; Singapore; Tokyo: Springer, 2000
(Springer desktop editions in chemistry)
ISBN 3-540-66970-1

Springer-Verlag is a company in the specialist publishing group BertelsmannSpringer
© Springer-Verlag Berlin Heidelberg 2000
Printed in Germany

Cover: design & production, Heidelberg
Typesetting: Fotosatz-Service Köhler OHG, Würzburg
Printed on acid-free paper   SPIN: 10720872   02/3020 hu - 5 4 3 2 1 0

Preface

Worldwide, industrial chemists are endeavoring to meet the criteria of a sustainable development. Ideally, an environmentally benign reaction utilizes nontoxic reagents and solvents for a quantitative conversion at the optimum level of product selectivity. Catalytic procedures are clearly the most economical means of effecting selective processes in organic synthesis. Therefore, Biocatalysis continues to attract considerable attention of the synthetic chemist's community due to its competitive "natural" advantages, particularly in its intrinsic capacity for the asymmetric synthesis of enantiomerically pure compounds. Despite the initial high promise, however, industrialization of biocatalytic processes so far has only been realized for a few large scale operations. Limited substrate tolerance of available enzymes, tedious and costly process development, and the need for an extended knowledge base across many scientific disciplines have often hampered a straightforward replacement of traditional chemical operations by the utilization of enzyme catalysis.

This book attempts to familiarize the synthetic organic community with a number of important new developments in the Biocatalysis arena in consideration of both enabling and able technologies. A number of leading contributors from the forefront of this exciting technology address the state of the art of biocatalysis in eight authoritative and timely reviews, from discovery through development to application. Recent landmark advances in molecular biology have the potential to profoundly alter the shape of the field of applied biocatalysis and the pace of its future progress at the beginning of the next millennium. The latest screening and selection technologies allow the rapid identification of enzyme activities that offer properties suitable for organic synthetic applications. Thus, unique enzymes with improved or with novel properties are becoming available from diverse, previously inaccessible sources, by ingenious DNA recombination techniques along an evolutionary approach, or by eliciting monoclonal antibodies with directable binding specificities. Besides appropriate techniques of protein handling and reaction engineering, for future successful biocatalytic processes it will be mandatory to have ready access to an extended database that precisely defines scope and limitations for each synthetically useful enzyme. Thus, half of the chapters illustrate the synthetic potential of recently emerging biocatalysts that have a capacity for the synthesis or modification of the most important classes of pharmaceutically interesting compounds, particularly in the phospholipid, epoxide, cyanohydrin, and oligosaccharide fields.

Unique new enzymes are now readily accessible in quantity with properties that are amenable to modification on demand. It is my firm belief that such fascinating possibilities not only open new playgrounds for creative minds but will also assist each practising scientist with effective tools for tackling the future challenges in Organic Synthesis, and it is my hope that this volume will actively support this goal.

Darmstadt, September 1998   Wolf-Dieter Fessner

Contents

Topics in Current Chemistry
Now Also Available Electronically

For all customers with a standing order for **Topics in Current Chemistry** we offer the electronic form via LINK **free of charge**. You will receive a password for free access to the full articles.
Please register at: **http://link.springer.de/series/tcc/reg_form.htm**

If you do not have a standing order you can nevertheless browse through the table of contents of the volumes and the abstracts of each article at:
http://link.springer.de/series/tcc

There you will also find information about the
- Editorial Board
- Aims and Scope
- Instructions for Authors

Screening for Novel Enzymes

David C. Demirjian[1] · Pratik C. Shah[2] · Francisco Morís-Varas[1]

[1] ThermoGen, Inc., 2225 W. Harrison, Chicago, IL 60612, USA. *E-mail: ddem@thermogen.com*
[2] NephRx Corporation, 2201 W. Campbell Park, Chicago, IL 60612, USA

The development of new biocatalysts as synthetic tools for chemists has been expanding rapidly over the last several years. It is now possible either to discover or engineer enzymes with unique substrate specificities and selectivities that are stable and robust for organic synthesis applications. This has been made possible by the application of the newest screening and selection technologies that allow rapid identification of enzyme activities from diverse sources. We focus on how to recognize and tackle important issues during the strategic design and implementation of screening for novel enzymes. We also review the approaches available for biocatalyst discovery and relate them to the isolation of thermostable enantioselective esterases and alcohol dehydrogenases for the purpose of illustration and discussion.

Keywords: Enzymes, Biocatalysts, Screening, Enzyme discovery, Biocatalysis, Selections, Biotransformation, Industrial enzymes.

1
Overview

1.1
A Bit of Biocatalyst History

The first enzymes were discovered in the 1830s: diastase by Payen and Persoz and pepsin by Schwann. The first asymmetric synthesis using an isolated enzyme was carried out by Emil Fischer in 1894 when he applied the cyanohydrin reaction to L-arabinose, establishing that enzymes have an extremely high degree of substrate specificity [1]. By the 1920s several different enzymes were known to exist. While the idea that enzymes could be used for a variety of commercial applications was always in the realm of possibility, it was only in the 1960s and early 1970s that commercial processes using enzymes were widely used. For example, carbohydrate-processing enzymes have been widely used in the food industry for the processing of corn, potato and other starches [2]. Also, the addition of proteases to detergents became an important industrial use of enzymes. While these applications have accounted for a majority of the bulk enzyme sales, an increasingly important application of enzymes has been as a catalytic tool in the synthesis of specialty organic chemicals [3–7].

For example, an early application of biocatalysis was a commercial process to make ascorbic acid starting with glucose, developed by Reichstein in 1934 (reviewed in [8]). The adoption of biocatalysis processes in the production of specialty chemicals has been slower than anticipated. Most synthetic chemistry biocatalysis applications currently in place have been developed using a limited set of commercially available enzyme tools such as lipases, esterases, or proteases. These enzymes were developed for other applications, such as those in the food and textile industries [9]. They were explored as biocatalytic solutions because they were readily available inexpensively and in large quantities. It is only within the last few years that there has been a concerted effort to discover, engineer, and develop enzymes specifically for chemical synthesis applications [10]. The development of new molecular screening techniques, recombinant DNA technology, and a new focus on the importance of biocatalysis has significantly advanced the field.

1.2
Tailoring Screening to Meet Biocatalysis Challenges

There are a number of challenges to the successful development of a commercial biocatalysis process. Most of these challenges can be addressed at the outset by properly setting up a screen in order to identify a novel catalyst which has the properties necessary to carry out a reaction that can be scaled-up. The most critical challenges include the following.

Every reaction may need a different enzyme. Even highly related molecules may need a different catalyst to optimally carry out a reaction. While the same enzyme can often recognize similar substrates, the differences in substrate specificity can be great enough to prevent a process from being economical. There are several approaches to expand the diversity of enzymes available to carry out biotransformations and develop new enzymes specifically tailored or discovered for chemical synthesis applications. Recently several groups, including our own, have begun to overcome this by developing families of enzymes such as hydrolases, oxidoreductases and others through enzyme discovery approaches. We have now developed specific families of enzymes capable of carrying out certain classes of reactions, giving the synthetic chemist a wider range of enzymes off-the-shelf. Another approach to diversify an enzyme collection is to use directed evolution to fine-tune an enzyme's activity [11]. This is even more effective when taken together with the enzymes available from biocatalyst discovery searches that now allow a greater range of starting templates for the reaction.

Enzymes are notoriously unstable and require special handling conditions. It is likely that no other factor has been more detrimental in keeping biocatalyst technologies from practical implementation on large scale than the inherent instability found in most enzymes. If enzymes are to play a significant role in large scale processing of chemicals, they must be able to endure the often harsh conditions associated with industry. Alpha-amylase and subtilisin are the most successful industrial enzymes primarily because of their thermostability and hardiness. In addition, while most enzymes lose a significant portion of their activity in organic solvents, thermostable enzymes are typically more tolerant to the denaturing conditions of many organic solvents [12, 13]. Other technologies such as immobilization [14, 15], cross-linked-enzyme-crystals (CLEC [16]), and directed evolution [11, 17] can also help to further stabilize enzymes, but by setting up an initial screen for stable enzymes, one can ensure an inherent stability in the enzyme that is developed.

Multidisciplinary science. Biocatalysis is truly a multidisciplinary science, incorporating microbiology, molecular biology, enzymology and biochemistry, synthetic chemistry, analytical chemistry, and chemical engineering. Effective screening for new enzymes for biocatalytic applications requires an understanding and integration of these sciences.

It takes a long time to develop a biotransformation process. Finally, the time it takes to develop a biotransformation can prevent the process from reaching a commercial scale. In the pharmaceutical industry time is money, and a delay in the development of a useful process will lead to the adoption of an alternative process – even if it is more costly. Biocatalyst discovery and screening represents one of the longest time factors involved. By pre-establishing a diverse library of enzyme activities, organisms and gene banks, biocatalyst discovery and engineering can be shortened considerably.

We will explore here some examples and how screens were used for developing these enzyme families, and discuss the importance of choosing the right conditions to allow for easier identification of an active enzyme that can readily be produced and scaled-up when needed. By implementing screening solutions that take these factors into account, one can exploit the vast diversity of nature's catalytic repertoire.

1.3
A Working Plan

Figure 1 outlines a general working plan for exploring a biocatalysis solution. The simplest solutions are tested first at the top of the flow-chart. The easiest solution is to find a commercially available enzyme for a given reaction. If one is

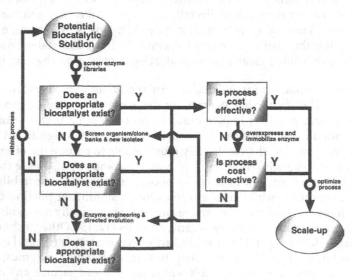

Fig. 1. Working plan for finding a biocatalytic solution

not found, then a more involved screening project needs to be undertaken. By setting up this screening project correctly and methodically, one can save time and effort in the discovery of a new catalyst. We usually screen organism banks, clone banks, and new isolates in parallel because it increases the probability of finding an appropriate enzyme of interest. It is difficult to determine ahead of time that a particular method is likely to lead to an enzyme with the desired activity, and all methods usually yield different results. Finally, if an ideal catalyst cannot be found, newer technologies such as directed evolution and other methods of protein engineering could be employed to customize an appropriate biocatalyst.

One of our activities is to develop new libraries of enzymes for certain classes of chemical reactions so that the speed of developing a new biocatalysis process is increased, thus expanding the options at the top of the diagram.

There are a number considerations which need to be taken into account when screening for new enzymes, and most of these can be generalized to a variety of screening projects.

2
Enzyme Sources

The first question that should be addressed is on the source of the enzymes that are to be screened. There are advantages and disadvantages to each source, and these have been compiled in Table 1.

2.1
Screening from Commercially Available Enzyme Libraries

By far the fastest and easiest route to finding a new enzyme is to find one that exists in a commercial library. This allows for instant access to the catalyst in sufficient quantities to take the project to the next step. For the most part there are still only a few sources. Traditional sources such as Sigma, Amano, Roche Molecular Biochemicals (formerly Boehringer-Mannheim), and Toyobo carry many of the classic enzymes that have been used in the past.

The biggest change over the last several years has been the development of larger commercial enzyme libraries to enhance and simplify biocatalyst discovery. We have been developing enzymes using a methodical plan to discover and develop useful new enzymes for synthetic chemists. As a result, the first new sets of thermostable enzyme families were discovered. These included esterases and lipases, and more recently alcohol dehydrogenases. Other groups have also taken similar approaches [19]. In addition, companies like Altus Biologics have assembled commonly used industrial catalysts for researchers and modified them with technology to stabilize the proteins [20].

Although larger commercial enzyme libraries are becoming available to the researcher, it is still not always possible to find an appropriate enzyme from a commercial source. In this case, a custom screening needs to be performed. This screening can be from a collection of microorganisms or from clone banks that have been generated from these organisms or isolated DNA.

Table 1. Enzyme sources

Source	Advantages	Disadvantages
Commercial enzymes[a]	• Available off the shelf and generally in large quantities • Fastest solution • Generally tested by multiple users	• Until recently only limited enzymes available • Generally not available for genetic engineering
	ThermoGen, Amano, Roche Molecular Biochemicals, Sigma-Aldrich, Fluka, Toyobo, Diversa, Altus Biologics, Biocatalysts	
Culture sources	• Source of new, undiscovered catalysts	• Need to clone for optimal enzyme expression • Findings limited to those which are expressed in the organism • Expensive and difficult to maintain collections
	International Culture Banks Proprietary banks	
Clone banks	• Source of new, undiscovered catalysts • Enzymes which are not expressed in original host can often be found • Generally easier to move into high volume production • Allows easier genetic engineering of enzyme	• Large libraries need to be maintained • Many proteins will not express in the cloning host
	Proprietary sources	
Clone banks from uncultured organisms	• Same advantages as clone banks plus source of new, undiscovered catalysts which have not been cultivated by other means	• All the disadvantages of clone banks • Possible redundancies in library • Possibly difficult to express evolutionarily distant genes
	Proprietary sources	

[a] For additional suppliers list see [18].

2.2
Culture Sources

Most enzymes of industrial importance developed in the past have been derived from species that are GRAS (Generally Regarded as Safe). These include bacterial species for *Bacillus* and *Lactobacillus*, and *Pseudomonas* and fungi from the *Ascomycota* and *Zygomycota* classes [21].

A partial listing including web addresses of some of the largest culture sources are given in Table 2, and many of these have links to most of the other available strain collections. Some of these include the American Type Culture Collection (ATCC) and the DSMZ German Collection of Microorganisms and Cell Cultures.

Access to a good proprietary collection of enzymes is also extremely helpful for finding an enzyme, especially if those collections have been logically assembled for applications of interest. If one knows which type of enzyme one is screening for, cultures can often be enriched for particular enzyme activities by standard methods [22].

Screening from culture sources has been successful in many cases [23]. There are, however, several challenges. First, since the media for different organisms will be diverse, the systematic screening of organism banks becomes more difficult. A general solution is to group like strains together for screening on different types of media and to find a media recipe that can support growth of the organism. Since the media composition ultimately effects which proteins are produced in the cell, the media choice for a particular enzyme screen can be critical to having the protein of interest expressed [24]. The cost of establishing and maintaining a proprietary strain collection can also be quite high. Several companies provide cost-effective access to their proprietary culture collections.

Strain redundancy is also another concern. When identifying new isolates, verification that a particular strain is unique can be accomplished by one of several methods. The first is a phenotypic characterization of traits such as colony morphology, classical strain typing such as the Bergey system [25], ribosome relationship data [26], or PCR-based comparison of strains [27]. One thing to note, however, is that very small differences in amino acid makeup of related enzymes can lead to significant differences in enzyme activity. This can often be overlooked if using a broader genomic comparison tool to eliminate duplicates.

Table 2. Partial list of online strain collections

Collection	www site
ATCC American Type Culture Collection	www.atcc.org
DSMZ German Collection of Microorg. and Cell Cultures	www.gbf.de/DSMZ
Microbial Information Network of China	sun.im.ac.cn
MSDN Microbial Strain Data Network	www.bdt.org.br/bdt/msdn
World Data Center for Microorganisms (Japan)	wdcm.nig.ac.jp
CGSC *E. coli* Genetic Stock Center	cgsc.biology.yale.edu/top.html

2.3
Screening from Clone Banks

Screening for new enzymes from clone-banks can be extremely rewarding. By setting up the clone banks in a unified or small set of host organisms (like *E. coli, Bacillus,* or yeast) only a limited number of different propagation methods need to be implemented, thus allowing systematic screening methods to be carried out more easily. If a cloned enzyme is discovered, it is generally easier to scale-up and produce in larger quantities. In addition, cloning is a prerequisite to most types of genetic modification of the gene including directed evolution. Additionally, in a clone bank the gene of interest is often removed from its regulatory elements that can repress expression and helps purify the gene away from isozymes and other competing activities.

There are a few disadvantages to screening clone banks. Removing a gene from its regulatory elements can turn the gene off instead of on, thus masking the activity. The gene may not express well in the host cloning strain (which typically includes *E. coli, Bacillus* and yeast) because of codon usage, nucleic acid structure, or lethality. The activity from enzymes that are post-translationally modified may be altered or destroyed [28]. In addition, the actual restriction sites used to construct the gene bank can affect whether any activity is observed or not, since the distance from the expression signals are altered. Finally, for each organism a clone bank is developed from, one needs to screen thousands to tens of thousands of clones for each organism to cover the entire genome of that organism. Several solutions have been devised, including automated screening and hierarchical screening methods, that will be discussed in more detail below.

The DNA used for cloning can originate from DNA prepared from cultured organisms or DNA prepared from uncultured organisms. It has been estimated that less than 1% of the world's organisms have been cultured [29]. By directly amplifying DNA from soil samples using PCR the DNA from these uncultured organisms can be isolated and used as a cloning source, thus allowing access to a greater diversity of enzymes [19]. There are a few concerns with DNA from uncultured organisms that need to be taken into account. First, it is difficult sometimes to clone DNA between closely related species. Generating DNA fragments from organisms that are less closely related presents even more problems. Second, by nature of the PCR reaction, certain DNA species will be amplified preferentially; thus there is no guarantee that if 100 esterases are isolated from one PCR reaction, they will be different.

One way of alleviating a number of these problems is actually to make several parallel clone banks using different expression vectors and strains, regulatable expression systems, and low copy-number vectors. Vectors choices [28] such as plasmids vs phagemids, high vs low copy number vectors, all can affect what is cloned, and there is not one solution which will satisfy every project. Furthermore, the cloning host also makes a difference. *E. coli* is usually the host of choice, but *Bacillus* can be preferred for secretable enzymes, and yeast for eukaryotic or post-transitionally modified enzymes.

2.4
Organizing the Enzyme Sources

If a commercial or pre-developed enzyme is not available from our library, we typically screen all the other sources simultaneously because one can never tell where a unique or useful activity is likely to be developed from first. In general, the use of formats such as microtiter plates helps to speed screening and creates an array that can be systematically screened, although one needs to take special care to avoid contamination problems. Clone banks can be stored in one of several ways. DNA libraries can be stored untransformed in tubes, or as transformed cultures. These transformed cultures can be stored as a pooled mixture or as an array of individual colonies. Depending on the insert size, a bank of thousands to tens of thousands of individual colonies needs to be stored to cover the complete genome of an organism. This should be done for every organism in the collection if one wants ultimate versatility in screening. The library from a single organism can then be arrayed and multiplexed into a single screening plate to save screening time. Once a positive candidate is identified, the sub-plates can then be tested quickly to determine which clone contains the appropriate activity.

There are advantages and disadvantages for each strategy. First, saving the clone banks as either DNA libraries or transformed pools provides the most flexibility and saves time and effort. The DNA can be transformed into any host that may be desired. Different hosts may be desired for different activities since

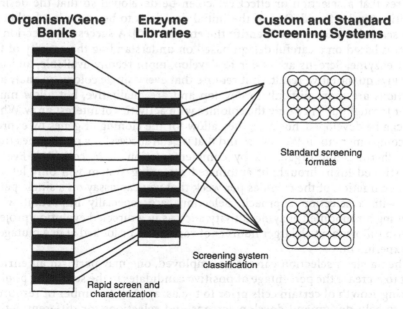

Fig. 2. Enzyme library organization. Enzymes discovered in organisms and gene banks are organized by activity type into enzyme libraries. These are characterized and can be subdivided into screening kits based on their substrate preferences and synthetic utility

host organisms often has some degree of background activities. Thus a proper host can make a big difference in implementing an effective screen or selection. Arraying all of the colonies can be extremely time-consuming, but is often important if one cannot develop an appropriate random plate screen or selection. In addition an arrayed system is somewhat easier to automate. These are all important issues and decisions that need to be made in setting up an appropriate screen or selection, and are discussed further in the next section.

Instead of re-culturing and re-cloning organisms and libraries each time a new enzyme is desired, strategies to centralize and organize the libraries and findings of previous experiments in-house can be implemented. Figure 2 shows one way to organize the results, making them useful for future screening projects both in-house and at other locations. Organism libraries and banks can be grouped based on media preferences and relatedness for easier screening.

3
Enzyme Screening

3.1
Some Strategic Issues

3.1.1
Screens vs Selections

Assaying a large number of sources, such as a large collection of clone banks, requires that a selection or effective screen be developed so that the desired activities can be found. One of the initial decisions to be made is whether to use a screen or selection to identify the enzyme [30]. A successful selection or screen is based on a careful design based on understanding the activity of the target enzyme. Screens are easier to develop, more readily available, and can often give quantitative results, but require that every single colony be analyzed. Selections are more difficult to develop and are qualitative, but allow much higher throughput since only the colonies with activities of interest grow. When they can be developed, however, they allow for the cloning of genes by expression complementation. We have found that we can successfully clone genes from many thermophilic organisms by complementation in E. coli [31]. Even a sophisticated high-throughput automated screening system will only let you look at a fraction of the colonies per week that you can assay on a single petri-plate with a selection approach. Selections are especially important when screening for modified enzyme activity, such as in a directed evolution project. Selection allows a much larger number of variants to be looked at in a mutagenesis experiment.

When a clean selection cannot be employed, one may perform an enrichment to increase the percentage of positive candidates in the screening pool by favoring growth of certain cells prior to the assay [22]. A number of resources exist to help design and develop screens and selections for different activities. LaRossa describes many selectable phenotypes that are available in E. coli [32].

However, even if a selection can not be developed in a reasonable period of time, a powerful screen can be extremely useful. The most convenient is a petri-plate-based screen that allows rapid visual identification of positive colonies or phage plaques. Generally a precipitable color reaction is the most desirable technique, since it allows visual and straightforward identification of active colonies. This is a result of color development upon enzymatic processing of the substrate (usually chromogenic) and the properties of the chromogen, which precipitates in place. If no diffusion occurs, the active colony can be easily identified. If a precipitable method cannot be developed, a more-tedious liquid-phase soluble substrate system often can be. An intermediate solution in which a non-precipitable chromogenic substrate is looked at in the solid phase is still possible with the right conditions, but diffusion can jeopardize the identification of the active species, especially when the high-throughput requires high density of colonies.

3.1.2
Substrate Selection

The most important consideration in finding a new enzyme is to set up screening conditions that mimic the target as much as possible. Optimally, the actual target substrate makes the best screening molecule, although this is often difficult to accommodate. Throughput considerations and other factors need to be taken into account and may require that a substrate analog be used. In this case, the closer the properties of the substrate analog are to the actual target substrate, the more likely it is to find an enzyme of interest. Substrate selection issues are addressed in more detail below. In the worst case scenario, if a suitable substrate or substrate analog cannot be used, a complicated method such as HPLC or GC must be used as an assay. This significantly limits throughput.

3.1.3
Screening Criteria

Setting up the proper screening criteria is also extremely important. The criteria will affect which candidates are turned up in the screening process as well as the physical and kinetic properties of the enzymes that are identified. By setting the wrong criteria, an enzyme might be found which catalyzes conversion of a particular substrate, but does it in such a way that would be economically ineffective or impossible to carry out on a large scale. These issues include pH, temperature, buffer, salt, cosolvents, and other conditions affecting the assay to be used and should closely mimic the end application as much as possible.

3.1.4
Plate vs Liquid Assay Systems

While more difficult and time consuming to develop, petri-plate-based screens or selections are highly desirable since they allow about 1000–10,000 colonies to be visually screened on a single petri with a random array of colonies. Potential positive candidates can then be isolated for further characterization. This allows the most flexibility and ease of use in screening. It allows non-ordered clone libraries to be rapidly screened as well. In order to develop a solid-phase screening system, one needs a good assay for visualizing positives. This is generally colorimetric assay based on a precipitable substrate, a pH-based assay, or a filter-bound assay [33].

If a solid-phase plate assay cannot be developed, the next best thing is a simple liquid-phase assay. Using soluble enzyme substrates, these assays are generally more quantitative than a simple colorimetric plate-assay. The liquid-based system is amenable to easy automation since many of the technologies for liquid handling, microtiter-plate handling, and microtiter-plate assays have been developed. The liquid-handling system requires that the colonies being screened are arrayed into microtiter plates initially. This limits throughput compared with the plate-based system even if a highly automated system is employed. Some researchers are now working on flow-cytometry assays to measure systematically individual reactions in microorganisms [34]. This will allow each individual cell in a screening application to be assayed without the need of first growing colonies and picking them manually or robotically into individual wells in a microtiter plate.

3.2
Different Methods of Screening

As the number of enzymes and the need to scan larger and larger libraries grows, methods for increasing throughput in screening become increasingly important. Two general approaches have been taken. These include automated high-throughput screening (HTS) and hierarchical structured screening approaches.

3.2.1
Hierarchical (Multi-Tiered) Screening Approaches (HMTS)

One can gain significant increases in throughput by implementing hierarchical screening approaches. In this type of approach several screening assays are often combined in series to narrow the scope of the screening project step by step. Using this method the easier, but perhaps less accurate, screens are carried out first. The more tedious but quantitative screens are then carried out on only a fractional subset of candidates which have been pre-validated as potentially useful isolates. We have often made use of this type of screening approach because it is rapid, useful, and cost effective. An example 3-level hierarchical screen is outlined in Fig. 3 and consists of the following strategy.

Level 1. Most General Screen – Fast and Simple. This type of screen eliminates the majority of negative candidates, and preferentially does not eliminate potential positives.

Level 2. Intermediate Screen(s). This step generally employs more specific substrates or semi-quantitative approaches.

Level 3. Specific Screen. Slowest and most accurate. This type of screen generally employs highly quantitative assays including HPLC, GC, or spectrophotometric or fluorogenic quantitative approaches.

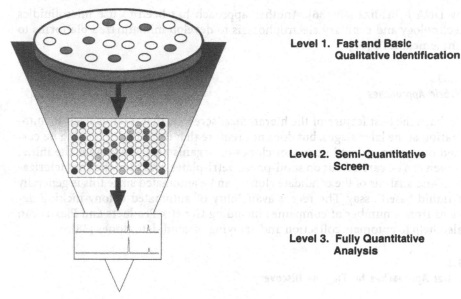

Level 1. Fast and Basic
          Qualitative Identification

Level 2. Semi-Quantitative
          Screen

Level 3. Fully Quantitative
          Analysis

Fig. 3. A schematic of a three-level hierarchical screen

This type of screening approach works because it is subtractive in nature. The first step acts to eliminate the majority of candidates and colonies that do not appear to have any desirable activity at all. As mentioned earlier, the use of substrate analogs almost assures that some potential candidates will be missed which act on a particular substrate of interest, or that some will be found that do not perform on the actual target substrate. For this reason it is important to try and pick substrates that give a good cross-section of activities from the library being screened. In the second step of the screen, expression levels and relative substrate specificities can be determined on a more quantitative basis, often using colorimetric candidates like nitrophenyl derivatives. Finally, actual candidate substrates are usually tested to determine which enzymes give the best activity for a particular application.

3.2.2
Automated High-Throughput Screening (HTS)

Recent progress in approaches for drug design and lead-optimization has led to an extensive development of new tools for high throughput screening (HTS) and ultrahigh throughput screening [35]. A number of commercial platforms for the development and implementation of HTS are now available that can allow for the screening of anywhere from 10,000 to over 100,000 wells per week. Another trend in implementing high throughput screening has been towards miniaturization and the use of microchips. The immobilization of DNA fragments on a microchip has allowed for the survey of large collections of genes for relatedness by DNA hybridization [36]. Another approach has been to use microfluidics technology and capillary electrophoresis to develop miniaturized platforms to implement HTS [37].

3.2.3
Hybrid Approaches

Perhaps the best feature of the hierarchical screen is that it lends itself to automation at the later stages, but does not require that the initial screening be carried out on an ordered library of clones or organisms. For example, the initial screen can be carried out on solid-phase petri-plates, and then the characterization and analysis of the candidate clones can be automated since this is generally a liquid-based assay. The recent availability of automated colony-picking devices from a number of companies including Genetix Products and Flexus can also help to automate collection and arraying of candidate clones [38].

3.3
Other Approaches for Enzyme Discovery

3.3.1
Screening for Gene Homology

Another approach to discover new enzymes is based on the sequence similarity across members of a class of enzymes. Rather than screening for activity based on an assay, this approach uses the sequence similarity between enzymes with similar properties. An analysis of conserved regions can be performed to design degenerate oligonucleotide primers. These primers are used to amplify the coding sequences of related enzymes using PCR from template libraries. For example, this approach has been used successfully for cellulases [39]. The diversity of enzymes that can be isolated using this approach would represent the diversity of ecological niches of the source organisms. Therefore, one can isolate enzymes with close matches of catalytic properties but different catalytic optima if source organisms are closely related but inhabit different environments.

3.3.2
Mining Genome Databases

With the wealth of genomics information that is now becoming available, the approach to using sequence similarity to discover new catalysts can be implemented with computer search tools, rather than carrying out actual experiments. This bioinformatics approach can allow for exploration of a vast number of genomes and organisms that span multiple phyla. A sequenced genome is first annotated using one of several methods such as BLAST, WIT, Magpie, and other programs to identify potential homologies with genes in genomic databanks [40–44]. Recent research is now focusing on methods to identify and assign gene functions to open reading frames that have no obvious homologs. Using these methods, we have discovered dozens of new, previously unidentified genes with potentially useful industrial properties from thermophilic organisms which are currently being studied and developed.

4
Implementing a Screen: Esterases and Dehydrogenases

4.1
Developing Screening Criteria

The considerations in the previous section need to be addressed and customized for every screening project. As an example, when we set out to develop a new library of esterases for synthetic chemistry use, we first needed to determine the criteria that would be used in the screening project. Esterases and lipases catalyze the hydrolysis of ester bonds as shown in Scheme 1 and are useful for reactions requiring different regioselectivities, chemoselectivities, and stereoselectivities depending on the enzyme's substrate specificities.

Our goals were to develop a series of esterases that offered a range of substrate specificities, enhanced enzyme stability, and were useful in a variety of synthetic chemistry applications on a large scale. At the onset of this project, the existing commercial enzymes did not fully address the need for different substrate specificities, improved stereoselectivity, stability under industrial reaction conditions, and availability in large quantities. We also wanted the enzymes to be produced at high levels and have high overall activity.

Within these parameters we thought it best to first develop a set of esterases that recognized short alkyl-chain esters. This was important since one of the primary applications of the esterases was in chiral resolution applications where the molecules being tested have generally small methyl or ethyl ester groups attached. We also chose to identify those enzymes that appeared to have high activity and/or production levels in the screening application. The temperatures at which the screening would be carried out were also critical. Thermostable enzymes are generally more stable under a variety of conditions including room temperature and in organic solvents.

Using high temperature screens we initially identified extremely stable enzymes such as E100 shown in Fig. 4. E100 was isolated from an extreme thermophile. While this enzyme is extremely stable, it had very low activity at room temperature, which made it practically ineffective below 50–60 °C. In fact, we were unable to identify its optimal temperature using p-nitrophenyl-propionate as a substrate since it spontaneously hydrolyzed at temperatures above 70 °C. While certain issues such as substrate and product solubility might make it desirable to have an enzyme that works at extremely high temperatures [45], most researchers are more interested in carrying out the reactions at or near ambient temperatures. In fact, it is a common misconception that an enzyme which functions at a high temperature is necessarily going to catalyze a reaction faster than its mesophilic counterpart at a lower temperature. Since the thermophilic enzyme evolved at a higher temperature, its optimal temperature is likely to be at the higher temperature. In contrast, the optimal catalytic rate is not necessarily

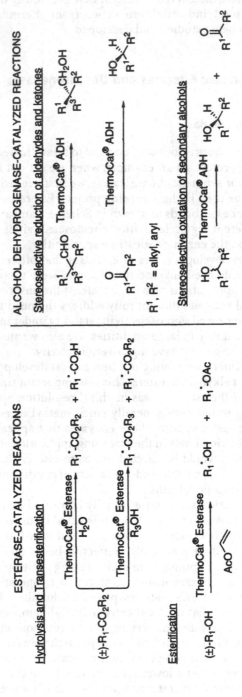

Scheme 1. Stereoselective outcome of esterase- and ADH-catalyzed reactions

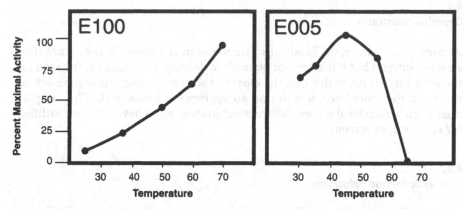

Fig. 4. Temperature vs rate plots for an enzyme from an extreme thermophile (E100) and from a moderate thermophile (E005). The substrate used in these experiments was p-nitrophenyl-propionate which was quantitated spectrophotometrically at O.D. 410. The data has been normalized to maximum activity. E100 has more thermostability and thermophilicity than E005, but does not retain much activity at room temperature

going to be superior to the mesophilic enzyme's optimal rate. One sees that enzymes from hyperthermophiles (optimal growth at 85 °C and above) will be nearly inactive at room temperature, enzymes from extreme thermophiles (optimal growth at approximately 60–85 °C) will be slightly more active at room temperature, and enzymes from moderate thermophiles (optimal growth at approximately 45–65 °C) will often have reasonably high activity at room temperature. For this reason, screening from the moderate and extreme thermophiles will often lead to better results than screening from the hyperthermophiles for certain applications. In addition, since the moderate and extreme thermophiles are more prevalent, the diversity of enzymes to be found among them is also generally greater, although an argument can be made that enzymes from hyperthermophiles will, on the whole, be more diverse from moderate and extreme thermophiles. Ultimately, a more rewarding approach to extending the temperature range of an enzyme is to take a moderately thermophilic (or mesophilic enzyme) and evolve it *in vitro* to have an increased thermostability, but not thermophilicity. We have developed a specialized system to do this *in vivo* in *Thermus* where selections for thermostability can be implemented [46].

Using a new set of screening criteria that called for subjecting the candidates to high temperature while performing the assay at low temperature, we were able to identify enzymes which were both thermostable and active at room temperature. Figure 4 shows a typical optimal activity plot for E005 in comparison to E100. This class of enzymes retains most of its stability, at least to about 60 °C, has optimal activities around 40 °C, and retains at least 25% of its full activity at room temperature. Similar criteria were put together for developing a new library of alcohol dehydrogenases. Again, we tailored the screen to identify enzymes that were active at room temperature, but stable at elevated temperature.

4.2
Choosing Substrates

As mentioned earlier, the ideal substrate to use in any screen is the actual substrate of interest but it is often not possible to develop a working assay based on the actual substrate. In this case the closer a substrate analog mimics the actual substrate, the more likely it is to find an appropriate biocatalyst. When implementing a hierarchical screen, different substrate analogs have different utilities for each level of screening.

4.2.1
Hydrolytic Enzyme Substrates

Scheme 2 lists three classes of general substrate analogs and their useful detection ranges for hydrolytic enzymes that we have used to isolate libraries of different types of hydrolases (esterases, proteases, lipases, phosphatases, glycosidases, and sulfatases). These are carboxylic acid, phosphate, or sulfate esters, amides and glycosides that contain a detectable probe, and a non-detectable moiety with the functionality of interest. The first class of substrate is the indigogenic substrates. This group includes substrates such as 5-bromo-4-chloro-3-indoxyl-galactopyranoside (X-gal) which is a classic substrate in the determination and quantification of β-galactosidase activity [47–49]. Since these substrates are precipible they are ideal for first-level hierarchical plate assays since the

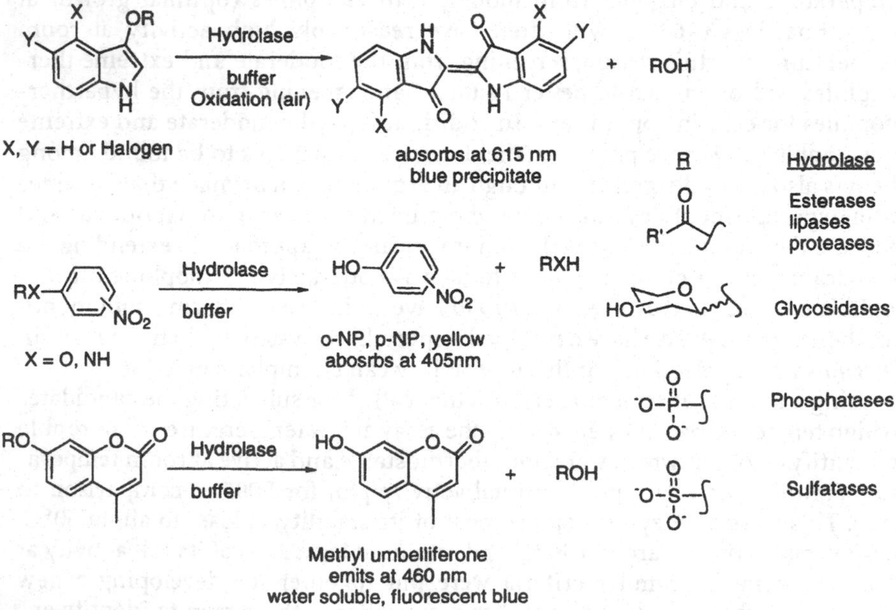

Scheme 2. Chromogenic, fluorogenic and precipitable reagents for screening of hydrolases

color develops and stays in the vicinity of the colony. The other two classes of substrates are both soluble substrates that are useful in second-level screens since they are quantitative. The chromogenic substrates based on nitrophenyl or nitroaniline can be used in a quantitative spectrophotometric liquid-assay, but are generally not useful in plate screens since they diffuse readily and are not sensitive enough. The fluorogenic substrates such as those based on umbelliferone or coumarin are at least one thousand times more sensitive than their chromogenic counterparts. Because of the high sensitivity, they can sometimes be used in plate assays if the reaction is fast enough and can be detected before significant hydrolysis and background appears, but still have their main application in a liquid-phase assay. Many of these substrate derivatives can be purchased from sources such as Sigma Chemicals or Biosynth International. In some cases we synthesize custom substrates to mimic the actual substrate more closely.

The solubility of the substrate analog can be a problem for quantitative studies. As a matter of fact, lipases exhibit better performance with hydrophobic substrates. When using analogs the additional hydrophobicity of the substrate makes detection more difficult. In these instances we have found the fluorogenic derivatives superior to the chromogenic ones, since the nanomolar detection requires very little substrate to be solubilized into the buffer, and small amounts of organic cosolvent can be used without severely affecting the activity of the target enzyme. Another possibility consists of amplification of the color yield of chromogenic substrates by coupling the release of the chromogen to an auxiliary system. This has been described for detection of enzyme catalyzed amidolysis using p-nitroanilides as substrates and p-dimethylaminocinnamaldehyde as a coupling agent. The resulting aldehyde forms a Schiff base with high molar absorbancy and the micromethod adaptation has been also described [50]. A color assay for lipases cleaving fatty acids has been described using Cu salts and chromogenic Au [51]. The method is sensitive and has been used in a semi-automated manner [52]. Another indirect method of activity detection relies on treatment with auxiliary enzymes. This includes luciferase to determine luminescence by detection of carboxylic acids [53] or the corresponding alcohol moiety [54], or with acyl CoA synthetase to form CoA esters which can be detected chromogenically by a coupled peroxidase reaction or quantitation of the AMP used during the synthetase reaction [55].

4.2.2
NAD(P)H-Alcohol Dehydrogenase Substrates

Redox reactions such as those catalyzed by alcohol dehydrogenases can be monitored colorimetrically. These biocatalysts generally require a cofactor to carry out the transfer of hydrogen. By combining the redox reaction with a dye-forming reaction, visualization of the conversion becomes possible. This is based on the transfer of a hydrogen atom from the reduced form of the coenzyme to a suitable dye. This is an indirect test requiring that an auxiliary enzyme or synthetic compound be applied. Tetrazolium salts are very useful because they can be reduced to formazans, which absorb light in the visible region. The most

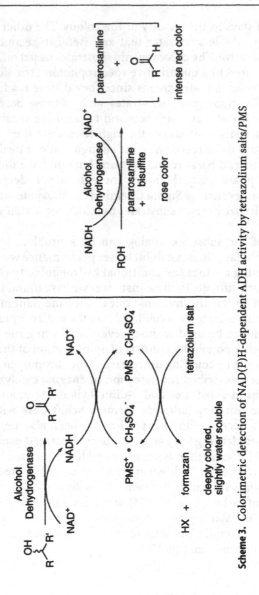

Scheme 3. Colorimetric detection of NAD(P)H-dependent ADH activity by tetrazolium salts/PMS

important advantages are the increase of the color during the reaction and the irreversibility of the reaction under biological conditions. The cascade of reactions leading to visualization is represented in Scheme 3, using phenazine methosulphate (PMS) as the non-enzymatic transferring agent [56]. The PMS can be substituted for an enzyme (diaphorase) or by another agent called Meldola Blue (8-dimethylamino-2-benzophenoxazine), which is less sensitive to light and even superior to PMS regarding proton rate transfer.

We have also used *p*-rosaniline as an alternative to the tetrazolium salts. This is even more convenient, since it allows a direct transfer without the need of either chemical or enzymatic transferring agents. The combination of p-rosaniline (also called p-fuchsine) with bisulfite converts the dye to the leuco form, which is rose-colored. When the oxidation of an alcohol takes place the combination of the resulting ketone or aldehyde with the p-fuchsin yields a red color which makes the identification of the alcohol dehydrogenase activities on agar plates possible allowing the rapid screening of thousands of colonies [57].

Another alternative is the use of oxygenases, which oxidize the reduced coenzymes and simultaneously incorporate molecular oxygen into the other substrate that is converted into a dye. The oxygenase reaction can be carried out in such a way that H_2O_2 is produced and determined by a variety of methods. A third method to produce color with NAD(P)H consuming enzymes is the reduction of nitroso compounds by ADH: the substrate is oxidized, the p-nitrosodimethylaniline is reduced, and the cofactor is recycled between both reactions [58].

One of the advantages of the dehydrogenase screens is that even though they rely on an indirect detection of the reaction through a linked assay, they are based on the reaction with the actual target substrate, thus reducing the number of hierarchical screening levels needed to identify an enzyme of interest. For example, enzymes can first be identified based on activity against the target compound with a linked colorimetric reaction, and then are characterized in a second level screening to quantitate traits like stereospecificity and substrate ranges.

4.3
Isolating and Characterizing Candidates

4.3.1
Isolating Candidates

Using indolyl-based substrate analogs such as 5-bromo-4-chloro-3-indolyl-acetate, butyrate, or propionate (X-acetate, X-butyrate, and X-propionate) for esterases and using the p-rosaniline assay coupled with ethanol, butanol, or propanol for alcohol dehydrogenases we were able to isolate several hundred candidates in the first level of a hierarchical screen. An intermediate screen was then used to characterize and compare the esterase candidates. We used nitrophenyl and methyl-umbelliferyl derivatives to analyze esterase candidates and their preference for chain length.

In order to determine if the new enzymes that were isolated met the criteria that we had set out with, approximately 20 esterases identified in the first and second level screening efforts were characterized. We carried out characterization such as molecular weight, optimal temperature, optimal pH, stereospecificity, and thermotolerance. Table 3 compares the activities of a representative range of the esterases that were selected for activity at room temperatures. As can be seen by the data, the enzymes span a wide range of optimal activities. Some of the enzymes function well at low pH, some function well at high pH, and others have a fairly broad pH range. All enzymes are extremely stable between room temperature and at least 40 °C and those that were isolated in room temperature assays generally retained at least 25 % of their activity at room tem-

Table 3. Useful activity conditions for some ThermoCat esterases

Biocatalyst	Temperature (°C)		pH		t 1/2 (40 °C)
	Optimum	Useful	Optimum	Useful	
E001	45	RT – 55	7.5	broad	–[a]
E004	45	RT – 60	6.5	< 6.0 – 8.0	–[a]
E005	45	RT – 60	7.0	broad	–[a]
E009	45	RT – 50	6.5 – 7.0	< 6.0 – 8.0	–[a]
E012	45	RT – 60	≤ 6.0	< 6.0 – 7.5	–[a]
E014	45	RT – 50	7.0	< 6.0 – 8.0	–[a]
E015	45	RT – 60	> 9.0	7.5 – 9.0	–[a]
E019	45	RT – 60	> 9.0	broad	nd
E100	> 75	45° – 85	> 8.5	8.0	–[a]
E101	> 75	45° – 85	> 8.5	8.0	–[a]

[a] No discernible decrease in activity over lifetime of experiment. Units defined as μmol/min for the hydrolysis of p-nitrophenyl-propionate.

perature. Many of the enzymes can also tolerate temperatures well above 60 °C. As expected most, but not all, of the esterases prefer the short-chain alkyl groups since they were isolated on those compounds.

The enantioselectivity of the esterases was examined by monitoring the hydrolysis of a racemic ester by chiral HPLC, phenethyl acetate. The enantiomeric excess achieved was > 99 % in some cases for the hydrolysis of the S-ester by esterases E003, E004, E007, E008, E009, and E013. Similar high ee values were observed for the R-ester hydrolysis with E002, E005, and E020 as shown in Table 4.

In addition, since esterases and lipases are known sometimes to catalyze hydrolysis of amide bonds [59, 60], we characterized the enzymes on p-nitroanilide compounds to observe if there was any activity. As shown in Fig. 5, many of the enzymes do indeed have activity on these substrates and may be useful in amide chemistry as well as the ester reactions mentioned earlier. This is an

Table 4. Enantioselective hydrolysis of (+)-phenethyl alcohol acetate by ThermoCat esterases

Enzyme	ee Alcohol (%)	ee Ester (%)[a]
E002	94.2	90.0
E003	92.8	95.0
E004	36.2	98.8
E005	95.0	82.8
E007	53.8	99.8
E008	29.4	99.0
E009	87.8	99.4
E013	43.0	98.2
E020	94.8	42.2

[a] Substrate: sec-phenyl-ethyl-acetate.

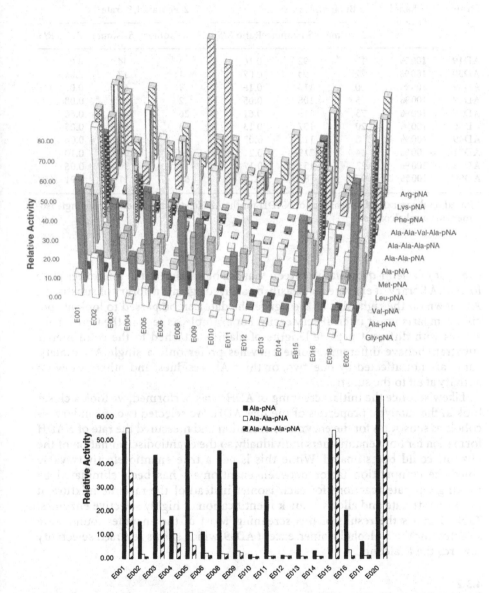

Fig. 5. Activity plots for a series of esterases tested against a library of p-nitroanilide (pNA) substrates. Many of the enzymes show significant activity against several pNA derivatives. These include amino acids such as glycine (Gly), alanine (Ala), valine (Val), Leucine (Leu), methionine (Met), phenylalanine (Phe), lysine (Lys), and arginine (Arg) attached alone and in combination to the pNA moiety. If a cross-section of the experiment is studied, one can see patterns in substrate preference. For example, using substrates with variable length peptide chains, some enzymes exhibit preference to single alanine and others show high activity on substrates consisting on di- or tripeptides of alanine

Table 5. Selectivity of new ThermoCat ADHs[a]

Strain	Ethanol % rate	2-Butanol, % rate			2-Pentanol, % rate		
		R-isomer	S-isomer	Ratio R/S	R-isomer	S-isomer	Ratio R/S
AD19	100%	15	92	0.16	3	44	0.06
AD30	100%	7	94	0.07	1	17	0.04
AD39	100%	20	113	0.18	4	61	0.07
AD49a	100%	5	108	0.05	2	21	0.08
AD49b	100%	73	118	0.61	26	40	0.66
AD55	100%	20	157	0.13	3	53	0.05
AD69	100%	8	112	0.07	1	21	0.04
AD71	100%	44	219	0.2	10	100	0.10
AD98	100%	27	153	0.17	4	70	0.06
AD99	100%	97	85	1.14	62	11	5.91

[a] Based on the % rate of appearance of NADH for the ADH catalyzed oxidation of single isomers of chiral alcohols.

example of a semi-quantitative screen that can be carried out in microtiter plate format. A library of enzymes can easily be assayed against a library of substrates. As shown on the right side of the figure, the data can be parsed to look at specific comparisons of interest very easily. For example, as seen in the figure, substrates with different peptide lengths of alanine attached to the colorimetric substrate behave differently. Some enzymes prefer only a single Ala moiety, some are not affected by one, two, or three Ala residues, and others show no activity at all to the substrate.

Likewise, once the initial screening of ADHs was performed, we took a closer look at the catalytic properties of the new ADH. We selected two secondary alcohols as substrates for the enzymatic oxidation and measured the rate of NADH formation for both enantiomers individually, so the enantiodiscrimination of the enzyme could be estimated. While this is not a true enantioselectivity value since the competition factor between enantiomers has been eliminated by making separate reactions for each isomer instead of the racemic mixture, it gives an estimate and allows a quick identification of highly selective enzymes. Table 5 shows the results of this screening. Most of the enzymes found were selective for the S-alcohol isomer, except AD99, which shows reversed selectivity towards the R-alcohol.

4.3.2
Stocking the Enzyme Library

These enzyme libraries represent the beginning of a new set of biocatalyst tools for the synthetic chemist and were the first thermostable enzyme libraries developed specifically for synthetic chemistry applications in each of these chemical classifications. We have since been expanding the esterase and dehydrogenase libraries using a variety of new and diverse substrate compounds. As these new enzymes are verified for their utility they will be added to a growing enzyme library, thus expanding the synthetic toolbox.

4.4
New Rapid Screening Technologies for Biocatalyst Discovery and Development

As enzyme libraries grow and high throughput screening methods develop, new types of assays which utilize the specific target substrates instead of substrate analogs are needed. We have been working on new screening methods that will allow for accelerated development of enzymes whether by screening enzyme libraries and organism/clone banks, or mutant populations for directed evolution.

One of the most promising new screening systems is the monitoring of a change in pH for hydrolase-catalyzed reactions that produce an acid such as esterases. Several research groups including our own have explored this type of assay at several levels of sophistication and for screening purposes. Scheme 4

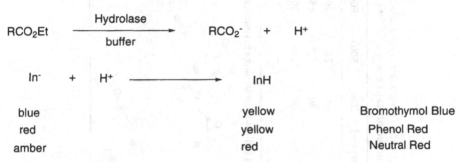

Scheme 4. pH-shift based assays for hydrolase catalyzed reactions

shows the principle behind the strategy: the acid produced during the course of the reaction will cause a drop in pH. This drop can be either large or small depending on, among other parameters, the buffer strength. If the buffer is chosen to cause a relatively large drop in pH the approach can be used qualitatively by visualizing the change in color if the initial pH and indicator are selected accordingly.

We studied a method for monitoring libraries of isolated esterases as a first-pass screening approach. The indicator used for this purpose was bromothymol blue at pH 7.25, which gives a blue color to begin with. As the reaction goes on and the pH drops below 7.0, a yellow color appears, facilitating the identification of positives from the library. This drop can be modulated by the buffer strength and the amount of enzyme to fit better in a convenient time-window. One of the most important features of the enzymes we are working on is their ability to perform stereoselective reactions, e.g., resolutions of racemic mixtures. In this case it is crucial to find enzymes with high enantioselectivity (E), a parameter well studied by Sih and coworkers [61]. This indicator-aided approach has been used in our company to evaluate enantioselectivities by reacting each single isomer

Table 6. The general strategy of a pH-based assay for enzyme activity and selectivity of ThermoGen esterases. If separate isomers are run side-by-side an indication of substrate preference can be developed for each enzyme

Substrate	Control (no enzyme)	E001	E003	E004	E005	E006	E016	E017	E018
Ethyl butyrate	–	60	240	20	<5	<5	–	–	15
Tributyrin	–	60	20	<5	<5	<5	<5	120	60
Phenethyl alchohol acetate	–	240	240	20	60	60	120	900	90
Control (no substrate)	–	–	–	–	–	–	–	–	–

Time (in min) needed to turn the color of an indicator containing enzymatic reaction

moderate selective
E002, E005, E006
E007, E017

non-selective
hydrolysis

highly selective
E002, E003, E005,
E006, E007, E012

separately with the same enzyme in parallel under the same conditions. The slow-reacting enantiomer will change the color of the solution slower compared to the fast-reacting one. Table 6 shows the results of such screening for three non-chiral esters and the evaluation of enantioselectivity for three pairs of single isomers. As stated before, the enantioselectivity value provides a reasonable approximation for screening purposes. In a hierarchical screening approach one can then identify the best results for analysis of the racemic mixture by HPLC or GC. In general, E values below 20 are of little use for the resolution of racemic mixtures [62], so only those clear results for the colorimetric assay will be considered further, thus minimizing the risk of enantioselectivity overrepresentation.

A quantitative enantioselectivity (E) assay has also been recently set up by Kazlauskas and coworkers [63]. They have developed a suitable system by choosing carefully the pair indicator-buffer so that the kinetics of the color change is linear and amenable for the kinetic study using a microplate reader. Despite the lack of competitive hydrolysis as explained above, the method is able to tell apart values of E < 5, and the sensitivity is ten times higher than the color based assay. In this assay, the quantitation relies on extremely small changes of the pH and there is no visible color change. The high throughput capability combined with the high sensitivity makes this method a good fit for directed evolution studies where most mutants (and wild-type isolates) will be active to some extent.

Indicator-based assays have also been described for the rapid characterization of serine proteases by their esterase activity using a microplate set-up [64]. A version of this assay that works for the screening of colonies on agar plates has been developed by Bornscheuer et al., using minimum media and glycerol esters of the substrate of interest, so the active colonies grow larger and are easier to detect by the color change on the plate [65].

5
Conclusions

Recent advances in molecular technology have allowed rapid and effective discovery and characterization of biocatalysts for chemical synthesis applications. New methods for screening the creation of diverse enzyme libraries, the development of diverse organism and clone banks, and new methods for assaying enzymes and evolving proteins of interest are helping to solve the challenges of biocatalysis. We have been able to develop a series of diverse catalysts that are stable and effective by developing and integrating a variety of these technologies using a multidisciplinary approach. Future developments in the field promise to make custom biocatalyst identification an attractive and cost effective possibility.

Acknowledgements. We would like to thank the National Heart Lung and Blood Institute and National Cancer Institute at the National Institutes of Health (CA63876, HL5773), the National Science Foundation (DMI-9561842), and the Department of the Army, US Department of Defense (DAA G55-97-C-0004) for their support. We gratefully acknowledge the work of Ms. Kelly Whitecotton, Sigma Chemical Company, St. Louis, MO in performing enantioselectivity experiments with the ThermoCat esterases.

6
References

1. Schoffers E, Golebiowski A, Johnson CR (1996) Tetrahedron 52:3769
2. Crabb WD, Mitchinson C (1997) Trends Biotechnol 15:349
3. Gerhartz W (1990) Enzymes in industry. VCH, Weinheim
4. Wong C-H, Whitesides GM (1994) Enzymes in synthetic organic chemistry. Pergamon, Oxford
5. Santaniello E, Ferraboschi P, Grisenti P, Manzocchi A (1992) Chem Rev 92:1071
6. Zaks A, Dodds DR (1997) Drug Discovery Today 2:513
7. Shimizu S, Ogawa J, Kataoka M, Kobayashi M (1997) Adv Biochem Eng Biotechnol 58:45
8. Turner MK (1995) Trends Biotechnol 13:173
9. Schwardt E (1990) Food Biotechnol 4:337
10. Roberts S (1997) Preparative biotransformations. Wiley, Chichester
11. Arnold FH, Moore JC (1997) Adv Biochem Eng Biotechnol 58:1
12. Owusu RK, Cowan DA (1990) Enzyme Microb Technol 12:374
13. Liao HH (1993) Enzyme Microb Technol 15:286
14. Tosa T, Shibatani T (1995) Ann N Y Acad Sci 750:364
15. Mazid MA (1993) Bio-Technology 11:690
16. St Clair NL, Navia MA (1992) J Am Chem Soc 114:7314
17. Kuchner O, Arnold FH (1997) Trends Biotechnol 15:523
18. Annual Product Guide Issue (1998) Nature Biotechnol 15:1107
19. Short JM (1997) Nature Biotechnol 15:1322
20. Lalonde J (1997) Chemtech 27:38
21. Dalboge H, Lange L (1998) Trends Biotechnol 16:265
22. Hummel W (1997) Exs 80:49
23. Chartrain M, Armstrong J, Katz L, King S, Reddy J, Shi YJ, Tschaen D, Greasham R (1996) Ann N Y Acad Sci 799:612
24. (1984) ATCC media handbook. American Type Culture Collection, Rockville, MD
25. Bergey DH, Holt JG (1993) Bergey's manual of determinative bacteriology. Williams & Wilkins, Baltimore
26. Giovannoni SJ, DeLong EF, Olsen GJ, Pace NR (1988) J Bacteriol 170:720
27. Busse HJ, Denner EBM, Lubitz W (1996) J Biotechnol 47:3
28. Sambrook J, Maniatis T, Fritsch EF (1989) Molecular cloning: a laboratory manual. Cold Spring Harbor Laboratory, Cold Spring Harbor, NY
29. Amann R, Ludwig W, Schliefer K-H (1995) Microbiol Rev 59:143
30. Fastrez J (1997) Mol Biotechnol 7:37
31. Vonstein V, Johnson SP, Yu H, Casadaban MJ, Pagratis NC, Weber JM, Demirjian DC (1995) J Bacteriol 177:4540
32. LaRossa RA (1996) In: Neidhart FC, Curtiss III R, Ingraham JL, Lin ECC, Low KB, Magasanik B, Reznikoff WS, Riley M, Schaechter M, Umbarger HE (eds) Escherichia coli and Salmonella, vol 2. ASM Press, Washington, DC, p 2527
33. Michal G, Möllering H, Siedel J (1983) In: Bergmeyer HU (ed) Methods of enzymatic analysis, vol I. Verlag Chemie, Weinheim p 197
34. Alvarez AM, Ibanez M, Rotger R (1993) Biotechniques 15:974
35. Houston JG, Banks M (1997) Curr Opin Biotechnol 8:734
36. Cheng J, Sheldon EL, Wu L, Uribe A, Gerrue LO, Carrino J, Heller MJ, Oconnell JP (1998) Nature Biotechnol 16:541
37. Hadd AG, Raymond DE, Halliwell JW, Jacobson SC, Ramsey JM (1997) Anal Chem 69:3407
38. Jones P, Watson A, Davies M, Stubbings S (1992) Nucl Acids Res 20:4599
39. Sheppard PO, Grant FJ, Oort PJ, Sprecher CA, Foster DC, Hagen FS, Upshall A, McKnight GL, O'Hara PJ (1994) Gene 150:163
40. Madden TL, Tatusov RL, Zhang J (1996) Methods Enzymol 266:131
41. Koonin EV, Tatusov RL, Rudd KE (1996) Methods Enzymol 266:295
42. Tatusov RL, Koonin EV, Lipman DJ (1997) Science 278:631

43. Gaasterland T, Sensen CW (1996) Biochimie 78:302
44. Dsouza M, Larsen N, Overbeek R (1997) Trends Genet 13:497
45. Govardhan CP, Margolin AL (1995) Chem Ind 17:689
46. Weber JM, Johnson SP, Vonstein V, Casadaban MJ, Demirjian DC (1995) Biotechnology (NY) 13:271
47. Casadaban MJ, Martinez-Arias A, Shapira SK, Chou J (1983) Methods Enzymol 100:293
48. Casadaban MJ, Chou J, Cohen SN (1980) J Bacteriol 143:971
49. Miller JH (1991) A short course in bacterial genetics: a laboratory manual and handbook for Escherichia coli and related bacteria. Cold Spring Harbor Laboratories Press, Cold Spring Harbor
50. Kwaan HC, Friedman RB, Szczecinski M (1978) Thromb Res 13:5
51. Hron WT, Menahan LA (1981) J Lipid Res 22:377
52. Bowyer DE, Cridland JS, King JP (1979) J Lipid Res 19:274
53. Ulitzer S, Heller M (1981) Methods Enzymol 72:338
54. Bjokhem I, Arner P, Thore A, Ostman J (1981) J Lipid Res 22:1142
55. Mizuno K, Toyosato M, Yabumoto S, Tauimizu I, Hirakawa H (1980) Anal Biochem 108:6
56. Sowerby JM, Ottaway JH (1961) Biochem J 79:21P
57. Conway T, Sewell GW, Osman YA, Ingram LO (1987) J Bacteriol 169:2591
58. Skursky L, Kovár J, Stachova M (1979) Anal Biochem 99:65
59. Quiros M, Sanchez VM, Brieva R, Rebolledo F, Gotor V (1993) Tetrahedron Asymmetry 4:1105
60. Gotor V, Brieva R, González C, Rebolledo F (1991) Tetrahedron 47:9207
61. Chen C-H, Fujimoto Y, Girdaukas G, Sih CJ (1982) J Am Chem Soc 104:104
62. Stecher H, Faber K (1997) Synthesis 1
63. Janes L, Löwendhal C, Kazlauskas R (1998) Eur J Chem (1998) 4:2317
64. Whittaker RG, Manthey MK, LeBrocque DS, Hayes PJ (1994) Anal Biochem 220:238
65. Bornscheuer UT, Enzelberger MM, Altenbuchner J, Meyer HH (1998) Stratagene Tech Bull 11:16

Superior Biocatalysts by Directed Evolution

Manfred T. Reetz[1] · Karl-Erich Jaeger[2]

[1] Max-Planck-Institut für Kohlenforschung, D-45470 Mülheim/Ruhr, Germany.
  E-mail: reetz@mpi-muelheim.mpg.de
[2] Lehrstuhl Biologie der Mikroorganismen, Ruhr-Universität, D-44780 Bochum, Germany.
  E-mail: karl-erich.jaeger@ruhr-uni-bochum.de

Useful biocatalysts for organic chemistry can be created by directed evolution. Mutations are introduced into genes encoding biocatalyst proteins of interest by error-prone PCR or other random mutagenesis methods. The mutated genes can be rearranged by recombinative processes like DNA shuffling, thereby significantly enhancing the efficiency with which genes can be evolved. These genes are expressed in suitable microbial hosts leading to the production of functional biocatalysts. Selection or screening procedures serve to identify in a large library of potential candidates the biocatalyst which possesses the desired properties. Examples of applications include subtilisin E with greatly improved catalytic activity and stability in organic solvent, an esterase with 50-fold higher activity in organic solvent, and a β-lactamase conferring a 32,000-fold increased antibiotic resistance. Furthermore, directed evolution of a bacterial lipase resulted in a significant increase in enantioselectivity, thereby demonstrating the enormous potential of this process for organic chemistry.

Keywords: Directed evolution, Random mutagenesis, DNA shuffling, Biocatalysis, Combinatorial libraries, Enantioselectivity.

Topics in Current Chemistry, Vol. 200
© Springer Verlag Berlin Heidelberg 1999

1
Introduction: Evolution in the Test Tube

It has been estimated that more than 80% of all industrial chemical processes are based on catalytic reactions [1]. The options available to industrial and academic chemists are homogeneous [2] and heterogeneous [3] chemical catalysts on the one hand, and biocatalysts [4] on the other. The decision as to which type of catalyst should be used in a given situation depends upon a number of factors, all of which ultimately relate to economic and ecological aspects, e.g. ease of catalyst production, its activity and selectivity, availability of specific building blocks, type and amount of solvent, and simplicity of workup. It is therefore not possible to provide general rules, although attempts to compare chemical catalysts with biocatalysts have occasionally been made [5]. In the final analysis the user will choose the option that works best in a given case. The number of industrial and academic reports concerning the use of enzymes in organic synthesis has increased dramatically during the last two decades [4]. The reasons behind this rapid development include increasing efforts in isolating and testing new enzymes, more extensive research in studying the performance of enzymes in organic solvents and growing investments in chemical engineering aspects of biotechnology.

The types of enzymes used by organic chemists vary widely and include such well-known biocatalysts as lipases, esterases, oxidoreductases, oxinitrilases, transferases and aldolases [4]. An example which illustrates the industrial application of a lipase concerns the kinetic resolution of a chiral epoxy ester used as the key intermediate in the synthesis of the calcium antagonist Diltiazem, a major therapeutic in the treatment of high blood pressure [6] (Fig. 1). In developing the industrial process for the production of this drug, many different lipases were screened, but only the bacterial lipase from *Serratia marescens* showed both a sufficiently high activity *and* enantioselectivity. The intermediate is produced industrially on a scale of 50 tons/year.

Although examples of this type are impressive, the use of enzymes is clearly restricted due to the typical phenomenon of substrate specificity. Being a product of millions of years of evolution, enzymes did not evolve to be efficient in every situation that organic chemists would like them to be. Catalyst activity and selectivity are unacceptable for numerous substrates of interest. In other cases selectivity may be high, but the process requires solvents and conditions which lead to low enzyme stability, activity and/or selectivity.

Theoretically, a way out of this dilemma is traditional protein engineering based on site-directed mutagenesis [7]. In a certain sense this is analogous to ligand tuning in homogeneous catalysis [2]. Accordingly, certain methods in molecular biology are utilised in order to substitute a specific amino acid for a different one at a defined position in the peptide chain of the enzyme, hoping that this will improve activity or selectivity (or both). Although this technique has been highly useful in studying structure/function relations of enzymes, practical applications are not as common as one would like, because in practice this form of "design" is even more difficult and laborious than ligand tuning in homogeneous catalysis. The practising chemist not only needs detailed infor-

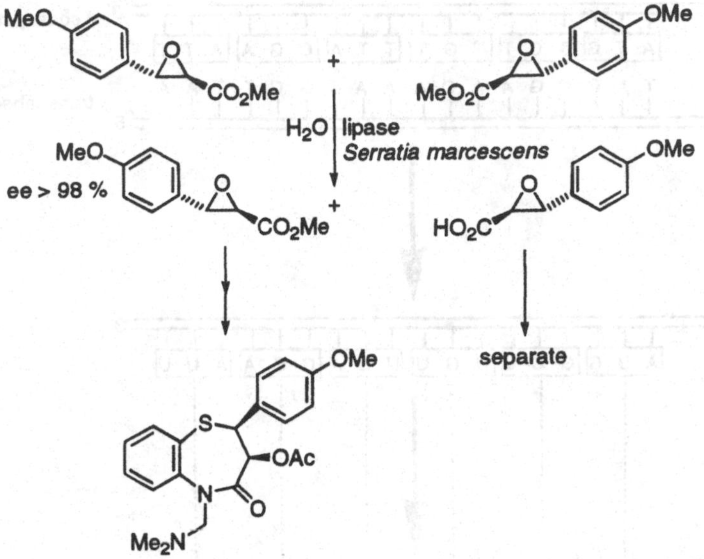

Fig. 1. Enzymatic kinetic resolution of an intermediate used in the synthesis of Diltiazem

mation concerning the three-dimensional structure of the enzyme and the mechanism of the catalysed reaction, but also predictive power with respect to the proper site of substitution and intuition concerning the optimal choice of the amino acid to be inserted [7]. Currently there is no reliable theory upon which such decisions can be based. The situation becomes even more difficult if more than one amino acid substitution needs to be carried out, which is generally the case.

In the late 1980s and early 1990s molecular biologists began to develop new and practical techniques for random mutagenesis [8–10]. It is important to remind the reader that mutagenesis is not performed on the enzyme itself, but on the gene (DNA segment) which encodes a particular enzyme. According to the instructions of a particular gene, Nature uses the 20 natural amino acids at its disposal to assemble a specific enzyme [11] (Fig. 2).

If mutations are randomly introduced into a gene, then, upon expression in a suitable bacterial system, mutant enzymes are produced. Thus, it is possible to create libraries of mutant enzymes in which the amino acids have been exchanged randomly. Moreover, the frequency of mutation and therefore the size of the enzyme library can be controlled in a simple manner [8–10]. By the mid-1990s it became obvious that the prospect of obtaining large numbers of new and improved enzymes was turning into reality, although the problem of identifying the best mutant enzyme in a given library remained to be solved in a general way. Irrespective of such challenges, the feasibility of the combinatorial principle in the development of biocatalysts was established [8–10, 12, 13].

Although such prospects are certainly exciting for organic chemists and other scientists in related fields, e.g. environmental chemists interested in enzymatic

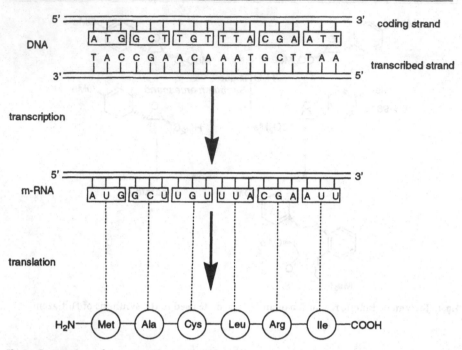

Fig. 2. Conversion of genetic information from DNA to protein via transcription and translation. The sequence of the coding DNA strand consists of nucleotide triplets called codons (*boxed*) each corresponding to an amino acids of the protein

detoxification of pollutants, the really novel quality in these efforts concerns an additional aspect which goes far beyond combinatorial chemistry, namely the evolutive process. If the library of enzymes contains a few improved but not yet optimal mutants, one or more of these can be identified and used as the starting point for the next cycle of mutagenesis and selection or screening, a process that can be repeated as often as needed. Thus, any number of generations of mutant enzymes can be created. On the basis of sequential cycles of mutation/replication/selection (screening), an enzyme having the desired properties can be evolved. Thus, what took Nature millions of years to develop for a specific purpose [14] can now in principle be performed in the test tube within months or weeks, namely the creation of an optimal catalyst for a reaction of interest to the practising organic chemist. This type of evolutive process requires no knowledge of the three-dimensional structure of the enzyme nor of the mechanism of the reaction which it catalyses, yet it is quite unlike a "lottery system" based on luck. Indeed, it can be viewed as the most rational way to develop catalysts. A particularly intriguing aspect is the prospect of creating enantioselective catalysts for application in asymmetric synthesis [15].

Before going into detail, we briefly address the question of protein sequence space, which is in fact huge. Consider, for example, an enzyme composed of 300 amino acids, which is fairly typical [16]. If all of the 20 natural amino acids were to be completely randomised in the chain, to include all combinations an astronomically high number of permutations would result, namely 20^{300} enzymes. In contrast, if we restrict ourselves to the smallest possible "surgical manipulation", namely a single amino acid exchange per enzyme molecule, this being performed randomly, then 5700 mutants are theoretically possible as calculated by the following algorithm:

$$N = 19^M X! / [(X-M)!M!] = 5700 \tag{1}$$

where N=number of mutants, M=number of exchanged amino acids per enzyme molecules, and X=number of amino acids per enzyme.

It should be possible to handle this relatively small number of mutant enzymes in an efficient assay system. However, it needs to be emphasised that the present methods of mutagenesis do not guarantee the creation of all 5700 mutants in a given mutagenesis experiment. There are several reasons for this, one being the fact that the probability of formation is not identical for all mutants. Relevant to the problem of exploring protein sequence space efficiently is the finding that only a fraction of the amino acid residues in an enzyme is actually critical for function, folding and stability (an insight that was obtained by mutagenesis experiments) [17]. In spite of these uncertainties, it is evident that the simplest and perhaps most effective strategy would be to choose a low mutation frequency with formation of small enzyme libraries, hoping that at least a few mutants with slightly improved properties would be generated in each generation, one or more of which would serve as a template for the next cycle of mutagenesis. Indeed, if M = 2 in the above example, the number of mutant enzymes N turns out to be about 16 million, which would be difficult to screen. This also indicates that the best tactics are to aim for small but significant sequential improvements until the desired catalytic properties have evolved. Moreover, the combination of various mutagenesis methods offers additional strategic possibilities for exploring protein sequence space efficiently [18].

The purpose of this review is to summarise the present status of this fascinating new area of endeavour, which has been termed "directed evolution" [8–10, 12, 13]. We begin with a description of the relevant molecular biology methods, follow up with typical examples from the field of applications and end with concluding remarks concerning future prospects.

2
Methods for Mutagenesis

Before specific methods for mutagenesis are discussed in detail, a few general remarks are in order. Mutations can be induced by many different methods, some of which may best be explained by considering an example: Suppose a gene *b* encoding a biocatalyst protein B consisting of 300 amino acids needs to be mutated in order to generate a variant protein B' which exhibits increased activity towards a given substrate. Then the following points should be considered:

1. If detailed knowledge is available concerning structure-function relationship and the specific role of particular amino acid residues, it is feasible to use

methods of *site-directed mutagenesis*. Assuming a catalytic site-residue of protein B is known, one could exchange this amino acid against another one predicted to result in improved substrate binding. Frequently, a particular amino acid residue can be identified as being important for enzymatic activity of the biocatalyst. However, it is impossible to predict which of the remaining 19 amino acids would result in an increased activity. In such a case, *saturation mutagenesis* would be performed resulting in exchange of the existing residue against all the remaining 19 amino acids.

2. If a particular region X of protein B has been identified as being involved in enzymatic activity, it might be useful to mutate this region by various techniques of *cassette mutagenesis* and to reinsert it into gene *b*. Subsequent selection and/or screening might result in a variant protein B' with an optimised region X.

3. If secondary and tertiary structural data are not available at all, methods of *directed enzyme evolution* should be applied to the gene *b* with two general approaches to be distinguished: (i) non-recombinative methods including *error-prone PCR* on the isolated gene or amplification of the gene in a so-called *bacterial mutator strain* which introduces random mutations, and (ii) recombinative methods including *DNA-shuffling* and *staggered extension processes* which start with multiple mutated or related genes. Usually, large libraries of mutated genes are created which must then be searched for the desired variant protein B'. At this stage, B' proteins are identified by appropriate screening assays which normally allow one to look at approximately 10^4 mutant proteins, or by genetic selection processes which may allow the experimenter to examine 10^8 individual variants, suggesting an increased chance of finding the desired biocatalyst. However, two important points must be emphasised: (i) protein B consists of 300 amino acids, making the total number of possible mutants 20^{300} which is an infinite number for practical considerations, and (ii) beneficial mutations occur very rarely, i.e. most mutations are deleterious. Therefore, it was suggested to start directed evolution experiments with a gene which encodes a protein exhibiting properties close to what is searched for [19]. The chance of finding the desired biocatalyst is far better when performing several small steps of evolutive improvement rather than one large step.

In the following sections we briefly discuss the biological basis of the most important methods of mutagenesis.

2.1
Site-Directed Mutagenesis

A large number of protocols have been established for site-directed mutagenesis [20] allowing one to exchange, insert or delete one or more defined nucleotides of a given gene, thereby resulting in desired amino acid substitutions. In principle, a short oligonucleotide primer is hybridised to a single-stranded DNA template of a circular plasmid containing the gene to be mutagenised. The oligonucleotide is completely complementary to a region of the template except for the mismatch carrying the mutation. The oligonucleotide primer is extended by DNA polymerase and the DNA fragment cloned and expressed in *E. coli* [21]. The most widely used mutagenesis protocols use PCR with mutagenic oligonucleotide primers. The principle of the popular overlap extension PCR [22] is schematically outlined in Fig. 3.

Two DNA fragments are separately amplified from a target gene using a matching and a mutant primer each. The resulting two DNA fragments share a small overlapping region which contains the same mutation in each strand. These strands are now mixed, melted and reannealed so that the two strands carrying the mutation can now act as primers on one another. In a final step, the full-length mutant DNA molecules are synthesised by DNA polymerase. Recently, we have developed a variation of this method eliminating the second PCR-amplification step which was therefore named one-step overlap extension PCR [23].

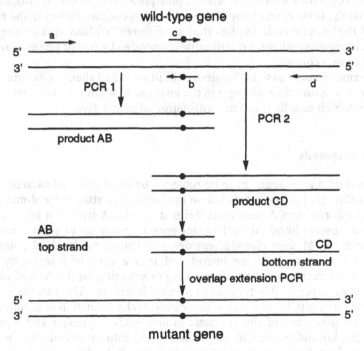

Fig. 3. Mutagenesis by overlap extension PCR [22]. PCR products are shown as two *paired strands*, primers are shown as horizontal *arrows*, and mutations in primers and products as *black dots*

2.2
Saturation Mutagenesis

Saturation mutagenesis is a generalised term pertaining to the substitution or insertion of codons encoding all possible amino acids at any predetermined position in a gene. A straightforward strategy is the application of a site-directed mutagenesis method for introducing the nucleotide exchanges necessary to obtain all the desired codons. However, a set of mutagenic oligonucleotide primers is needed and all of the resulting mutagenised DNAs have to be sequenced in order to confirm the presence of the mutations. One of the available methods named codon cassette mutagenesis [24] uses a set of eleven mutagenic codon cassettes. The target gene is cleaved with a restriction endonuclease which generates blunt ends at an appropriate position. The mutagenic cassettes are inserted at these positions and positive clones are identified by restriction analysis and DNA sequencing. Upon digestion of the target DNA with an appropriate restriction enzyme, three base-cohesive ends are formed representing the mutagenic codon which result in recircularisation of the plasmid. Another method which is easy to perform uses PCR [25] to introduce site-directed mutations. In this particular case, oligonucleotide primers degenerated at one codon position are used which are synthesised employing equimolar concentrations of nucleoside phosphoramidites dA, dC, dG, and dT. This allows the formation of any of the existing 64 codons. However, the redundancy of the genetic code implies that six different codons encode the same amino acid arginine whereas methionine is encoded by one or phenylalanine is encoded by two codons.

Furthermore, some base exchanges result in so-called silent mutations which do not cause amino acid exchanges in the enzyme. In addition, nonsense codons also occur which usually result in termination of translation.

2.3
Cassette Mutagenesis

In this method a gene segment to be replaced by combinatorial cassette mutagenesis (CCM) [26] is first identified on the basis of existing three-dimensional structural information. A cassette is defined as a DNA-fragment consisting of three up to several hundred nucleotides encoding one up to several hundred amino acids. CCM uses oligonucleotides containing randomised codons as mutagenic cassettes which are introduced into a gene of interest by PCR-methods. The next step consists of screening or selection for the desired variant. An elegant example of this type was recently described [27] with the enzyme chorismate mutase (CM) of E. coli which catalyses the formation of prephenate, a biosynthesis precursor of the aromatic amino acids L-tyrosine and L-phenylalanine. The homodimeric CM was transformed into an enzymatically active monomeric form by insertion of a six-residue interhelical turn which was optimised by directed evolution. Selection was performed by growing the bacteria in minimal medium lacking the aromatic amino acids tyrosine and phenylalanine. Other examples of structure-based evolutionary protein design relate to the optimisation of binding of antibody fragments to particular ligands or probing and improving enzyme active sites [28]. An interesting extension of the method described above is called combinatorial multiple cassette mutagenesis and allows the creation of a mutant library in which each of the multiple cassettes contains a mixture of wild-type and randomised sequences; this results in a complete permutation of all wild-type and mutant cassettes [29].

2.4
Directed Enzyme Evolution by Non-Recombinative Methods

The powerful new technique of directed enzyme evolution aims at mimicking the process of enzyme improvement by natural evolution, differing in two major aspects: (i) researchers define the properties to be optimised, and (ii) the evolutive process takes place in a relatively short period of time (weeks to months). Generally, the process of directed evolution consists of the following reiterative steps: (1) creation of a library consisting of mutated genes, (2) functional expression of these genes in suitable hosts, and (3) identification of enzyme variants with improved properties by selection or screening. In the following sections we first discuss non-recombinative and recombinative methods needed for the creation of mutant libraries of enzymes. Selected examples for gene expression systems are then presented.

2.4.1
Error-Prone PCR

A PCR reaction is normally carried out in order to amplify any given DNA with high fidelity. The intrinsic $3' \rightarrow 5'$ exonuclease activity (also called proof-reading activity) of DNA polymerases ensures that DNA amplification proceeds in an accurate manner. A standard PCR protocol reads as follows: 1.5 mmol l^{-1} $MgCl_2$, 50 mmol l^{-1} KCl, 10 mmol l^{-1} Tris-HCl, pH 8.3, 0.2 mmol l^{-1} each of dNTP's, 0.3 µmol l^{-1} of each primer, and 2.5 units of thermostable DNA polymerase in a total volume of 100 µl. This reaction mixture is incubated for 30 PCR cycles consisting of 1 min at 94 °C, 1 min at 45 °C, and 1 min at 72 °C in a thermal cycler. Occasionally, wrong nucleotides are incorporated during amplification, leading to mutations with a frequency of $0.1 - 2 \times 10^{-4}$ per nucleotide for the thermostable DNA polymerase from *Thermus aquaticus* (*Taq*-polymerase) [30]. This very small error rate can be increased to 7×10^{-3} per nucleotide by: (1) increasing the concentration of $MgCl_2$ to 7 mmol l^{-1}, (2) addition of 0.5 mmol l^{-1} $MnCl_2$, (3) increasing the concentration of dCTP and dTTP to 1 mmol l^{-1}, and (4) increasing the amount of *Taq* polymerase to 5 units [31]. For the purpose of directed enzyme evolution, these conditions should be modified so as to achieve an average mutation frequency of one to two nucleotide exchanges per gene, leading to an average exchange of one amino acid per mutant enzyme. Such conditions will typically result in a mutant library having a screenable size (e.g. 5700 mutants for a protein consisting of 300 amino acids; see above).

2.4.2
Bacterial Mutator Strains

Wild-type *Escherichia coli* strains exhibit a spontaneous mutation frequency of about 2.5×10^{-4} mutations per 1000 nucleotides of DNA propagated on a pBlue-script-like plasmid after 30 generations of growth. Strains which contain mutations in various DNA repair pathways show a considerable increase in the spontaneous mutation rate which is 5- to 100-fold higher than that of a wild-type strain [32]. Recently, an *E. coli* strain named XL1-Red was constructed and made commercially available (Stratagene, La Jolla, USA); it contains mutations in three independent DNA repair pathways, thereby exhibiting a spontaneous mutation frequency of about 0.5 mutations per 1000 nucleotides of DNA under the conditions described above [33]. In order to generate a mutant library, a gene encoding a biocatalyst protein can be cloned into an appropriate plasmid, transformed into *E. coli* XL1-Red and the strain grown overnight. However, if the target DNA is of small size (< 100 bp) or if multiple mutations are required, the number of generations needed for propagation of the plasmid DNA will become impractically high. In addition, due to the lack of DNA repair pathways, the strain is fairly unstable and cannot be propagated for prolonged periods [34].

2.5
Directed Enzyme Evolution by Recombinative Methods

Recombination in a genetic sense means the breaking and rejoining of DNA in new combinations. The genetic information associated with one DNA molecule may become associated with a different DNA molecule, or the order of genetic information on a DNA molecule may be altered. This process is very important for all living species because it significantly speeds up the process of evolution. Inside a living cell, both homologous recombination (occurring between identical or very similar DNA sequences) and non-homologous recombination (occurring between distinct DNA sequences) require the presence of a set of recombination enzymes ensuring breakage and rejoining of DNA molecules. Various approaches have been developed to mimic Nature's recombination strategy in the test tube.

2.5.1
DNA-Shuffling

The most prominent novel method emphasizing recombination in DNA shuffling [10, 35] (Fig. 4). One or more closely related genes are digested with DNaseI to yield double-stranded oligonucleotide fragments of 10–50 bp which are purified and used in a PCR-like reaction. Repeated cycles of strand separation and reannealing in the presence of DNA polymerase and a final PCR-amplification step result in reassembling of a full-length gene. Recombination occurs by template switching, i.e. fragments from one copy of the gene prime on another copy. The combination of this technique with those described above to introduce point mutations at a low and controlled rate [36] allows the simultaneous permutation of both single mutations and large blocks of DNA sequences. Mutants with improved properties can be identified and the genes can then be reshuffled against each other or against the wild-type gene which would correspond to the backcross-technique known from classical genetics. This process may eventually lead to an elimination of neutral and deleterious mutations from the gene pool. DNA shuffling can be performed with mutants of the same gene, but also with homologous genes from different species

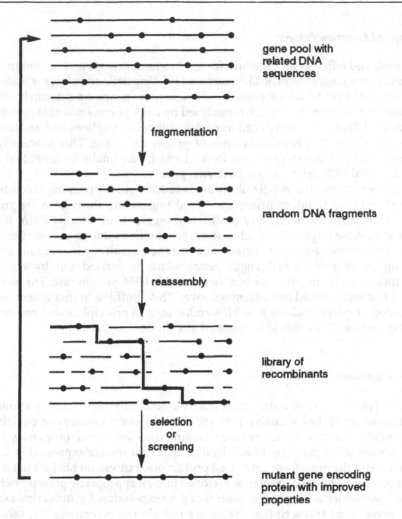

gene pool with
related DNA
sequences

fragmentation

random DNA fragments

reassembly

library of
recombinants

selection
or
screening

mutant gene encoding
protein with improved
properties

Fig. 4. DNA shuffling method [10, 35]. Related genes with different beneficial mutations (*black dots*) are randomly fragmented with DNaseI. During reassembly of fragments recombination occurs yielding in progeny genes with improved positive mutations which can serve as starting points for another round of mutation and recombination

(called family shuffling) which results in sparse sampling of a large sequence space, thereby significantly accelerating the rate of functional enzyme improvement [37]. The enormous potential of this novel technique to evolve both single genes encoding β-lactamase, β-galactosidase, green-fluorescent protein, alkyl transferase, benzyl esterase, and t-RNA synthetase or whole operons encoding enzymes for arsenate- or atrazine degradation has been demonstrated [38]. Selected examples will be discussed below in more detail (Sect. 3.1).

2.5.2
Staggered Extension Process

A simple and efficient new method for in vitro mutagenesis and recombination is based on a staggered PCR-like reaction [39]. One defined primer is added to the template DNA which may consist of two or more genes. An extremely abbreviated primer extension reaction catalysed by DNA polymerase then produces short DNA fragments which can anneal to different templates and are further extended during the next short cycle of primer extension. This process is repeated until full-length genes are formed which can finally be amplified in a conventional PCR reaction using external primers.

Another alternative to DNA shuffling is called random-priming recombination (RPR) [40]. Random primers are used to generate short DNA fragments (50–500 bases) complementary to different segments of a target DNA. In the next step, these fragments which also carry mutations due to mispriming and base misincorporation can prime one another, resulting in recombination during reassembly to full-length genes which is carried out by repeated thermocycling in the presence of a thermostable DNA polymerase. The authors claim that this method has advantages over DNA shuffling in that it uses single stranded templates including m-RNA and, at least in principle, every nucleotide of the template DNA should be mutated at a similar frequency [40].

2.6
Gene Expression Systems

Directed evolution of biocatalysts requires the functional expression in a suitable host organism of the biocatalyst genes to be tested as a necessary prerequisite to successfully identifying an enzyme variant with improved properties [41]. Today, most of the genes optimised by directed evolution are expressed in E. coli which is still the most frequently used prokaryotic expression host for heterologous proteins. The gene of interest is cloned into an appropriate plasmid behind a promoter which allows a tight control of gene expression, i.e. induction as well as repression of transcription. Examples include the promoters lac, tac, and T7-lac which can be induced by addition of synthetic chemicals (isopropyl-β-D-thiogalactopyranoside) and regulated by appropriate repressors (lacI^Q). A number of additional factors including transcriptional terminators, elements affecting translation, and codon usage have to be controlled to ensure optimal expression levels [42]. Many of the biocatalysts used in organic chemistry belong to the group of secreted enzymes with lipases being the most prominent examples. At least in some cases it was demonstrated that expression of a lipase gene is not sufficient to obtain enzymatically active protein. Indeed, the process of protein secretion is also required to ensure correct folding into an enzymatically active conformation. This is the reason why a biocatalyst protein originating from P. aeruginosa will exhibit very low activity or remain enzymatically inactive upon overexpression in E. coli [43]. Some of the protein secretion mechanisms are highly specific processes involving up to 25 different proteins and normally work only in the homologous host. Extensive engineering of host bacteria will be necessary to construct appropriate "secretor" strains which may also contain additional folding catalysts [44].

Other bacterial expression hosts are Gram-positive bacteria belonging to the genera *Bacillus*, *Clostridium*, *Lactococcus* and *Staphylococcus*. These bacteria have a cellular architecture different from *E. coli* in that they contain only one instead of two membranes allowing for direct secretion of proteins into the fermentation medium. Important characteristics of expression systems working in these organisms and their control as well as new developments have recently been reviewed [45]. Methylotrophic yeasts such as *Hansenula polymorpha* and *Pichia pastoris* represent eukaryotic expression hosts of emerging importance. They have been used to express enzymes including plant α-amylases and bacterial β-galactosidases [46]. In summary, it should be emphasised that the choice of an appropriate expression system is a prerequisite in the design of a successful strategy for directed evolution of a biocatalyst.

2.7
Selection and Screening

Large libraries of up to 10^{10} mutant genes can easily be created by using one of the methods described above. Therefore, the development of appropriate search systems is of key importance to the whole process of directed enzyme evolution. Novel strategies of selection and screening have recently been reviewed [28, 47, 48], although there is a need to develop further methods for most real applications. The terms "selection" and "screening" are often mistakenly used. We therefore reiterate the proper definitions. Screening is the process of identifying a desired member of a library in the presence of all other members. If selection is applied only the desired member of a library appears, e.g. as a viable microbial clone. This so-called in vivo selection is usually very efficient in that it allows only those microorganisms to grow which express a gene encoding a particular enzyme necessary to survive. However, it can be tedious to develop such a selection system, because microbial cells are extremely versatile with respect to circumventing restrictions imposed by a certain selection system. In vitro selection by phage display [48] is a powerful approach to identify peptide ligands and receptors which are displayed on the surface of a bacteriophage [48b]. Those phages displaying the desired peptide can be selected from a phage library by "biopanning", i.e. binding to a ligand immobilised on appropriate column matrices or microtiter plates. Phages with high affinity for the ligand can be enriched, amplified and further characterised. Efficient screening procedures have to handle every single member of a given library which usually requires automation. Reactions which take place in microtiter plates allowing for spectrophotometric detection of the products are still the method of choice in the development of high-throughput screening methods. A thermographic method was recently developed to detect the black-body radiation produced by the reaction of (R)- and (S)-configured enantiomers of a chiral alcohol. The substrate was acylated by an enantioselective lipase in a microtiter plate and the reaction was monitored in a time-resolved manner by using an IR-camera [49]. It remains to be seen if this method can be refined to allow quantification of enzyme activities and enantioselectivities.

3
Applications of Directed Evolution Methods

3.1
Enzymes with Higher Stability and Activity

When using enzymes as catalysts in organic synthesis it is important to consider enzyme stability, even if the reaction to be catalysed occurs in an aqueous medium which may be similar to in vivo conditions. Moreover, the actual conditions under which enzymes are employed as catalysts in organic transformations often differ considerably from the natural environment, which means that the question of stability becomes even more important [4]. The necessity to perform reactions at higher temperatures and/or in a non-aqueous medium requires increased thermostability and/or sufficient stability and activity in organic solvents. Although several approaches to increasing enzyme thermostability have been described, none are completely general. These include certain types of enzyme immobilisation [50, 51], the formation of cross-linked enzyme crystals [52], and chemical modification by the introduction of disulfide bonds, salt bridges or chemical cross-links in the enzyme [13, 53].

An alternative to these approaches is traditional protein engineering based on the use of site-directed mutagenesis [7]. Many examples outlining the use of this technique to obtain enzymes with higher thermostability and activity have been described. Enhanced thermostability is based on improved electrostatic, hydrophobic and hydrogen bonding interactions and on the introduction of disulfide bonds, which means that, inter alia, an intimate knowledge of the three-dimensional structure of the enzyme is required. A case in point is the attempt to increase the thermostability and activity of subtilisin BPN', a catalytic esterase used by organic chemists in the synthesis of peptides, regioselective acylation of carbohydrates and enantioselective transformation of chiral alcohols, acids and amines [54]. Following six site-specific mutations, a subtilisin mutant (8350) was obtained which was found to be 100 times more stable than the wild-type enzyme in aqueous solution at room temperature and 50 times more stable than the wild-type in anhydrous dimethylformamide. This extensive and successful study included X-ray crystal structures of the wild-type enzyme and of the engineered mutant, which allowed conclusions regarding structure, stability and activity [54].

Although higher temperatures were actually not tested in the above example, increasing the temperature in a given reaction may be necessary in actual industrial or academic applications due to higher rates, increased substrate solubility and decreased viscosity of the medium. Thermostability is often correlated with the melting temperature (T_m) of the enzyme. Using early and/or more recent techniques of random mutagenesis, a number of authors have shown that single beneficial mutations increase T_m of an enzyme by $1-2°C$, although larger changes are sometimes possible [19, 41, 55, 56].

A strategy that was first introduced in early mutagenesis experiments allows in some cases the direct identification of thermally stable mutants [57]. It is based on biological selection rather than screening for activity or selectivity, which means that the method is restricted to functional enzymes which are necessary for the survival and growth of the respective host. For example, the gene coding for a given enzyme from a mesophilic organism was cloned and introduced into a thermophile, e.g. *Bacillus stearothermophilus*, and selection for increased enzymatic activity was then performed on the basis of the growth rate of the host organism at higher temperatures [57]. This technique was applied to kanamycin nucleotidyltransferase (KNTase), an enzyme which confers resistance to the antibiotic kanamycin. Variants of this enzyme displaying kanamycin resistance at 63 °C were produced by two sequential rounds of mutagenesis resulting in two amino acid substitutions. Although related strategies have been reported for other model systems [19, 41, 55, 56], it is currently not clear how general these are. Although biological selection is elegant, many enzymatic systems of interest to organic chemists may not be amenable to this technique. The only other detection method is screening [15, 49, 58, 59].

An early example of in vitro evolution based on error-prone PCR concerns an attempt to enhance the catalytic activity of subtilisin E (an enzyme composed of 275 amino acids) in a model reaction carried out in polar organic medium [12]. It was of interest to see if enzyme activity in the hydrolysis of a short peptide suc-Ala-Ala-Pro-Pro-*p*-nitroanilide (sAAPF-pna) in mixtures of H_2O/DMF could be improved:

$$\text{suc-Ala-Ala-Pro-Pro-}p\text{-nitroanilide} \xrightarrow[\text{mutant subtilisin E}]{H_2O} \text{components}$$

In doing so, a randomly mutated mini-DNA library of subtilisin E was first created by inserting a PCR-amplified gene fragment into *B. subtilis* expression vector pKWC [12]. The mutant bacteria were first blotted on a nitrocellulose filter and transferred to DMF-containing agar plates without damaging the *B. subtilis* host. A relatively crude screening system was developed on these plates containing DMF and casein (a milk protein). The latter was used for the purpose of detecting positive mutants. Enhanced hydrolytic activity was identified by simple visual discrimination of halos produced around individual colonies on the agar plates containing casein. A total number of 300 *B. subtilis* clones were produced, and of these 72% showed visible halos. Twenty-seven clones which produced halos larger than that of the wild-type were scrutinised more closely for proteolytic activity on the specific substrate sAAPF-pna. One mutant exhibited significantly higher activity than the wild-type enzyme in both aqueous buffer and in the presence of 10% DMF. DNA sequencing revealed a substitution at nucleotide position 762 (A by G), which means that at amino acid position 103 of the wild-type enzyme glutamine was replaced by arginine. Another base substitution in the coding region led to a "silent" mutation (replacement of A by C), producing no change in the amino acid sequence. The corresponding gene was then used as the template for a second round of mutagenesis by error-prone PCR. The resulting variants were screened by the same procedure, leading to the identification of a new mutant exhibiting similar activity in aqueous buffer, but higher activity in the presence of DMF. Specifically, it turned out to be 38 times more active than wild-type subtilisin E in 85% DMF. Sequencing of the gene showed that an aspartate to asparagine substitution had occurred at position 60 of the enzyme. Although the information obtained by sequencing was not actually exploited in the efforts to generate higher enzyme activity, it was of interest with respect to structure/activity relations. Indeed, the substitutions are located near the substrate binding pocket or near the active site. However, one should not draw the general conclusion that positive mutants are always the result of amino acid substitutions near the active site of an enzyme. Indeed, many examples are now known in which such substitutions occur on the surface of the enzyme [19, 41, 55]. Importantly, this even applies to mutant enzymes displaying enhanced enantioselectivity, an observation that is of substantial theoretical interest [18] (cf. Sect. 3.2).

In a follow-up study more extensive mutagenesis experiments were performed [60]. Following each mutagenesis and screening step, the mutant producing the largest halo on the DMF/casein plates was isolated and used in the next random mutagenesis reaction. Starting with a mutant containing four effective amino acid substitutions produced by random mutagenesis, six additional mutations were brought about during three further sequential rounds of mutagenesis screening, a process which generated about 4000 colonies (mutants). The best mutant (PC3) was found to hydrolyse the model peptide substrate sAAPM-pna about 256 times more efficiently than wild-type subtilisin E in 60% DMF, i. e. $k_{cat}/K_m = 256$. In 85% DMF a value of $k_{cat}/K_m = 131$ was determined. The wild-type subtilisin E and several of the mutants were also tested as catalysts in the synthesis of poly(L-methionine) from L-methionine methyl ester in 70% DMF (Fig. 5). Whereas under these conditions the wild-type enzyme produced no detectable amount of polymer, the PC3 mutant led to significant amounts of product.

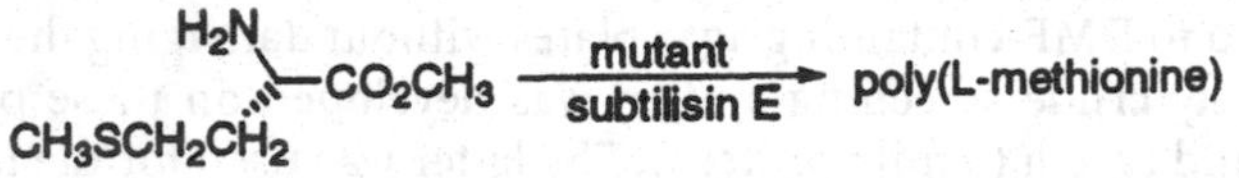

$$ \underset{CH_3SCH_2CH_2}{\overset{H_2N}{\diagdown}}\!\!-CO_2CH_3 \xrightarrow[\text{subtilisin E}]{\text{mutant}} poly(L\text{-methionine}) $$

Fig. 5. Polymerizing amide formation catalyzed by mutant subtilisin E

Even more active subtilisin mutants were obtained upon further rounds of mutagenesis [60]. The final 'product' was an enzyme that is 500-fold more active in 60% DMF than the wild-type subtilisin E. The overall process needed to accomplish this impressive result involved about 10.000 mutants, all of which were screened.

Although the success of these studies cannot and need not be disputed, the simple screening system actually used in the identification of the best mutants is based on the assumption that enzyme activity in casein hydrolysis parallels activity in the model reaction(s). However, screening is more precise and efficient if performed directly on the substrate of interest. This was accomplished in subsequent studies using other substrates. One such substrate concerns a cephalosporin-derived antibiotic [61]. During the synthesis of certain derivatives required in the synthetic sequence leading to the final drug, p-nitrobenzyl alcohol (pNB-OH) is routinely used to protect carboxylic acid functionality. Large amounts of zinc salts are employed in the final deprotection step [62]. In order to develop an economically attractive alternative, the possibility of enzymatic deprotection based on an esterase was explored [62] (Fig. 6).

Extensive searching of microorganisms led to the identification of "p-nitro-benzyl esterase" (p-NB esterase) from B. subtilis, which resulted in high yields of product in the deprotection step [62]. Nevertheless, the rate of the reaction was too slow, among other factors, to compete effectively with the traditional chemical method of deprotection.

Therefore, an attractive goal was to create a more active pNB esterase by directed evolution. Since the protected form of the antibiotic is poorly soluble in water, it was clear that the mutant enzyme would need to function well in a polar organic solvent. Unfortunately, the pNB esterase is very sensitive to organic solvents. Since the wild-type esterase is not secreted by the E. coli cells in which it is produced, the screening strategy used in the previous evolutionary experiments with subtilisin E could not be applied. A colorimetric signal detected by UV-visible spectroscopy would be ideal, since such techniques are well-known in biochemistry and combinatorial chemistry [63]. However, starting material and products (Fig. 6) do not differ significantly in their absorption properties. Thus, a compromise was made in that the analogous p-nitrophenol ester was prepared and used in the screening system, hoping that positive mutants would also be effective in the hydrolysis of the actual substrate employed in the industrial process (Fig. 7) [61]. In doing so, a commercially available microtiter plate with 96 wells was used, p-nitrophenol being the compound detected on the basis of its UV-visible absorption band.

In the first round of mutagenesis about 1000 colonies were produced and screened for activity on the p-NPA substrate in 20% DMF. About 33 mutants appeared to be better than the wild-type and of these the 3 best were studied more closely. Only one of them showed a higher total activity than the wild-type

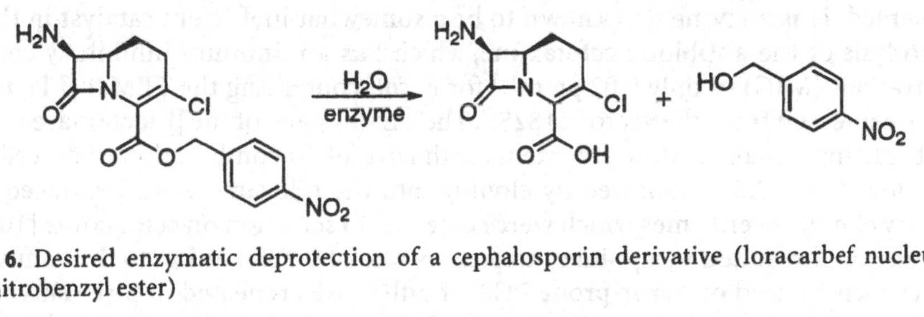

Fig. 6. Desired enzymatic deprotection of a cephalosporin derivative (loracarbef nucleus-p-nitrobenzyl ester)

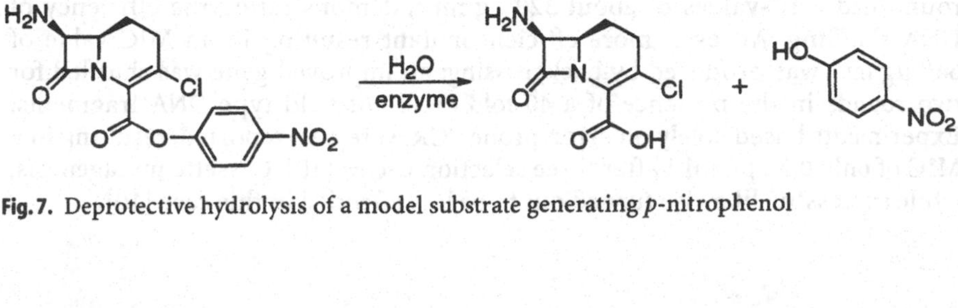

Fig. 7. Deprotective hydrolysis of a model substrate generating p-nitrophenol

on the actual p-NB substrate. The gene corresponding to this mutant was then used as the starting point for the second cycle mutagenesis in which 2800 mutants were studied. A third and a fourth generation of mutants (1500 and 7400 members respectively) were then produced and screened on the p-NP substrate. Sixty-four positive clones were tested on the actual p-NB substrate. Five of the mutants showed significant increases in activity. The best variant (4–54B9) displayed a 15-fold activity enhancement over the wild-type [61].

Further dramatic improvements were achieved upon applying DNA-shuffling to the genes of the five best mutants [64]. They were collected in the test tube and fragmented with an enzyme that cuts the DNA randomly (Sect. 2.5.1). Following reassembly the combined genes were inserted back into the plasmid and expressed in E. coli. Only 400 colonies were screened, resulting in the identification of 8 mutants showing significantly enhanced performance relative to the 5 immediate parents. The total activity of the best mutant was measured to be more than 50 times as high as that of the wild-type enzyme. It is interesting to note that the "yield" of positive mutants in the DNA shuffling experiments is about 20-fold higher than that obtained by screening the mutants with point mutations which were generated by error-prone PCR. Thus, DNA shuffling can enhance the efficiency of directed evolution by making use of the information present in mutant genes resulting from error-prone PCR. In this thorough study, the information obtained by sequencing was discussed in detail [64]. In contrast to the subtilisin case [60], none of the amino acid substitutions occur in segments of the esterase in which interaction with the bound substrate is predicted. It is therefore clear that application of site-directed mutagenesis would have been much more laborious, success being less than certain.

The principle of DNA shuffling in the evolution of improved enzymes was first illustrated in a seminal paper in 1994 [10]. To organic chemists this publication might appear to be of little relevance, but in fact it has ramifications for organic chemistry as well. In this paper a model study relating to β-lactamase is presented. This enzyme was known to be a somewhat inefficient catalyst in the hydrolysis of the antibiotic cefotaxime, which has a minimum inhibitory concentrations (MIC) of only 0.02 μg ml^{-1} for E. coli containing the TEM-1 β-lactamase expressed from the vector p182Sfi. The TEM-1 gene of the β-lactamase was first split into small random fragments with DNaseI. Recombination as described above (Sect. 2.5.1) followed by cloning into the relevant vector produced a library of mutant enzymes which were subjected to selection on cefotaxime [10]. This method resulted in a point mutagenesis rate of 0.7%, similar to the control experiments based on error-prone PCR. Shuffling was repeated in three successive rounds. After each round mutants with improved resistance were selected by plating on increasing levels of cefotaxime. Several colonies from the third round had MIC-values of about 320 μg ml^{-1}, demonstrating the efficiency of DNA shuffling. An even more efficient mutant resulting in an MIC value of 640 μg ml^{-1} was produced by backcrossing; an improved gene was shuffled for two rounds in the presence of a 40-fold excess of wild-type DNA fragments. Experiments based solely on error-prone PCR were also reported, resulting in a MIC of only 0.32 μg ml^{-1} after three selection cycles [10]. Cassette mutagenesis, while successful, likewise turned out to be less effective in this case [10].

The potential application of DNA shuffling in environmental chemistry [65] (enzymatic degradation of pollutants) was illustrated in a study concerning the molecular evolution of an arsenate detoxification pathway [66]. Accordingly, the activity of arsenate reductase was increased 40-fold. The detection method was again based on a crude but effective biological selection test, not on enzyme screening. Specifically, culture turbidity as measured in Klett units was recorded as a function of arsenate concentration [66]. Although the usefulness of the final product of DNA shuffling was not demonstrated in a real application [66], it is likely that the method will have a significant impact on environmental biotechnology [65]. Indeed, the task of selecting genes for a given purpose will be easier in the future due to the increasing availability of relevant information on microbial catabolic reactions, e. g., on the World Wide Web [67]. An example is the University of Minnesota Biocatalysis/Biodegradation Database [68].

DNA shuffling has also been applied in other systems [35, 37, 69], e. g. in improving whole cell fluorescence by the green fluorescent protein (GFP) [70]. This protein is often employed as a reporter for gene expression and regulation and is therefore of importance in developmental and cell biology, drug screening and diagnostic assays [71]. In this particular study the improved variants were identified by simple visual screening based on intrinsic fluorescence of the mutant proteins under UV illumination [70]. Essentially the same screening system was also used in related investigations directed towards improving the properties of GFP, including suppressed thermosensitivity [72]. In these studies error-prone PCR or combinatorial mutagenesis was used. It is also interesting to note that DNA shuffling has been reported to be successful in the evolution of a fucosidase from a galactosidase [69b].

A number of other reports concerning in vitro evolution of enzymes have appeared [56, 73], but again application in organic synthesis was seldom the goal. The same applies to studies based on cassette mutagenesis [74]. Nevertheless, these studies are important in other respects, e.g. in the development of methods which may actually be of direct or indirect use to organic chemists. The StEP-method of mutagenesis has been applied successfully in the evolution of an esterase [39].

3.2
Enzymes with Improved Enantioselectivity

As outlined in the Introduction (Sect. 1), enantioselective enzymes offer genuine opportunities to organic chemists, the major problem being the low degrees of enantioselectivity in many specific cases of interest [4]. Of course, the alternative is homogeneous transition metal catalysis [2, 5, 75], but this option is also in no way general. Principally, in vitro evolution of enantioselective biocatalysts constitutes a potential way out of the dilemma, at least for those reaction types which are known to be catalysed by enzymes. Nevertheless, the challenge involved is particularly difficult, because enantioselectivity is not a simple parameter to deal with, especially if high levels are striven for. Moreover, at the time when chemists began to apply in vitro evolution to the creation of enantioselective biocatalysts, there were no screening systems available. Certainly, the first law of in vitro evolution of enzymes, "you get what you screen for" [19, 41, 55, 58] applies all the more to enantioselectivity [76]. It is a painful reminder that a great deal of research in this particular area of analytical chemistry is necessary before significant advances in actually obtaining enantioselective enzymes can be expected.

In a classical study the lipase-catalysed enantioselective hydrolysis of racemic p-nitrophenyl-2-methyldecanoate was chosen as the test reaction [15] (Fig. 8). The p-nitrophenyl ester was employed in the kinetic resolution instead of the methyl or ethyl ester, in order to make screening possible [76] (see below). The lipase from the bacterium *Pseudomonas aeruginosa* PA01 [77] was used as the enzyme [15]. The wild-type enzyme shows an enantioselectivity (*ee*) of only 2 % in favour of the (*S*)-configured 2-methyldecanoic acid, which means that the enzyme had essentially no preference for either of the enantiomeric forms.

The enzyme in its mature form is composed of 285 amino acids [77]. Upon applying the algorithm described in Sect. 1, it can be calculated that the number of mutant lipases in which one amino acid per enzyme molecule is substituted by one of the remaining 19 amino acids is 5415 [15]. It was of great interest to see whether such small libraries (or even smaller ones) would contain lipases with significantly improved enantioselectivity in the model reaction [15, 18, 43, 76].

The lipase gene consisting of 933 base pairs was subjected to random mutagenesis using error-prone PCR [15]. In doing so, a relatively low mutation frequency was chosen so that statistically only one to two base substitutions per lipase gene were introduced. The mutated genes were then ligated into a suitable expression vector, amplified in *E. coli* and transformed into *P. aeruginosa*. Thereafter, the bacterial colonies (clones) were removed individually from the agar plates using toothpicks and grown in the wells of microtiter plates containing nutrient broth. This process turned out to be the slow step in the overall screening system and was therefore later automated. In the first cycle of mutagenesis about one thousand such colonies were harvested [15]. In order to screen these potential catalysts for enantioselectivity in the above asymmetric model reaction within a reasonable period of time, an efficient screening system based on the UV-visible absorption of the p-nitrophenolate anion at 410 nm was developed. Since the origin of this yellow-coloured anion is not defined when hydrolysing the racemate, the number of screening events was doubled. Thus, the 96 wells of commercially available microtiter plates were loaded with the culture

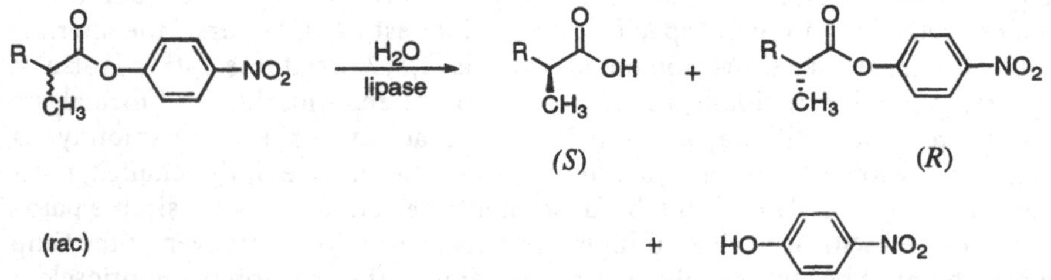

Fig. 8. Lipase-catalyzed kinetic resolution of a chiral ester

supernatants of the lipase mutants originating from the bacterial colonies together with 0.01 mol l^{-1} Tris/HCl buffer (pH 7.5). Thereafter, samples of enantiomerically pure (R)- and (S)-substrate dissolved separately in DMF were added pairwise. In this way 48 mutants per microtiter plate were screened. The enzyme-catalysed hydrolysis of each (R)/(S) pair was monitored by measuring the absorption of the p-nitrophenolate anion as a function of time. This required about 8–10 min per microtiter plate. About 500–600 mutants could be screened in one day by one person. Figure 9 shows the reaction profiles produced by the original wild-type enzymes and an improved mutant. Obviously, if the slopes of the lines corresponding to the (R)- and (S)-enantiomers are different, then this indicates that the enzyme prefers one of the enantiomers [15, 78]. In the worst situation the two lines coincide, which in fact was found for the vast majority of the mutants. As expected most of the mutants are the result of deleterious amino acid substitutions. Of about 1000 mutants screened in the first generation, 12 showed enhanced enantioselectivity. The exact enantioselectivity was then determined by hydrolysing the racemate in the presence of the corresponding mutant lipases and analysing the reaction products by conventional chiral gas chromatography. The best mutant in the first generation showed an *ee*-value of 31%.

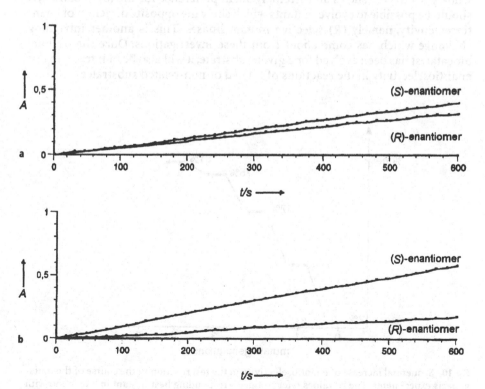

Fig. 9 a, b. Course of the lipase-catalyzed hydrolysis of the (R)- and (S)-ester as a function of time: **a** wild-type lipase from *P. aeruginosa*; **b** improved mutant in the first generation

The gene corresponding to the best mutant in the library of 1000 members was then chosen as the template for the second mutagenesis cycle, a process that was repeated until 4 generations of mutant lipases, each 1000 to 2400 in number, had been produced. The results summarised in Fig. 10 are remarkable especially in view of the fact that only four consecutive cycles of mutagenesis were traversed and only relatively few mutants were actually screened [15].

It was also interesting to observe that the mutant lipase which shows an *ee* of 81% in the model reaction of the *p*-nitrophenyl ester displays similar performance if the corresponding ethyl ester is used (80% *ee*) [76]. Nevertheless, it was not clear at this stage whether the degree of enantioselectivity would continue to climb in further mutagenesis experiments.

Ideas concerning further increases in enantioselectivity revolved around the following strategies [18, 79]: (1) more comprehensive screening, (2) production of further generations, (3) DNA-shuffling as an alternative mutagenesis method, and (4) saturation mutagenesis. Recent work indicates that combinations of these methods may well constitute the strategy of choice. Indeed, *ee*-values of >90% have been observed for the model reaction [18, 79]. More efforts are necessary to define the optimal strategy for the most efficient exploration of protein sequence space in obtaining highly enantioselective enzymes. Since the wild-type enzyme shows an extremely small preference for the (S)-substrate, it should be possible to evolve mutants which show the opposite direction of enantioselectivity, namely (R)-selective mutant lipases. This is another intriguing challenge which has come about from these investigations. Once the optimal biocatalyst has been evolved for a given substrate, it will also be of interest to test enantioselectivity in the reactions of related or non-related substrates.

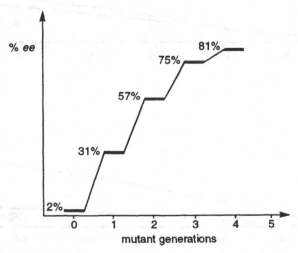

Fig. 10. Sequential increase of enantioselectivity in the test reaction in the course of the mutagenesis experiments. The *ee* values refer to the corresponding best mutant in a reaction with a conversion range of 20–30%; corresponding *E* values: 1.00, 2.10, 4.40, 9.40, and 11.3, respectively

Following the publication of the above study [15], another paper appeared reporting preliminary results concerning the directed evolution of an enantio-selective lipase [80]. In this case a lipase from *Pseudomonas fluorescens* was used in a different hydrolysis reaction. An *ee*-value of 25% was obtained. It is currently not clear whether an enhancement is possible [80], and if so, by which strategy.

In a brief report, random mutagenesis was successfully used to improve the enantioselectivity of a transaminase [81]. Transaminases are enzymes which are known to catalyse the conversion of ketones to the corresponding primary amines, the *ee*-values often being >95% [4, 82]. However, in the particular case of β-tetralone, an (*S*)-selective transaminase was found to show an *ee* of only 65%. Since the product is of commercial interest, a library of mutants was created, in which an enzyme showing an enantioselectivity of 98% in favour of the (*S*)-enantiomer was identified (Fig. 11) [81]. Unfortunately, details concerning the molecular biology and screening system in this interesting study were not reported.

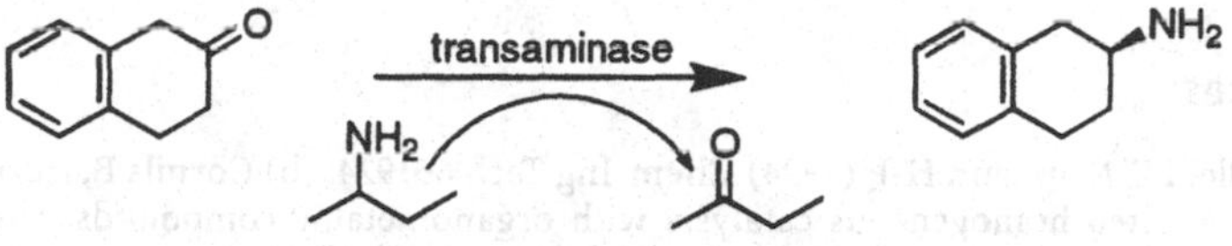

Fig. 11. Enantioselective transaminase-catalysed formation of a chiral amine

The studies relating to in vitro evolution of enantioselective enzymes, although limited in number, clearly point to enormous opportunities for organic chemists. It is also evident that in order to generalise, intensive research is necessary. The optimal strategy for exploring protein sequence space with respect to enantioselectivity still needs to be defined. However, it is already obvious that the solution to a given problem, as defined by the attainment of an arbitrary *ee*-value in a given reaction, is not unique. This means that we are not searching for a specific enzyme, i.e. for a single mutant enzyme having a certain amino acid sequence. Rather, more than one or perhaps even many mutants may fulfil the requirements specified by the organic chemist.

Chemists, biochemists and biologists can also learn a great deal from studies directed towards obtaining highly enantioselective enzymes, especially with respect to structure/activity/selectivity. For example, preliminary investigations show that amino acid substitutions on the surface of the mutant lipases from *P. aeruginosa*, i.e. at positions far removed from the active site, can lead to substantial improvements in enantioselectivity [18, 79].

4
Conclusions and Future Prospects

Nature synthesises biomolecules of amazing complexity by enzyme-catalysed reactions. Over evolutionary time, random mutations have occurred in DNA and were distributed by genetic recombination. Natural selection has created an impressive number of different proteins, all of them perfectly suited to fulfill their respective biological functions. Organic chemists have long recognised that these biocatalysts can also solve some of their problems. However, until now it had been a matter of trial and error to find and identify the right ones. Some biocatalysts are principally able to catalyse a given reaction, but they often lack a sufficient degree of stability, activity, substrate specificity, or enantioselectivity. The techniques of directed evolution now provide a tool to improve existing biocatalysts or create those with novel properties in a relatively short period of time. The basic strategy includes random mutagenesis to create libraries of mutated genes, exchange and reassembly of mutated gene fragments by recombinative methods like DNA-shuffling, and screening or selection to identify the best biocatalyst. With ever more applications appearing [79, 83], no serious doubt remains that applied molecular evolution technology will bring biological diversity in the form of whole families of catalysts into the test tubes of organic chemists.

5
References

1. (a) Gallei EF, Neumann H-P (1994) Chem-Ing Tech 66:924; (b) Cornils B, Herrmann WA (1996) Applied homogeneous catalysis with organometallic compounds, vols 1 and 2. VCH, Weinheim
2. Parshall GW, Ittel SD (1992) Homogeneous catalysis. Wiley, New York
3. Thomas JM, Thomas WJ (1997) Principles and practice of heterogeneous catalysis. VCH, Weinheim
4. (a) Davies HG, Green RH, Kelly DR, Roberts SM (1989) Biotransformations in preparative organic chemistry: the use of isolated enzymes and whole cell systems in synthesis. Academic Press, London; (b) Wong CH, Whitesides GM (1994) Enzymes in synthetic organic chemistry. Pergamon, Oxford; (c) Drauz K, Waldmann H (eds) (1995) Enzyme catalysis in organic synthesis. VCH, Weinheim; (d) Faber K (1997) Biotransformations in organic chemistry, 3rd edn. Springer, Berlin Heidelberg New York; (e) Johnson CR, Wells GW (1998) Curr Opin Chem Biol 2:70; (f) Fessner WD (1998) Curr Opin Chem Biol 2:85
5. Jacobsen EN, Finney NS (1994) Chem Biol 1:85
6. (a) Brandt S, Rossi RF, Dodds DR, Lopez JL (1990) WO-A9004643 A1 (Chem Abstr 113: 76620); (b) Schmid RD, Verger R (1998) Angew Chem Int Ed Engl 37:1608
7. (a) Fersht AR (1987) Biochemistry 26:8031; (b) Bryan PN (1987) Biotechnol Adv 5:221; (c) Gerlt GA (1987) Chem Rev 87:1079; (d) Knowles JR (1987) Science 236:1252; (e) Benkovic SJ, Fierke CA, Naylor AM (1988) Science 239:1105; (f) Wells JA, Estell DA (1988) Trends Biochem Sci 291; (g) Pantoliano MW, Whitlow M, Wood JF, Finzel BC, Gilliland GL, Poulos TL, Rollence ML, Bryan PN (1988) Biochemistry 27:8311; (h) Russell AJ, Fersht AR (1987) Nature (London) 328:496; (i) Holmquist M, Clausen IG, Patkar S, Svendsen A, Hult K (1995) J Protein Chem 14:217; (j) Beer HD, Wohlfahrt G, McCarthy JEG, Schomburg D, Schmid RD (1996) Protein Eng 9:507; (k) Hirose Y, Kariya K, Nakanishi Y, Kurono Y, Achiwa K (1995) Tetrahedron Lett 36:1063; (l) Haas MJ, Joerger RD, King G, Klein RR (1996) Ann N Y Acad Sci 799:115
8. Leung DW, Chen E, Goeddel DV (1989) Technique (Philadelphia) 1:11
9. (a) Eckert KA, Kunkel TA (1991) PCR Methods Appl 1:17; (b) Cadwell RC, Joyce GF (1994) PCR Methods Appl 3:136
10. Stemmer WPC (1994) Nature (London) 370:389
11. For example see: (a) Lewin B (1997) Genes VI. Oxford University Press and Cell Press, Oxford; (b) Stryer L (1995) Biochemistry. WH Freeman, New York

12. Chen K, Arnold FH (1991) Biotechnol 9:1073
13. Rubingh DN (1997) Curr Opin Biotechnol 8:417
14. Enzymes in "nature" can evolve much faster under certain conditions, e.g. in contaminated soils. An example is a phosphotriesterase, an enzyme discovered in a soil bacterium, which degrades certain pesticides. It is believed that this enzyme evolved during the last 50 years from a homologous sequence in the *E. coli* bacterium. Scanlan TS, Reid RC (1995) Chem Biol 2:71
15. Reetz MT, Zonta A, Schimossek K, Liebeton K, Jaeger K-E (1997) Angew Chem Int Ed Engl 36:2830
16. Arnold FH (1996) Chem Bioeng Sci 23:5091
17. (a) Heinz DW, Baase WA, Matthews BW (1992) Proc Natl Acad Sci USA 89:3751; (b) Poteete AR, Rennel D, Bouvier SE (1992) Proteins Struct Funct Genet 13:38
18. Liebeton K (1998) PhD Thesis, Ruhr-Universität Bochum
19. Arnold FH (1998) Acc Chem Res 31:125
20. Trower MK (ed) (1996) In vitro mutagenesis protocols. Humana Press, New Jersey
21. O'Donohue MJ, Kneale GG (1996) Mol Biotechnol 6:179
22. (a) Higuchi R, Krummel B, Saiki RK (1988) Nucleic Acids Res 16:7351; (b) Ho S, Hunt HD, Horton RM, Pullen JK, Pease LR (1989) Gene 77:51
23. Urban A, Neukirchen S, Jaeger KE (1997) Nucleic Acids Res 25:2227
24. Kegler-Ebo DM, Docktor CM, DiMaio D (1994) Nucleic Acids Res 22:1593
25. (a) Barettino D, Feigenbutz M, Valcárel R, Stunnenberg HG (1994) Nucleic Acids Res 22:541; (b) Barik S (1995) Mol Biotechnol 3:1
26. Reidhaar-Olson JF, Sauer RT (1988) Science 241:53
27. (a) MacBeath G, Kast P, Hilvert D (1998) Science 279:1958; (b) MacBeath G, Kast P, Hilvert D (1998) Protein Sci 7:1757
28. Kast P, Hilvert D (1997) Curr Opin Struct Biol 7:470
29. Crameri A, Stemmer WPC (1995) BioTechniques 18:194
30. Eckert KA, Kunkel TA (1990) Nucleic Acids Res 18:3739
31. Cadwell RC, Joyce GF (1995) Mutagenic PCR. In: Dieffenbach CW, Dveksler GS (eds) PCR Primer: a laboratory manual. CSHL Press, Cold Spring Harbor, p 583
32. Miller JH (1992) A short course in bacterial genetics. CSHL Press, Cold Spring Harbor
33. Greener A, Callahan M, Jerpseth B (1996) An efficient random mutagenesis technique using an *E. coli* mutator strain. In: Trower MK (ed) In vitro mutagenesis protocols. Humana Press, New Jersey
34. Greener A, Callahan M (1994) Strategies 7:32
35. Stemmer WPC (1994) Proc Natl Acad Sci 91:10,747
36. Zhao H, Arnold FH (1997) Nucleic Acids Res 25:1307
37. Crameri A, Raillard SA, Bermudez E, Stemmer WPC (1998) Nature 391:288
38. Patten PA, Howard RJ, Stemmer WPC (1997) Curr Opin Biotechnol 8:724
39. Zhao H, Giver L, Shao Z, Affholter JA, Arnold FH (1998) Nature Biotechnol 16:258
40. Shao Z, Zhao H, Giver L, Arnold FH (1998) Nucleic Acids Res 26:681
41. Kuchner O, Arnold FH (1997) Trends Biotechnol 15:523
42. Hannig G, Makrides SC (1998) Trends Biotechnol 16:54
43. Reetz MT, Jaeger KE (1998) Chem Phys Lipids 93:3
44. Sandkvist M, Bagdasarian M (1996) Curr Opin Biotchnol 7:505
45. de Vos WM, Kleerebezem M, Kuipers OP (1997) Curr Opin Biotchnol 8:547
46. Hollenberg CP, Gellissen G (1997) Curr Opin Biotchnol 8:554
47. Zhao H, Arnold FH (1997) Curr Opin Struct Biol 7:480
48. (a) Fastrez J (1997) Mol Biotechnol 7:37; (b) O'Neil KT, Hoess RH (1995) Curr Opin Struct Biol 5:443 (c) Smith GP, Petrenko VA (1997) Chem Rev 97:391
49. Reetz MT, Becker M, Kühling K, Holzwarth A, Angew Chem (in press)
50. For example see: (a) Mosbach K (ed) (1987/88) Methods in enzymology: immobilized enzymes and cells, parts B–D, vols 135–137. Academic Press, San Diego; (b) Tanaka A, Tosa T, Kobayashi (eds) (1993) Industrial application of immobilized biocatalysts, vol 16. Marcel Decker, New York; (c) Kennedy JF, Melo EHM, Jumel K (1990) Chem Eng Prog 7:81;

(d) Okahata Y, Mori T (1997) Trends Biotechnol 15:50; (e) Parthasarathy RV, Martin CR (1994) Nature 369:298
51. (a) Reetz MT (1997) Adv Mater 9:943; (b) Reetz MT, Zonta A, Simpelkamp J (1995) Angew Chem Int Ed Engl 34:301
52. Khalaf N, Govardhan CP, Lalonde JJ, Persichetti RA, Wang Y-F, Margolin AL (1996) J Am Chem Soc 118:5494
53. (a) Reiter Y, Brinkmann U, Jung S-H, Pastan I, Byungkook L (1995) Protein Eng 8:1323; (b) Noda Y, Fukada Y, Segawa S (1997) Biopolymers 41:131; (c) Fairman R, Chao H-G, Lavoie TB, Villafranc JJ, Matsueda GR, Novotny J (1996) Biochemistry 35:2824
54. Wong C-H, Chen S-T, Hennen WJ, Bibbs JA, Wang Y-F, Liu JL-C, Pantoliano MW, Whitlow M, Bryan PN (1990) J Am Chem Soc 112:945
55. Shao Z, Arnold FH (1996) Curr Opin Struct Biol 6:513
56. (a) Matsumura M, Aiba S (1985) J Biol Chem 260:15,298; (b) Bryan PN, Rollence ML, Pantoliano MW, Wood J, Finzel BC, Gilliland GL, Howard AJ, Poulos TL (1986) Proteins Struct Funct Genet 1:326; (c) Wells JA (1990) Biochemistry 29:8509; (d) Joyet P, Declerck N, Gaillardin C (1992) Biotechnol 10:1579; (e) Haruki M, Noguchi E, Akasako A, Oobatake M, Itaya M, Kanaya S (1994) J Biol Chem 269:26,904; (f) Rellos P, Scopes RK (1994) Protein Expres Purif 5:270; (g) Pjura P, Matsumura M, Baase WA, Matthews BW (1993) Protein Sci 2:2217; (h) Kotsuka T, Akanuma S, Tomuro M, Yamagishi A, Oshima T (1996) J Bacteriol 178:723; (i) Okada Y (1995) Biosci Biotechnol Biochem 59:1152
57. Liao H, McKenzie T, Hageman R (1986) Proc Natl Acad Sci USA 83:576
58. Zhao H, Arnold FH (1997) Curr Opin Struct Biol 7:480
59. Janes LE, Kazlauskas RJ (1997) J Org Chem 62:4560
60. (a) Chen K, Arnold FH (1993) Proc Natl Acad Sci USA 90:5618; (b) You L, Arnold FH (1996) Protein Eng 9:77
61. Moore JC, Arnold FH (1996) Nature Biotechnol 14:458
62. (a) Brannon DR, Mabe JA, Fukuda DS (1976) J Antibiotics 29:121; (b) US Patent (1975) 3,725,359; (c) Zock J, Cantwell C, Swartling J, Hodges R, Pohl T, Sutton K, Rosteck P Jr, McGilvray D, Queener S (1994) Gene 151:37
63. (a) Devlin JP (1997) High throughput screening. Marcel Decker, New York; (b) Czarnik AW, DeWitt SH (1997) A practical guide to combinatorial chemistry. ACS, Washington, DC
64. (a) Zhao H, Arnold FH (1997) Proc Natl Acad Sci USA 94:7997; (b) Moore JC, Jin H-M, Kuchner O, Arnold FH (1997) J Mol Biol 272:336
65. Minshull J (1995) Chem Biol 2:775
66. Crameri A, Dawes G, Rodriguez E Jr, Silver S, Stemmer WPC (1997) Nat Biotechnol 15:436
67. Wackett LP (1997) Nat Biotechnol 15:415
68. Ellis LBM, Wackett LP The University of Minnesota Biocatalysis/Biodegradation Database on the WWW: http://dragon.labmed.umn.edu/~lynda/index.html
69. (a) Stemmer WPC (1995) Bio/Technology 13:549; (b) Zhang J, Dawes G, Stemmer WPC (1997) Proc Natl Acad Sci USA 94:4504
70. Crameri A, Whitehorn E, Tate E, Stemmer WPC (1996) Nat Biotech 14:315
71. (a) Cubitt AB, Heim R, Adams SR, Boyd AE, Gross LA, Tsien TY (1995) Trends Biochem Sci 20:448; (b) Coxon A, Bestor TH (1995) Chem Biol 2:119
72. (a) Delagrave S, Hawtin RE, Silvia CM, Yang MM, Youvan DC (1995) Bio/Technology 13:151; (b) Heim R, Tsien RY (1996) Curr Biol 6:178; (c) Ehrig T, O'kane DJ, Prendergast FG (1995) FEBS Lett 367:163
73. (a) Gulick AM, Fahl WE (1995) Proc Natl Acad Sci USA 92:8140; (b) Shinkai A, Hirano A, Aisaka K (1996) J Biochem 120:915; (c) Strausberg SL, Alexander PA, Gallagher DT, Gilliland GL, Barnett BL, Bryan PN (1995) Biotechnol 13:669; (d) Black ME, Newcomb TG, Wilson HMP, Loeb LA (1996) Proc Natl Acad Sci USA 93:3525; (e) Tamakoshi M, Yamagishi A, Oshima T (1995) Mol Microbiol 16:1031; (f) Kano H, Taguchi S, Momose H (1997) Appl Microbiol Biotechnol 47:46; (g) Beuve A, Danchin A (1992) J Mol Biol 225:933; (h) Hawrani AS, Sessions RB, Moreton KM, Holbrook JJ (1996) J Mol Biol 264:97; (i) Sidhu SS, Borgford TJ (1996) J Mol Biol 257:233; (j) Ohnuma S-I, Nakazawa T, Hemmi H, Hallberg A-M, Koyama T, Ogura K, Nishino T (1996) J Biol Chem 271:10,087; (k) Sousa R, Padilla R

(1995) EMBO J 14:4609; (l) Widersten M, Mannervik B (1995) J Mol Biol 250:115; (m) Gaskin DJH, Bovagnet AH, Turner NA, Vulfson EN (1997) Biochem Soc Trans 25:15S

74. (a) Oliphant AR, Nussbaum AL, Struhl K (1986) Gene 44:177; (b) Derbyshire KM, Salvo JJ, Grindley NDF (1986) Gene 46:145
75. Berrisford DJ, Bolm C, Sharpless KB (1995) Angew Chem Int Ed Engl 34:1059
76. Schimossek K (1998) PhD Thesis, Ruhr-Universität Bochum
77. (a) Jaeger K-E, Schneidinger B, Liebeton K, Haas D, Reetz MT, Philippou S, Gerritse G, Ransac S, Dijkstra BW (1996) In: Nakazawa T, Furukawa K, Haas D, Silver S (eds) Molecular biology of Pseudomonads. ASM Press, Washington, p 319; (b) Jaeger K-E, Liebeton K, Zonta A, Schimossek K, Reetz MT (1996) Appl Microbiol Biotechnol 46:99
78. (a) Jaeger K-E, Reetz MT (1998) Trends Biotechnol 16:396; (b) Reetz MT, Zonta A, Schimossek K, Liebeton K, Jaeger K-E: patent application DEA19731990.4
79. Jaeger K-E, Liebeton K, Zonta A, Schimossek K, Reetz MT (submitted)
80. Bornscheuer UT, Altenbuchner J, Meyer HH (1998) Biotech Bioeng 58:554
81. Zhu X, Lewis CM, Haley MC, Bhatia MB, Pannuri S, Kamat S, Wu W, Bowen ARSTG (1997) IBC's 2nd Annual Symposium on Exploiting Enzyme Technology for Industrial Applications, Feb 20–21 1997, San Diego USA
82. Matcham GW, Bowen ARSTG (1996) CHIM OGGI 14:20
83. (a) Kazlauskas RJ, Weber HK (1998) Curr Opin Chem Biol 2:121; (b) Arnold FH (1998) Nat Biotechnol 16:617; (c) Tawfik DS, Griffith AD (1998) Nature Biotechnol 16:652; (d) Buchholz F, Angrand P-O, Stewart, AF (1998) Nature Biotechnol 16:657; (e) Kumamaru T, Suenaga H, Mitsuoka M, Watanabe T, Furukawa K (1998) Nat Biotechnol 16:663

Catalytic Antibodies for Organic Synthesis

Jean-Louis Reymond

Department of Chemistry & Biochemistry, University of Bern, Freiestrasse 3, CH-3012 Bern, Switzerland. E-mail: jean-louis.reymond@ioc.unibe.ch

Monoclonal antibodies with strong binding affinities tailored to any molecular target can be produced from immunized mice using hybridoma technology. Due to their directable binding specificities and ease of preparation, monoclonal antibodies are used in the daily practice of biological research as well as in many diagnostic applications. Since 1986 monoclonal antibodies with catalytic properties have been prepared by immunizing mice with stable transition state analogs of chemical reactions, thereby providing a versatile source of novel biocatalysts. This review discusses catalytic antibodies with respect to organic synthetic reactions.

Keywords: Catalytic antibodies, Biotransformations, Enantioselective reactions, Transition state analogs, High throughput screening.

Topics in Current Chemistry, Vol. 200
© Springer Verlag Berlin Heidelberg 1999

1
Introduction

Catalytic antibodies were first reported in 1986 by Lerner and Schultz for the hydrolysis of simple esters. Since then over 100 different reactions have been catalyzed by antibodies, from enantioselective preparative reactions to prodrug activations in vivo [1]. New methods have been developed, including specific immunization protocols, direct screening assays for catalysis, and manipulations with recombinant antibodies via phage display. This article gives an overview of work done towards applying catalytic antibodies for synthetic organic reactions by our and other laboratories.

1.1
Antibodies

Antibodies, also called immunoglobulins, are homodimeric globular proteins with a molecular weight of approximately 150 KDa (~ 1330 amino acids). Each unit consists of a heavy chain with domains VH (genetically variable domain), CH1, CH2, and CH3 (constant domains) and a light chain with domains VL (genetically variable domain) and CL (constant domain). Disulfide bridges link covalently heavy and light chain and both HL-pairs together. Each individual domain contains approximately 110 aminoacids and is folded according to a common structural motif known as the immunoglobin fold, which contains two characteristic anti-parallel β-sheets and an intra-domain disulfide bridge (Fig. 1).

Antibodies are produced by the immune system, an ensemble of cells found in all vertebrate animals, as part of a molecular defense mechanism against pathogens such as viruses and bacteria [2]. They possess two identical binding sites of approximately 700 Å^2, which are formed by each pair of variable domains VH-VL. These binding sites allow antibodies to bind tightly ($10^6 - 10^{12}$ M^{-1}) to structural elements (antigenic determinants) present on the pathogen, which thereby becomes tagged for destruction by the immune system.

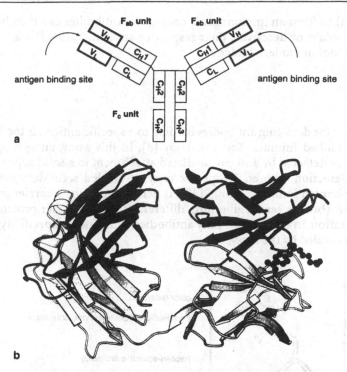

Fig. 1. a Structure of antibodies. V_H and V_L, C_L and C_H1 form a so called Fab fragment. The C_H2-C_H3 dimer forms the Fc fragment. F_c is normally glycosylated (not shown). **b** Ribbon diagram of Fab-fragment of catalytic antibody 48G7 with bound hapten (see Sect. 4.1). Heavy chain domains C_H1 and V_H *dark gray*, light chain domains C_L and V_L *light gray*, hapten in *ball-and-sticks model*. Produced from X-ray structure coordinates in pdb

Importantly, binding specificity and affinity depends on the amino-acid sequence of the antibody. As a consequence of somatic mutations, each anti-body-producing immune cell (B-cell) is genetically distinct and produces an antibody with unique amino-acid sequence in the variable domains VH and VL encoding the antibody combining site. Hence each B-cell produces an antibody with unique binding specificity. Their total number is estimated at 10^9 in the primary repertoire of B-cells. During the primary immune response B-cells producing antigen-specific antibodies are activated and multiply. A second contact with the antigen induces a process of maturation, whereby activated B-cells undergo further random mutations, which may provide an additional 10^4 structural variations for each antibody. In turn daughter cells with mutations increasing antibody-antigen binding are selected.

Antibodies binding to proteins, viruses, or cells with high specificity and affinity can be readily prepared by immunization, which simply consists of injecting the antigen of interest into an experimental animal. Antibodies binding to small organic molecules, called haptens, are also accessible by co-valently attaching haptens to a carrier protein, usually KLH (keyhole limpet hemocyanin) to form an immunogenic conjugate. Antibodies can then be used to identify, isolate, or deactivate their respective antigen. Antibodies are an indispensable tool in modern biology.

1.2
ELISA

The basic tool for detecting antibodies binding to a specific antigen is the ELISA, or *Enzyme Linked Immuno-Sorbent Assay* [3]. In this assay, antigen-specific antibodies are detected by antigen-mediated attachment to a solid support and secondary detection with an Fc-specific enzyme-labeled secondary antibody (Fig. 2). For hapten immunization, ELISA is performed using a carrier protein, typically BSA (bovine serum albumin), different from the carrier protein used for immunization. In this manner only antibodies with binding specificity to the hapten are revealed by the assay.

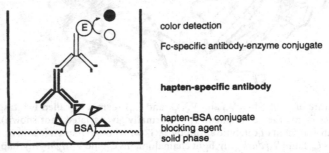

color detection

Fc-specific antibody-enzyme conjugate

hapten-specific antibody

hapten-BSA conjugate
blocking agent
solid phase

Fig. 2. Principle of ELISA. Antibodies elicited against a hapten-KLH conjugate are tested against a hapten-BSA conjugate. Only hapten-specific antibodies are detected. A blocking agent is used to prevent direct binding of antibodies to the solid support

1.3
Practical Considerations Regarding Antibody Production

Antigen-specific antibodies produced by B-cells are secreted and appear in the blood serum of an immunized animal, from where they can be isolated as polyclonal mixtures, usually from rabbits. Alternatively, isolation of B-cells from the spleen of an immunized mouse and fusion with myeloma (tumor cells) yields "hybridoma", which are antibody-producing cells that can be propagated in vitro [4]. Each hybridoma cell line secreting an antigen-specific antibody can be isolated and provides a source of a single monoclonal antibody. Polyclonal and monoclonal antibodies to a variety of proteins are commercially available, and many industrial and academic laboratories offer to produce polyclonal or monoclonal antibodies.

Despite their variable binding specificities determined by the structure of their binding sites, antibodies produced from hybridoma or rabbit serum are relatively homogeneous in their physico-chemical properties. They are stable at

37 °C for prolonged periods in aqueous physiological buffer, and can resist large variations of pH (3 – 11). A limited compatibility with organic solvents is observed. In our experience, straight, non-optimized in vitro culture of hybridoma followed by a single chromatographic separation on an antibody-specific protein-G column yields between 5 and 30 mg of pure antibody per liter of cell culture. Specialized reactors are commercially available that will yield up to 40 mg l^{-1} day^{-1} of antibody. These advantages of stability and ease of large scale production make antibodies extremely attractive to work with, and have contributed to making experimentation with catalytic antibodies easily accessible to chemists.

1.4
Catalytic Antibodies

Linus Pauling in 1945 recognized that antibodies and enzymes may be distinguished according to affinity, the former binding preferentially to stable ground state molecules, and the latter to transition states of reactions [5]. Thus, an enzyme active site could be viewed as a binding pocket for the reaction's transition state, leading to the speculation that this pocket might also bind stable compounds if those were structurally closely related to this transition state. This insight lead to the development of stable transition state analogs as enzyme inhibitors. For example, phosphonates as stable analogs of the tetrahedral intermediate in ester and amide hydrolysis provided inhibitors for esterases and proteases.

Applying this reasoning in reverse, Jencks predicted that antibodies with high binding affinity for a stable transition state analog of a reaction should then catalyze the reaction like enzymes [6]. Indeed, antibodies raised against a particular antigen generally do not show absolute binding specificity for the antigen but may sometimes bind to related structures. In 1986 Lerner et al. and Schultz et al. reported independently that immunization against phosphonate haptens yielded monoclonal antibodies which catalyzed the corresponding ester and carbonate hydrolysis [7]. The phosphonate group acted as a stable analog of the high energy tetrahedral intermediate involved in ester hydrolysis (Fig. 3).

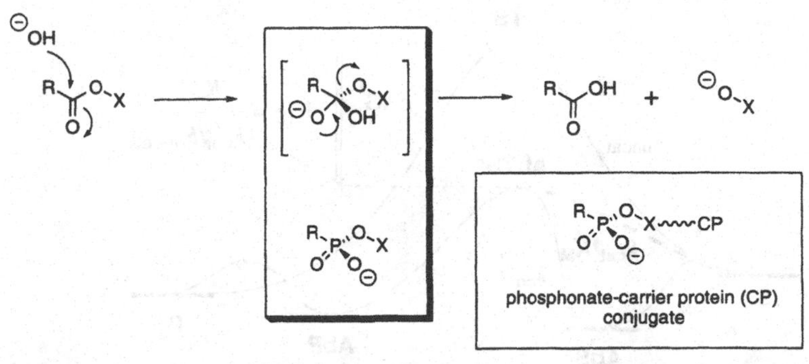

Fig. 3. Mechanism of ester hydrolysis involves a high energy tetrahedral intermediate. Phosphonates accurately mimic this intermediate and can be used as haptens to induce catalytic antibodies for ester hydrolysis

These experiments represented a dramatic breakthrough and established immunization with stable transition state analogs as a new paradigm in enzyme design [8]. They were followed by many others, encompassing enzyme-like as well as genuinely non-enzymatic processes. Research has focused on extending the scope of reactions accessible to catalytic antibodies, as well as on improving the methods by which these can be prepared.

1.5
K_{TS} as a Measure of Catalytic Efficiency

The concept of transition state binding applies generally to chemical catalysis. K_{TS} is defined as the dissociation constant of the complex between catalyst and a reaction's transition state. It expresses Pauling's postulate in quantitative terms, and relates directly to the catalytic effect $\Delta\Delta G^{\#}$ (Fig. 4).

This analysis can be applied to enzymatic as well as to simple chemical transformations [9–11], for uni- and multi-substrate [12] reactions according to Eqs. (1) and (2). $\Pi_N K_M$ denotes the product of Michealis-Menten constants for all substrates. In this analysis one assumes that kinetics follow the Michealis-Menten model, which is the case for most antibody-catalyzed processes discussed below. The k_{cat} denotes the rate constant for reaction of the antibody-substrate complex, K_M its dissociation constant, and k_{uncat} the rate constant for reaction in the medium without catalytic antibody or when the antibody is quantitatively inhibited by addition of its hapten. In several examples given below there is virtually no uncatalyzed reaction. This of course represents the best case.

$$K_{TS} = k_{uncat}/(k_{cat}/K_M) \tag{1}$$

$$K_{TS} = k_{uncat}/(k_{cat}/\Pi_N K_M) \tag{2}$$

An efficient catalyst binds the transition state very tightly, and therefore K_{TS} is very small. This number may be directly compared to K_D, the dissociation con-

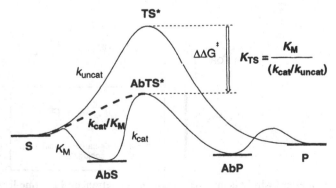

Fig. 4. Free energy diagram for catalysis. The dissociation constant K_{TS} relates quantitatively to the magnitude of the catalytic effect $\Delta\Delta G^{\#}$

stant of antibody-hapten complexes. Since this analysis applies to both uni- and multi-substrate reactions, it provides a convenient tool to compare widely different systems such as hydrolytic and bimolecular reactions. K_{TS} must be very small to have an interesting catalyst; for catalytic antibodies this is typically $K_{TS} < 10^{-7}$ mol l^{-1}. For each substrate i, the rate enhancement is given as

$$k_{cat}/k_{uncat}(i) = K_M(i)/K_{TS} \tag{3}$$

A complete assessment of catalyst performance must in addition address product inhibition and catalyst inactivation. For preparative applications, the final criterion concerns the absolute value of k_{cat}, which expresses how much product can be obtained from the catalysts per unit of time. As will be seen below, several catalytic antibodies indeed turn out to perform excellently in all respects.

2
Catalytic Antibodies for Enzyme-Like Reactions

2.1
Esterase Antibodies

Applications of enzymes in preparative chemistry is largely dominated by kinetic resolutions using esterases and lipases. Due to the ease of preparation of esterase catalytic antibodies from immunization with phosphonate transition state analogs, these represent the vast majority of catalytic antibodies prepared to date, even if it is not yet clear whether they will be competitive with currently available enzymes.

Hydrolysis of activated aryl esters, in particular nitrophenyl esters, has been used extensively to demonstrate the principles of antibody catalysis. This type of substrate is advantageous for model studies because formation of the tetrahedral intermediate is rate limiting, so that the issue of guiding decomposition of this intermediate towards hydrolysis does not need to be addressed in design. Nevertheless, hydrolyses of non-activated esters are also catalyzed by antibodies, the most notable example being that of R- and S-selective esterolyses of 2 by antibodies raised against hapten 1 (Scheme 1) [13]. Another anti-1 antibody was shown to promote enantioselective acylation of alcohols such as (S)-4 using vinyl ester 3 as acyl donor [14].

Mechanistic and structural investigations have revealed that while some esterase antibodies promote direct attack by hydroxide on the ester function [15], others unexpectedly may also use a covalent mechanism involving a serine-histidine dyad strongly reminiscent of the catalytic triad of serine proteases [16].

One of the limiting factors in obtaining efficient esterase antibodies is the observation of product inhibition by the carboxylate product. Non-phosphonate haptens have been used to induce esterase activity in antibodies [17]. Alternatively, taking carbonates as substrates instead of esters provides an elegant solution to the problem of product inhibition, as demonstrated in a kinetic resolution experiment [18]. This might be the most practical approach to obtain preparatively useful esterase catalytic antibodies in the future for kinetic resolu-

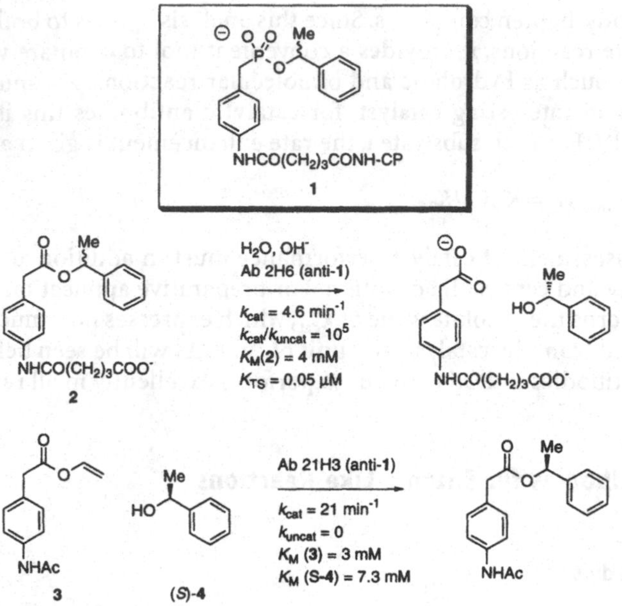

Scheme 1. Enantioselective lipase antibody

tion of racemic alcohols. Regioselective deprotection of acylated carbohydrates has been catalyzed by an antibody, demonstrating the usefulness of these catalysts for regioselective transformations [19].

2.2
Amidase and Ligase Antibodies

Cleaving or forming the native peptide bond, if possible with directable sequence specificity, is one of the main goals of catalytic antibody research. Several different p-nitroanilides have been cleaved using catalytic antibodies raised against phosphonates or related haptens [20], although these examples have little relevance to peptide bond cleavage because nitroanilides are highly activated. There is one report of an antibody-catalyzed hydrolysis of a carboxamide ($CONH_2$). The antibody was prepared using a methyl phosphinate hapten for immunization and an efficient direct screening protocol for catalysis to identify activity [21].

The most interesting report with respect to preparative applications concerns a ligase antibody obtained by immunization against phosphodiester hapten 5 (Scheme 2) [22]. This antibody catalyzes the coupling of an N-acylated amino-acid 4-nitrophenylester such as 6 with an L-tryptophan ester or amide such as 7 with useful selectivity and rate.

Inspired by enzymatic mechanisms, several groups have attempted to obtain amidase antibodies by recruiting a metal cofactor for catalysis. Introduction of

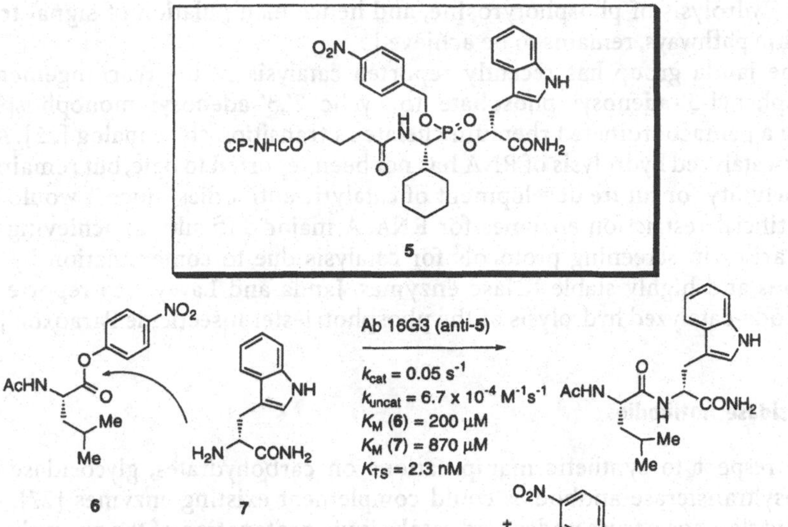

Scheme 2. A peptide ligase catalytic antibody

histidine residues in antibody sequences by genetic engineering has been used to install metal ions such as copper into an antibody binding site [23], although without success regarding esterase or amidase activity to date. Due to the enormous rate enhancement achievable with metal cofactors, there is little doubt that this strategy will ultimately succeed in producing useful amidase antibodies.

2.3
Phosphatase, Phosphodiesterase and Phosphotriesterase Antibodies

In analogy to phosphonates and phosphates providing optimal transition state analogs for inducing esterase activity, a stable pentacoordinated anionic group should allow one to prepare phosphatase and phosphodiesterase antibodies. An elegant solution has been demonstrated by Scanlan et al. who used α-hydroxy-phosphonate hapten 8 to induce a nitrophenyl phosphatase antibody (Scheme 3)

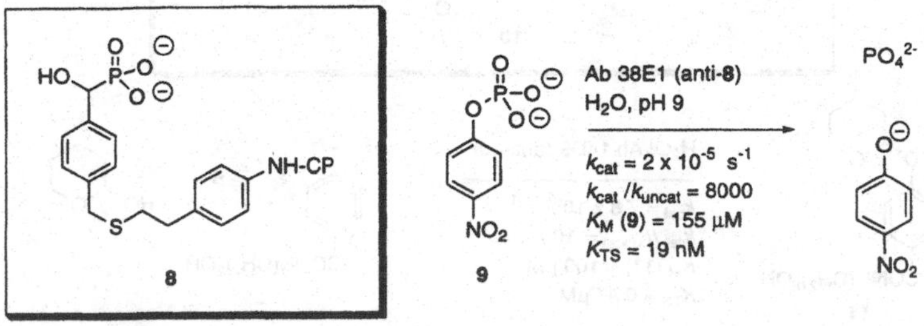

Scheme 3. A phosphatase antibody

[24]. Hydrolysis of phosphotyrosine, and hence manipulation of signal transduction pathways, remains to be achieved.

The Janda group has recently reported catalysis of the rearrangement of nitrophenyl-3'-adenosyl phosphate to cyclic 2',3'-adenosyl monophosphate using a pentacoordinated rhenium chelate as transition state analog [25]. Antibody-catalyzed hydrolysis of RNA has not been reported to date, but remains an opportunity for future development of catalytic antibodies, since it would lead to artificial restriction enzymes for RNA. A major difficulty in achieving that goal arises in screening protocols for catalysis due to contamination by ubiquitous and highly stable RNase enzymes. Janda and Lavey also reported an antibody-catalyzed hydrolysis of the phosphotriester insecticide Paraoxon [26].

2.4
Glycosidase Antibodies

With respect to synthetic manipulations on carbohydrates, glycosidase and glycosyltransferase antibodies could complement existing enzymes [27]. Glycosidic cleavage occurs under acid catalysis via protonation of the exocyclic C-1 oxygen followed by bond breaking to form a high energy oxocarbonium cation intermediate (Scheme 4). The reaction is marvelously achieved by glycosidase

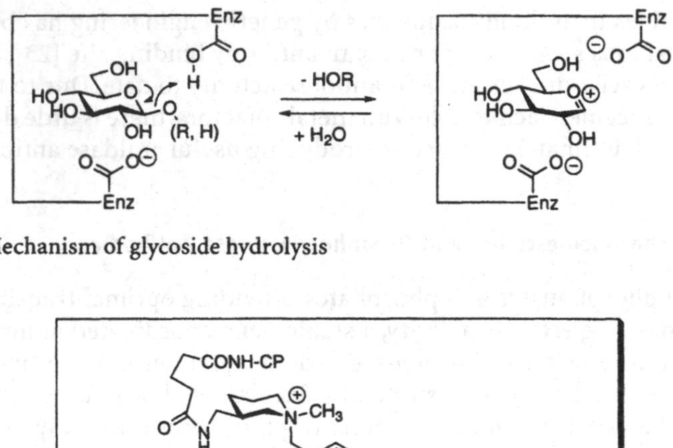

Scheme 4. Mechanism of glycoside hydrolysis

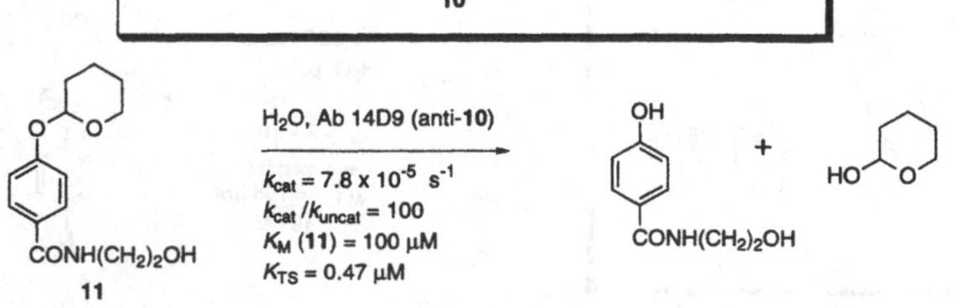

10

H2O, Ab 14D9 (anti-**10**)

$k_{cat} = 7.8 \times 10^{-5}$ s^{-1}
$k_{cat}/k_{uncat} = 100$
K_M (**11**) = 100 µM
$K_{TS} = 0.47$ µM

11

Scheme 5. A model glycosidase antibody

enzymes with the help of two carboxylate groups, one of which may be removed if provided on the substrate leaving group [28]. Carboxylate groups alone however are insufficient for catalysis to occur, and non-covalent interactions with the carbohydrate inducing strain towards the half-chair conformation of the oxocarbonium ion intermediate play a decisive role in enabling enzymatic catalysis. Indeed, cationic mimics of this intermediate are efficient inhibitors of glycosidases [29].

The mechanistically related hydrolysis of aryl-tetrahydropyranyl ether 11 [30] is catalyzed by antibody 14D9 prepared by immunization with cationic transition state analog 10, which features a positive charge in position C1 and a leaving group in axial position according to stereoelectronic rules (Scheme 5). This transition state analog design has yielded nanomolar inhibitors for glycosidases [31]. A similar model system has been studied by the Schultz group [32].

A glycosidase antibody for the hydrolysis of 13 has been reported by Masamune et al., using half-chair cationic transition state analog 12 as hapten [33]. Recently, a Scripps team has isolated a recombinant galactosidase catalytic antibody using immunization against pyrrolidine transition state analog 14 combined with an elaborate genetic chain shuffling and covalent selection experiment on phage-displayed antibodies (see Sect. 4.2) [34] (Scheme 6).

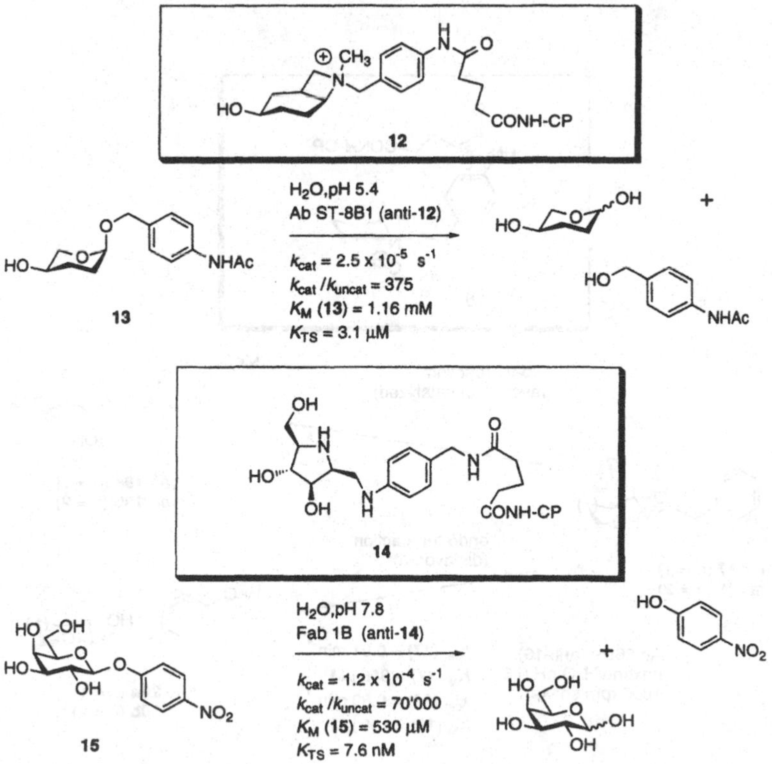

H_2O, pH 5.4
Ab ST-8B1 (anti-12)

$k_{cat} = 2.5 \times 10^{-5}$ s^{-1}
$k_{cat}/k_{uncat} = 375$
K_M (13) = 1.16 mM
$K_{TS} = 3.1$ μM

H_2O, pH 7.8
Fab 1B (anti-14)

$k_{cat} = 1.2 \times 10^{-4}$ s^{-1}
$k_{cat}/k_{uncat} = 70'000$
K_M (15) = 530 μM
$K_{TS} = 7.6$ nM

Scheme 6. Glycosidase antibodies

As for esterase catalysis, an important issue for glycosidic bond cleavage concerns the nature of the leaving group. While hydrolysis of aryl glycosides can serve preparative purposes as with glycosidase enzymes [35], the reaction is chemically biased due to the nature of the leaving group. A native glycosidic bond normally features a non-activated alkyl leaving group. We have recently shown in experiments with unactivated acetals that antibody-catalyzed cleavage is possible without using activated leaving groups using electrostatic catalysis in the antibody binding pocket [36].

2.5
Epoxide Hydrolase Antibodies

Epoxide hydrolysis is a valuable synthetic transformation. A number of epoxide hydrolase enzymes are known [37]. One of the most interesting and original transformations described using catalytic antibodies is the intramolecular cyclization of racemic hydroxyepoxide **17** (Scheme 7). This compound normally yields tetrahydrofuran **19a** following Baldwin's rule. By contrast a single enantiomer of the disfavored product tetrahydropyran **20a** is obtained using a catalytic antibody raised against *N*-oxide hapten **16** [38]. The same antibody also catalyzes cyclization of hydroxyepoxide **18** to yield optically enriched oxepane

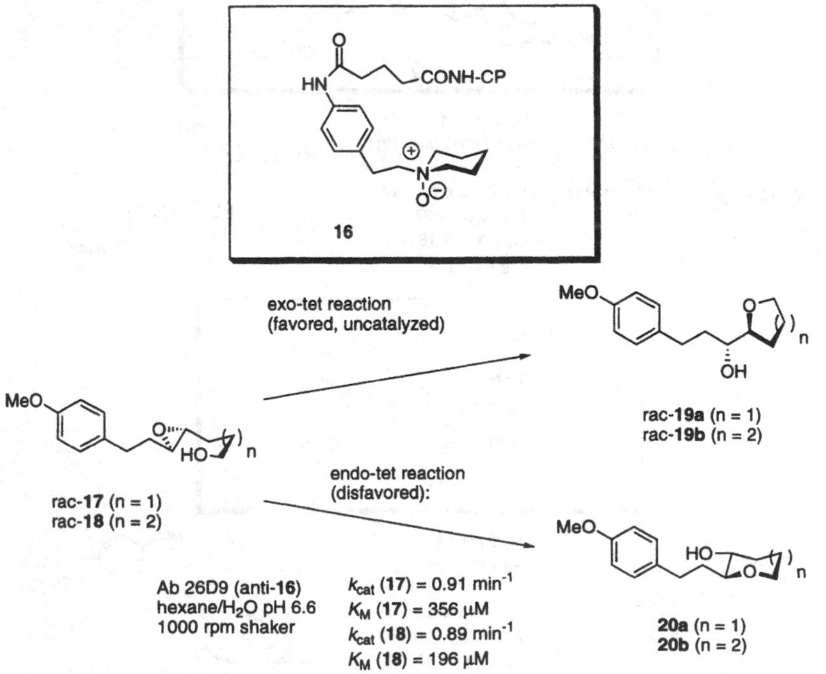

Scheme 7. "Anti-baldwinase" antibody

20b (78% *ee*) along the same disfavored endo-tet pathway [39]. These transformations are carried out in a biphasic hexane/water system. This "anti-Baldwinase" antibody operates by polarizing one side of the epoxide for nucleophilic opening [40].

Following a similar principle for hapten design, we have shown that antibody 14D9 obtained by immunization against **10** catalyzes the hydrolysis of a single enantiomer of epoxide **21** to yield diol **22**, evidently by participation of a carboxyl side-chain acting as a general acid (Scheme 8) [41]. The spontaneous epoxide opening by chloride from the medium to give chlorohydrin **23** is not catalyzed by the antibody. As for the anti-Baldwin selectivity described above, this chemoselectivity may be ascribed to a selective polarization of the homobenzylic epoxide bond within the antibody binding pocket.

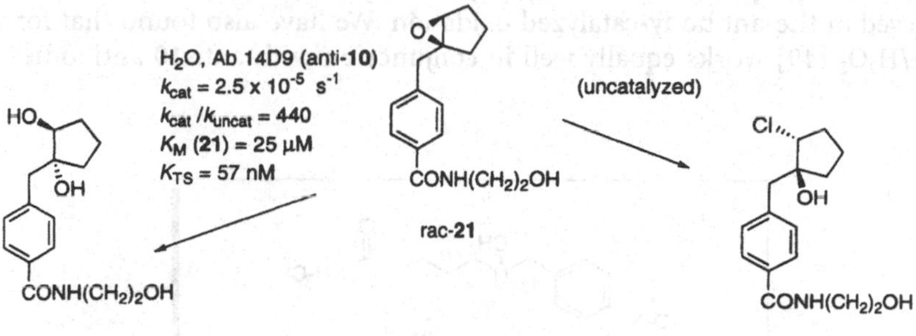

H_2O, Ab 14D9 (anti-**10**)
$k_{cat} = 2.5 \times 10^{-5}$ s^{-1}
$k_{cat}/k_{uncat} = 440$
K_M (**21**) = 25 µM
$K_{TS} = 57$ nM

(uncatalyzed)

Scheme 8. Epoxide hydrolase antibody

Epoxide hydrolysis in water has a low activation energy and can be readily triggered by general acid-base catalysis, as we have shown for hydrolysis of epoxide **21** by catalytic antibody 14D9. In addition the epoxide hydrolysis product is not charged and can be expected to show negligible product inhibition for a catalytic antibody raised against charged transition state analogs, e.g., **10**. These properties make epoxide hydrolysis a very favorable reaction for catalytic antibodies. A practical biphasic system has been described by Janda et al. for the preparation of optically pure tetrahydropyran **20a** on the gram scale starting with hydroxyepoxide **17** [42]. This suggests favorable future developments of epoxide reactions using catalytic antibodies for asymmetric synthesis.

2.6
Oxido-Reductase Antibodies

Redox catalysis with antibodies was first described for the reduction of resazurin using sulfite [43] and for heme-based peroxidase activity [44]. Concerning reactions relevant to preparative synthesis, Schultz and coworkers have described an elegant regio- and enantio-selective reduction of ketone **25a** with cyanoborohydride giving enantioselectively keto-alcohol **26** using an antibody raised against *N*-oxide transition state analog **24** (Scheme 9) [45]. Phosphonates were also found to induce catalytic antibodies for the same reaction [46]. Antibody catalysis of hydride transfer from an alcohol and an NAD-like cofactor has not yet been reported. An elaborate transition state analog of suitable design must be addressed to meet this challenge [47].

A number of oxidation reactions have been catalyzed by antibodies. Unfunctionalized olefins such as **27** can be oxidized enantioselectively to the corresponding epoxide **28** using antibody 20B11 directed against hapten **10** (Scheme 10) [48]. The oxidant used for this reaction is H_2O_2/CH_3CN, which generates $CH_3C(NH)OOH$ as the oxidizing species in situ. Peracids such as magnesium monoperphthalate oxidize antibodies quantitatively. The H_2O_2/CH_3CN reagent is milder and does not affect the antibody catalyst, yet is strong enough to attack di- and trisubstituted olefins. Although the rate enhancements observed are too low to lead to preparative chemistry, high enantioselectivities of epoxidation are observed in the antibody-catalyzed oxidation. We have also found that formamide/H_2O_2 [49] works equally well in conjunction with anti-**10** antibodies for

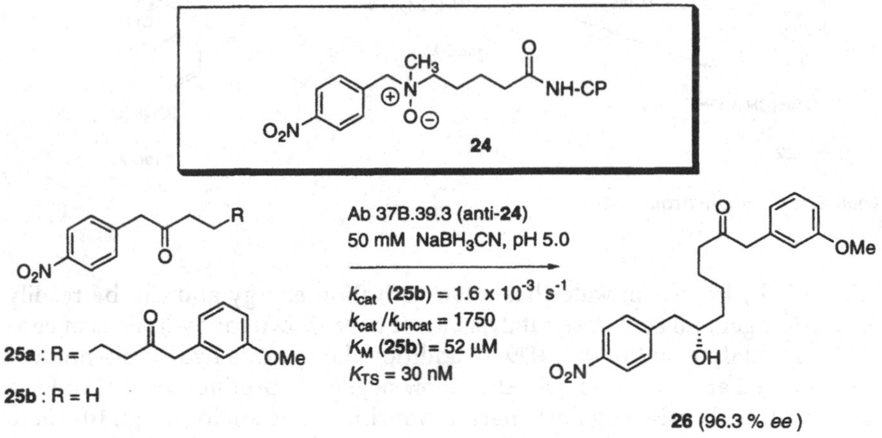

Scheme 9. Regio- and enantio-selective reduction catalyzed by an antibody

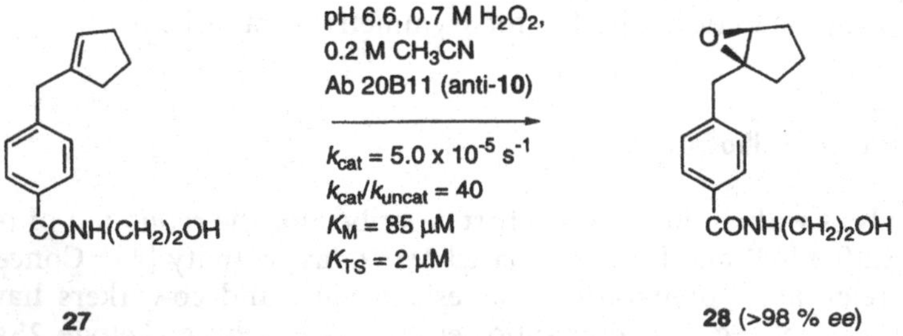

Scheme 10. Antibody-catalyzed enantioselective epoxidation

these oxidations. Either *R*- or *S*-epoxides can be obtained enantioselectively from olefin **27** and close analogs using a series of anti-**10** catalytic antibodies [50]. These results suggest that antibodies might become excellent enantioselective catalysts for epoxidations, although practical rate enhancement remains to be achieved. Oxidation of sulfide to sulfoxide by periodate has been catalyzed by antibodies raised against a phosphonate transition state analog [51].

2.7
Dehydratase Antibodies

Most antibody-catalyzed reactions are hydrolytic processes, which are highly exergonic in aqueous environment. β-Elimination of β-hydroxyketones to form α,β-unsaturated ketones is a rare case of a dehydration reaction proceeding exothermically in aqueous environment. Catalytic antibodies promoting β-elimination of β-hydroxy and β-halo-carbonyl compounds have been described [52]. The most interesting from a synthetic viewpoint concerns compound **30**, which can be dehydrofluorinated to *Z*-olefin **32** using catalytic antibody 1D4 raised against hapten **29**, while the uncatalyzed reaction exclusively yields the more stable *E*-olefin **31** (Scheme 11) [53]. Dehydratase activity also arises in aldolase antibodies and will be discussed below in that context.

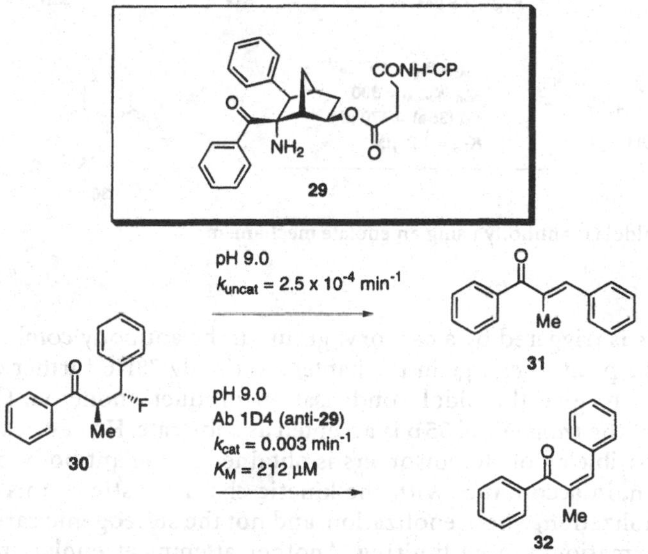

Scheme 11. Antibody-catalyzed disfavored β-fluoro elimination to a *Z*-olefin

2.8
Aldolase Antibodies

Carbon-carbon bond forming processes are highly valuable in synthetic chemistry. The aldol reaction, which generates up to two chiral centers upon coupling of two carbonyl compounds to form a β-hydroxycarbonyl, or aldol, product, is one of the most impressive of such processes. In aqueous medium, aldolization can be achieved either via enolates or via enamines as activated nucleophiles.

We have shown that enolate-mediated aldolization of keto-aldehyde 34 to give aldols 35 a/b is catalyzed by an antibody raised against hapten 33 (Scheme 12)

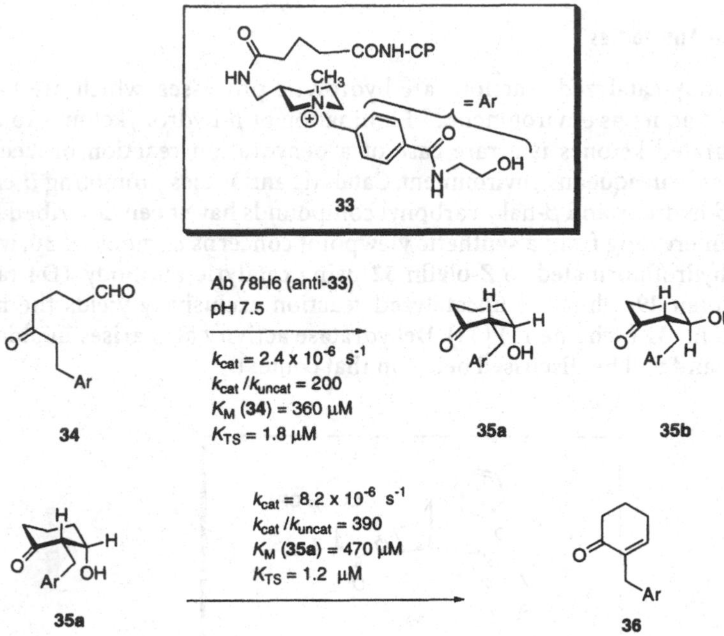

Scheme 12. An aldolase antibody using an enolate mechanism

[54]. Catalysis is triggered by a carboxyl group in the antibody combining site induced by the positive charge in the hapten. Antibody 78H6 further catalyzes dehydratation to give the aldol condensation product enone 36. Only one enantiomer of the *trans*-aldol 35 b is accepted as substrate. However, a mixture of all four possible aldol stereoisomers is obtained in the antibody-catalyzed aldol addition, in accordance with the kinetic characteristic of this enolate-mediated aldolization, where enolization, and not the stereogenic carbon-carbon bond formation, is rate limiting. Another attempt at enolate-mediated aldolization by the Schultz group yielded a catalytic antibody for a retro-Henry reaction [55].

Enamine-mediated aldolizations offer much better prospects for a stereo-controlled process. The famous enantioselective proline-catalyzed triketone cyclization to the Wieland-Miescher ketone 43 [56], as well as the chemistry of type I aldolase enzymes [57], provide ample precedents for stereo- and enantio-selective enamine-mediated reactions.

The aldol addition of aldehyde **38** and acetone to give aldols **39** is an excellent test reaction for a catalytic aldolization. We found that anti-**33** antibodies combine with primary amine **37** to form an aldolase antibody which catalyzes the reaction with good enantioselectivities (Scheme 13) [58]. Analysis of the reverse aldolization process shows that (S)-aldols undergo retroaldolization with the

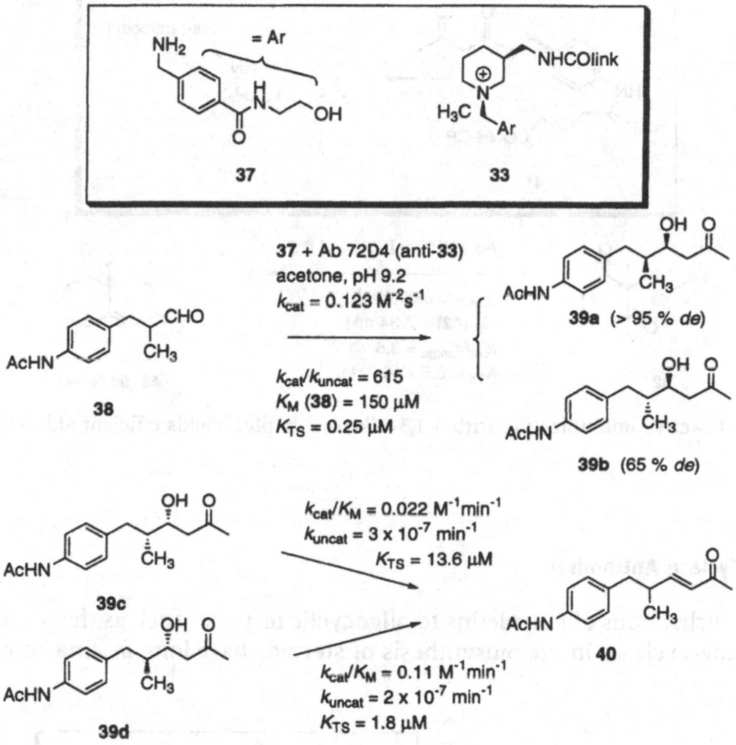

$37 + $ Ab 72D4 (anti-**33**)
acetone, pH 9.2
$k_{cat} = 0.123$ M^{-2}s^{-1}

$k_{cat}/k_{uncat} = 615$
K_M (**38**) $= 150$ µM
$K_{TS} = 0.25$ µM

39a (> 95 % de)

39b (65 % de)

$k_{cat}/K_M = 0.022$ M^{-1}min^{-1}
$k_{uncat} = 3 \times 10^{-7}$ min^{-1}
$K_{TS} = 13.6$ µM

$k_{cat}/K_M = 0.11$ M^{-1}min^{-1}
$k_{uncat} = 2 \times 10^{-7}$ min^{-1}
$K_{TS} = 1.8$ µM

38

39c

39d

40

Scheme 13. An enantioselective aldolase antibody using a primary amine cofactor

same stereoselectivites, as expected from the principle of microscopic reversibility. Most interestingly, (R)-aldols, which are not formed in the aldolization process, undergo β-elimination to enone **40** under catalysis by the antibody, which leads to a possible transition state model for the reaction [59].

A breakthrough experiment has come from the group of Barbas, who catalyzed the same reaction using a catalytic antibody raised against 1,3-diketone **41** (Scheme 14) [60]. This hapten selects for a lysine residue in the antibody combining site due to the formation of a stable vinilogous amide during immunization. While stereoselectivities are similar to antibody 72D4 for the aldolization of **37**, antibody 33F12 is a much more efficient catalyst. A beautiful series of synthetic applications has been reported using the similar catalytic antibody 38C2 (anti-**41**), which displays a surprising substrate versatility [61]. Many aldolizations proceed with very high enantioselectivities and can be carried out on a preparative scale; among many striking examples is the synthesis of optically pure Wieland-Miescher ketone 43 from prochiral diketone 42 [62]. This catalytic antibody is the first generally applicable preparative catalytic antibody, and is available commercially [63].

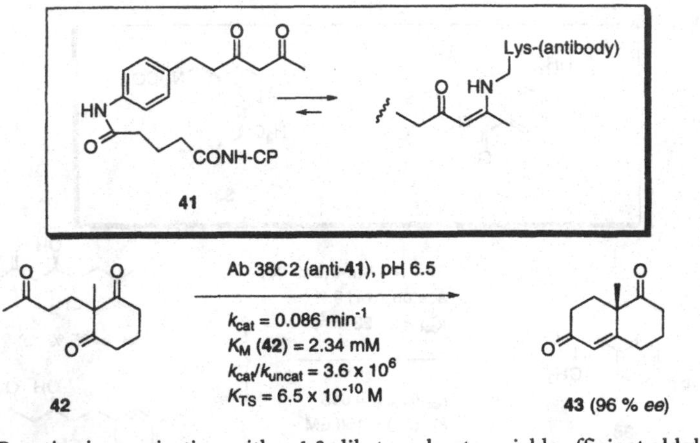

Scheme 14. Reactive immunization with a 1,3-diketone hapten yields efficient aldolase antibodies

2.9
Terpene Cyclase Antibodies

Cationic cyclizations of polyolefins to oligocyclic terpenes such as that catalyzed by squalene-cyclases in the biosynthesis of steroids have long fascinated chem-

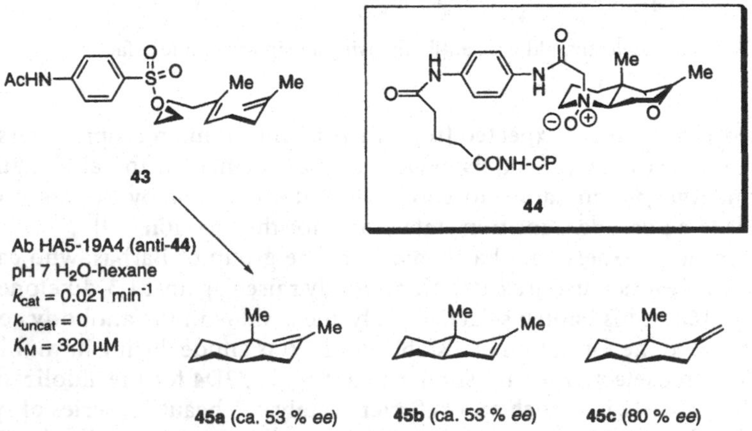

Scheme 15. Terpene cyclase antibodies

Catalytic Antibodies for Organic Synthesis 77

ists [64]. A number of protocols have been developed to effect these cyclizations stereo- and enantioselectively in solution [65]. Janda et al. have reported a series of catalytic antibodies that effect cyclization of olefinic tosylates in a biphasic hexane-buffer mixture at neutral pH [66]. Cyclization products are obtained whose structure depends both on starting material and catalyst [67]. Most remarkably, the products observed are not those obtained when the cyclization is run without catalyst, demonstrating the dramatic influence of the antibody on the outcome of the reaction. In one example a cyclopropane product is obtained in the reaction [68]. In a striking recent report antibody-catalyzed bicyclization of olefinic tosylate 43 to decalins 45a–c was observed using an antibody against hapten 44 (Scheme 15) [69].

3
Catalytic Antibodies for Non-Enzymatic Processes

Since the activity of catalytic antibodies is derived from the very general principle of transition state binding, there should be no limitations as to the types of reactions that can be catalyzed by such catalysts. Indeed, a number of non-enzymatic reactions have been catalyzed by antibodies.

3.1
Diels-Alderase Antibodies

Hilvert et al. reported the first catalytic antibody for a Diels-Alder cycloaddition [70]. Further examples were reported later using similar principles for the design of transition state analogs [71]. In one instance, cycloaddition catalysis was coupled to an esterase activity [72]. The most interesting example for asymmetric synthesis is the preparation of enantiomerically pure *exo*- and *endo*-cycloaddition products 49a/b from diene 48 and dimethyl-acrylamide 50 using catalytic antibodies raised against transition state analogs 46 and 47, respectively (Scheme 16) [73]. Similar activities were also detected in antibodies raised against a ferrocene derivative [74].

The crystal structure of two Diels Alderase antibodies has recently been reported [75], which suggests that catalytic activity not only derives from the entropic effect of binding both substrates in a reactive conformation, but also from activating hydrogen bonding interactions between antibody and dienophile.

3.2
Antibody-Catalyzed Electrocyclic Processes

Catalytic antibodies have been obtained for sigmatropic rearrangements, as first demonstrated for the biotransformation of chorismate to prephenate by two chorismate mutase antibodies developed in parallel by Hilvert et al. and Schultz et al. [76]. The Schultz group also described catalysis of an oxy-Cope rearrangement [77]. Schultz et al. and Hilvert et al. have prepared catalytic antibodies for concerted eliminations of aminoxide 53 and selenoxide 54 from immunization

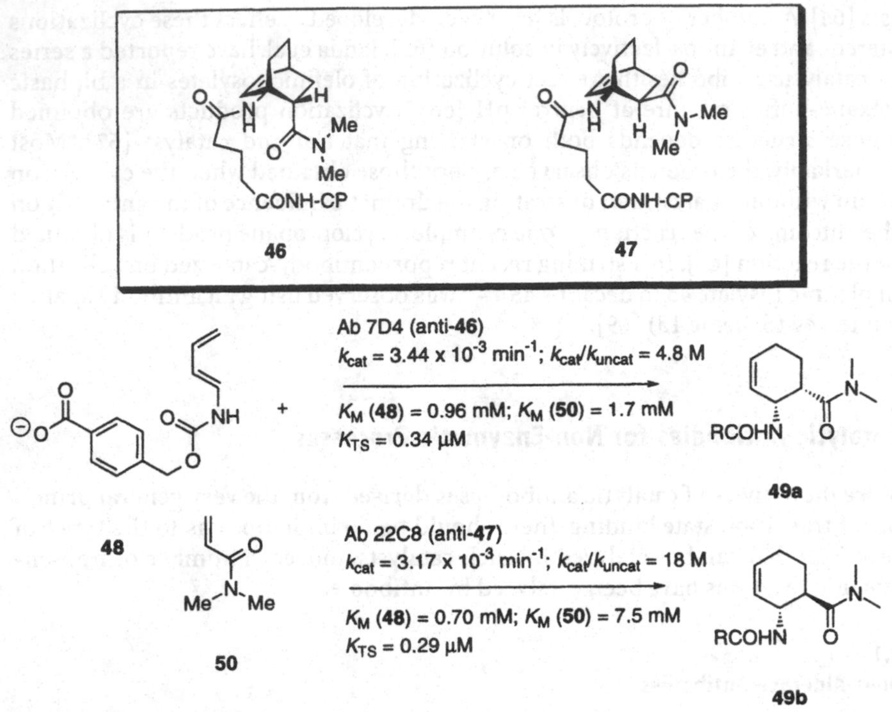

Scheme 16. *Exo-* and *endo*-selective enantioselective Diels-Alderase antibodies

with similar five-membered ring transition state analogs **51** and **52**, respectively (Scheme 17) [78]. An antibody raised against the tertiary amine analog of hapten **33** catalyzes an acid-promoted 1,2-dienone-phenol rearrangement [79].

3.3
Enantioselective Protonation with Catalytic Antibodies

Enantioselective protonation of prochiral enols or enolates, which provides synthetic access to optically active carbonyl compounds, is an elegantly simple test reaction for enantioselective reagents and catalysts, for which a number of examples have been described [80]. The only reaction described with alkyl enol ethers concerns the highly enantioselective protonation of enol ethers such as **55** by catalytic antibody 14D9, an antibody raised against hapten **10** [81]. Antibody 14D9 has a practical turnover of $k_{cat} = 0.4$ s^{-1} for substrate **55** and produces (S)-configured ketone **56** (Scheme 18). The antibody-catalyzed reaction is completely enantioselective (>99%) [82].

Catalytic antibody 14D9 derives its activity in large part from a carboxyl side chain induced in the antibody combining site by the positive charge in hapten **10** [83]. This side chain acts as a general acid catalyst in the rate-limiting protonation step. Hydrophobic interactions with alkyl substituents of the enol ether

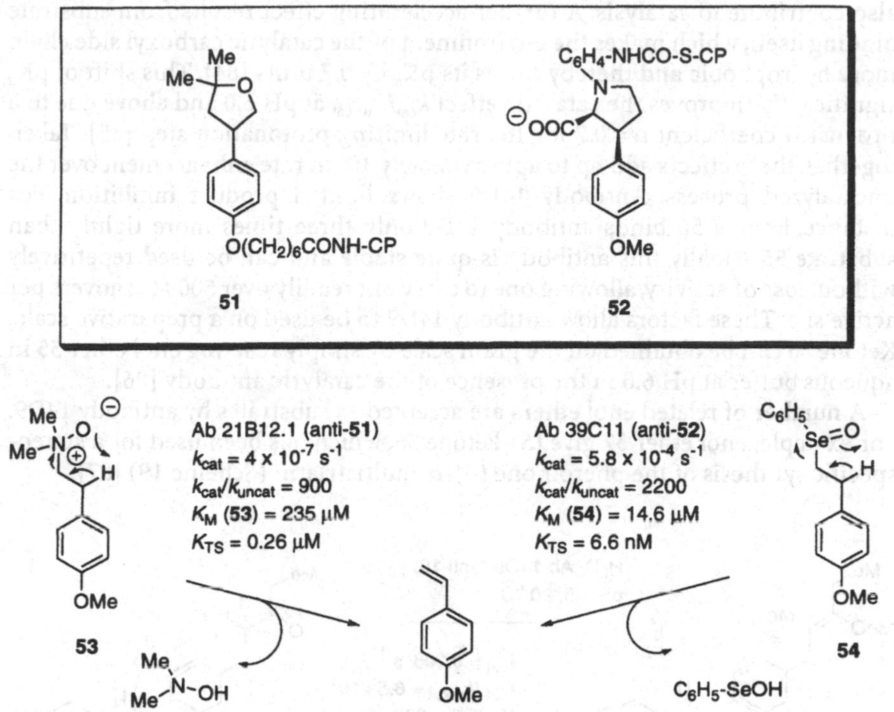
Me
Me
O
O(CH2)6CONH-CP
51
C6H4-NHCO-S-CP
N
OOC
OMe
52
Me
Me
N
O
H
OMe
53
Ab 21B12.1 (anti-51)
kcat = 4 x 10-7 s-1
kcat/kuncat = 900
KM (53) = 235 μM
KTS = 0.26 μM
Ab 39C11 (anti-52)
kcat = 5.8 x 10-4 s-1
kcat/kuncat = 2200
KM (54) = 14.6 μM
KTS = 6.6 nM
C6H5
Se=O
H
OMe
54
Me
N-OH
Me
OMe
C6H5-SeOH

Scheme 17. Antibody-catalyzed aminoxide and selenoxide eliminations

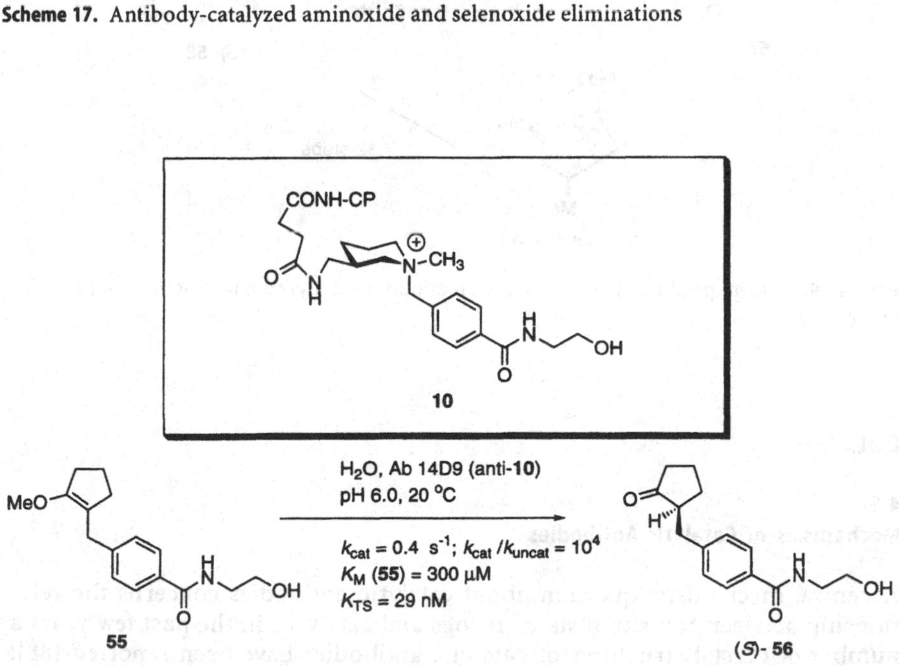
CONH-CP
O
N
H
N-CH3
H
N
OH
O
10
MeO
H
N
OH
O
55
H2O, Ab 14D9 (anti-10)
pH 6.0, 20 °C
kcat = 0.4 s-1; kcat /kuncat = 10^4
KM (55) = 300 μM
KTS = 29 nM
O
H
H
N
OH
O
(S)- 56

Scheme 18. Antibody-catalyzed enantioselective protonation

also contribute to catalysis. A further accelerating effect results from substrate binding itself, which makes the environment of the catalytic carboxyl side chain more hydrophobic and thereby raises its pK_a by 0.7 units [84]. This shift of pK_a significantly improves the catalytic effect k_{cat}/k_{uncat} at pH 6.0 and above due to a Brønsted coefficient $\alpha = 0.7$ for the rate limiting protonation step [85]. Taken together, these effects add up to approximately 10^4 in rate enhancement over the uncatalyzed process. Antibody 14D9 shows limited product inhibition. For instance, ketone 56 binds antibody 14D9 only three times more tightly than substrate 55. Finally this antibody is quite stable and can be used repetitively without loss of activity, allowing one to carry out readily over 5000 turnovers per active site. These factors allow antibody 14D9 to be used on a preparative scale. Ketone 56 can be obtained on the gram scale by simply reacting enol ether 55 in aqueous buffer at pH 6.0 in the presence of the catalytic antibody [86].

A number of related enol ethers are accepted as substrates by antibody 14D9. For example, enol ether 57 give (S)-ketone 58, which has been used for a stereo-specific synthesis of the pheromone (–)-α-multistriatin (Scheme 19) [87].

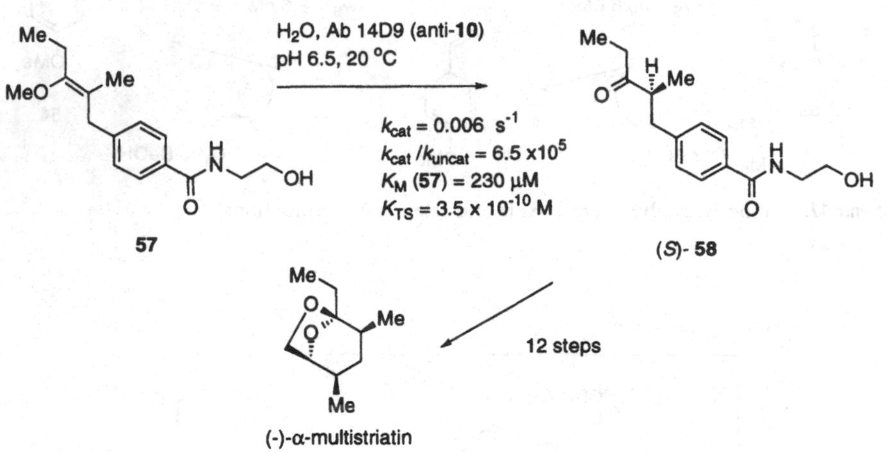

Scheme 19. Natural product synthesis using an antibody-catalyzed enantioselective protonation as key step

4
Outlook

4.1
Mechanisms of Catalytic Antibodies

A central mechanistic question about catalytic antibodies concerns the relationship between transition state analogs and catalysis. In the past few years a number of crystal structures of catalytic antibodies have been reported [88]. Identification of the active site is usually straightforward since antibodies are cocrystallized with the transition state analog used for immunization. It is reasonable to assume that the antibody-catalyzed reactions proceed according to the observed transition state analog's positions within the active sites, and an interpretation along these lines provides excellent insights into the relationship between antibody structure and catalysis.

Structural data provide an excellent starting point for site-directed mutagenesis experiments to investigate the role of individual amino-acid residues in catalysis. For example, antibody 48G7 raised against a nitrophenyl phosphonate catalyzes the hydrolysis of nitrophenyl acetate [89]. Crystal structure data reveal that three residues make direct hydrogen bonding contact with the negatively charged phosphonate moiety in the hapten, namely Arg^{96L} (light chain), Tyr^{33H} and His^{35H} (heavy chain) (Fig. 5). These residues are believed to stabilize the tetrahedral intermediate during hydrolysis. Single mutations of these key residues reduce catalysis (as measured by k_{cat}) by a factor of 11 ($Arg^{96L} \rightarrow Asn^{96L}$), 3.2 ($Tyr^{33H} \rightarrow Phe^{33H}$), and 1.7 ($His^{35H} \rightarrow Gln^{35H}$). His^{35H} is a conserved residue also found to play a critical role in a catalytic Ser-His dyad in esterase antibody 17E8 [90], where it acts as a general base. However, its role in 48G7 might be that of stabilizing the tetrahedral intermediate. The most profound mutational effect is found upon replacement of this histidine by a negatively charged glutamate, which reduces catalysis by a factor of 28 ($His^{35H} \rightarrow Glu^{35H}$). Interestingly, none of the nine residues mutated in the course of affinity maturation of the germline antibody to antibody 48G7 is in direct contact with the phosphonate hapten. Nevertheless, affinity maturation caused a 10^4-fold increase in binding affinity for this hapten and an 82-fold increase in catalytic efficiency [91].

Crystallographic analysis may provide only limited information when the hapten is either weakly bound or is a poor transition state analog of the reaction,

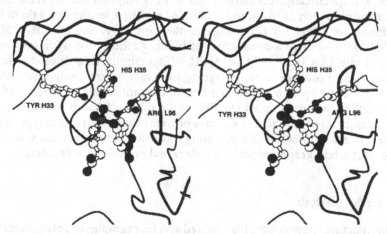

Fig. 5. Stereo-view of hapten binding site in catalytic antibody 48G7 with nitrophenyl phosphonate hapten and residues Arg^{96L} (*light chain*), Tyr^{33H} and His^{35H} (*heavy chain*). Heavy chain backbone as *dark gray ribbon* and light chain backbone as *light gray ribbon*. H-bonds shown as *lines*. Additional hapten–antibody contacts include three additional Tyrosine residues (not shown). Data from pdb files. See also Fig. 1b

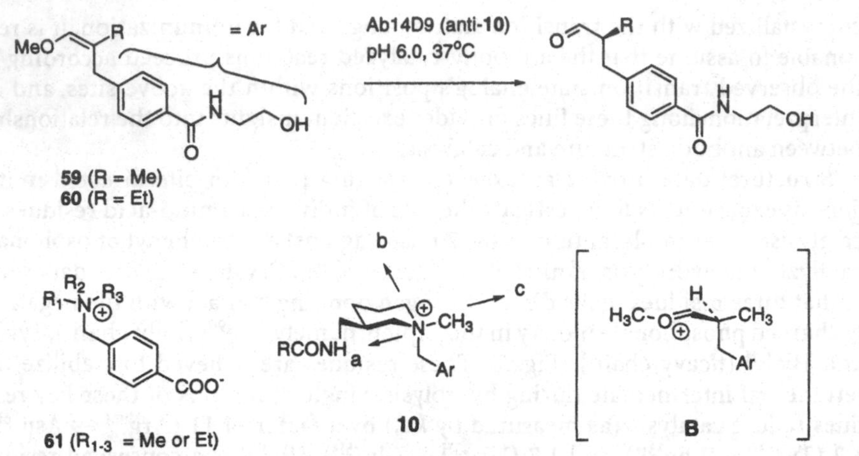

Scheme 20. Transition state structure of antibody-14D9 catalyzed enantioselective protonation

which is true for many highly efficient catalytic antibodies. In such cases kinetic experiments with substrate and hapten analogs can provide detailed information about the mechanism. In the case of antibody 14D9 (Scheme 20) hapten **10** is bound with nanomolar affinity, so that this hapten, even if not a very convincing transition state analog of the reaction, can be considered as a good negative image of the antibody's active site. Kinetic experiments using substrate **59** and hapten analogs show that enantioselective protonation occurs with the enol ether being in conformation B, with a proton being delivered from direction b in the rate limiting step, this relative to the hapten as defining the antibody binding pocket [92]. Specifically, successively substituting ethyl for methyl at the quaternary ammonium center in hapten analogs **61 a – d** first results in a gain of binding affinity for 14D9 for occupancy of the piperidine site ($ArCH_2N^+Me_3$ to $ArCH_2N^+MeEt_2$) but in a drastic loss of binding affinity for occupancy of the N-methyl site ($ArCH_2N^+MeEt_2$ to $ArCH_2N^+Et_3$). Similarly, a dramatic drop in transition state binding is measured going from methyl substituted enol ether **59** to its ethyl counterpart **60**. Furthermore, both enantiomers of product **56** bind equally well to the antibody, which discounts a simple chiral discrimination between enantiomeric products as a source of enantioselectivity. This type of kinetic analysis is very powerful for revealing transition state geometries, and complements crystallography which can only reveal ground state geometries.

4.2
Screening for Catalysis

Recent structural papers have highlighted several examples of antibody-catalyzed reactions for which a small catalytic effect was already present in a "germline" antibody, one that originated from the original library of 10^9 structures present in the unimmunized animal. In the case of an oxy-Cope rearrangement, catalysis was more efficient in the germline antibody compared with its "improved" counterpart, which had been optimized for binding during maturation [93]. Other spontaneous catalytic activities have been reported in antibodies, such as peptide- and DNA-hydrolyzing activities in autoimmune antibodies [94].

These findings suggest that immunization against transition state analogs might sometimes select an already existing catalytic activity from the germline library of antibodies, rather than creating it during antibody maturation. This points to the fact that efficient assays for catalysis are equally important in discovering catalytic antibodies as are transition state analogs. More practically, rapidly identifying potentially catalytic clones out of many hundreds of hapten binding clones from an immunization with a transition state analog, and concentrating cloning efforts on those selected clones, keeps work and material costs at a manageable level. Being able to conclude early on the presence or absence of a catalytic antibody also potentially allows feedback to hapten design, thereby increasing chances of success for a given target.

4.2.1
Cat-ELISA

Numerous methodologies have been developed to assay catalytic activity in cell culture media in high throughput format. Green et al. and Hilvert and MacBeath have developed a "cat-ELISA" similar to the ELISA for binding affinity. A substrate-carrier protein conjugate is surface-bound in wells of a 96-well plate. The reaction is allowed to proceed in the presence of a test solution. Formation of the surface bound product is quantitated using ELISA detection with a product specific polyclonal rabbit antibody. The presence of product indicates a possible catalyst in the test solution [95]. Similar systems applied to surface-bound substrates have been described using either biotinylated or DNA-tagged substrates [96].

The best method involves competitive ELISA between the reaction product and a product analog covalently tagged with acetylcholinesterase (AChE), which are allowed to compete for a surface bound, product-specific polyclonal rabbit antibody [97]. In the absence of product, AChE-tagged product binds to the well surface. After washing unbound reagents, surface-bound AChE is revealed using a chromogenic AChE substrate. If reaction has occurred, the untagged product binds to the surface and no AChE-tagged product remains bound after washing. Thus, no coloration is observed (Fig. 6).

4.2.2
Acridone Tags

We have described an efficient general protocol for assaying catalysis using substrates labeled with the fluorescent tag acridone [98]. In this method, reactions are simply analyzed by thin layer chromatography (TLC). Acridone is a highly photoresistant and intensely fluorescent group whose derivatives behave well on TLC. As little as 1 picomole of an acridone-tagged compound is visually detected under illumination with a simple UV lamp, a detection limit that is 10–100 times lower than that for fluorescein or dansyl derivatives. This method is easi-

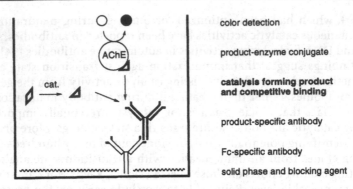

color detection

product-enzyme conjugate

**catalysis forming product
and competitive binding**

product-specific antibody

Fc-specific antibody

solid phase and blocking agent

Fig. 6. Principle of competitive cat-ELISA assay

ly adapted to practically any reaction and can be readily applied in high throughput format. A typical example is the epoxidation of (S)-citronellol derivative **61** to two TLC-separable stereoisomers **62 a, b**, which provides a simple high throughput screening assay for enantioselective epoxidation (Scheme 21). While **61** is not water soluble, related acridone derivatives with functional groups such as hydroxy, carbonyl or amide groups are well soluble at 100 µmol l^{-1} concentration in aqueous buffers.

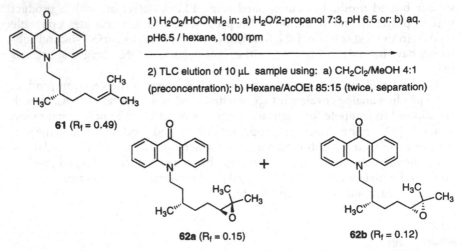

1) H_2O_2/HCONH$_2$ in: a) H_2O/2-propanol 7:3, pH 6.5 or: b) aq. pH6.5 / hexane, 1000 rpm

2) TLC elution of 10 µL sample using: a) CH$_2$Cl$_2$/MeOH 4:1 (preconcentration); b) Hexane/AcOEt 85:15 (twice, separation)

61 (R$_f$ = 0.49)

62a (R$_f$ = 0.15)   **62b** (R$_f$ = 0.12)

Scheme 21. Acridone tags allow highly sensitive and selective catalysis assays by TLC

4.2.3
Fluorogenic Substrates

In our hands, the most practicable assays are those involving fluorogenic substrates [98]. For example, non-fluorescent compound **63** undergoes a retro-

Diels-Alder reaction releasing, besides nitroxyl which can be converted to nitric oxide, a strongly blue fluorescent anthracene product 64 (Scheme 22) [99]. This reaction can therefore be followed sensitively in cell culture supernatants. Using early screening we have isolated seven different catalytic antibodies for the reaction out of approximately 12,000 individual cell culture wells resulting from fusions with ten different immunized mice. Due to early screening, the experiment was completed in a matter of weeks and comprised cloning of only a handful of antibodies [100].

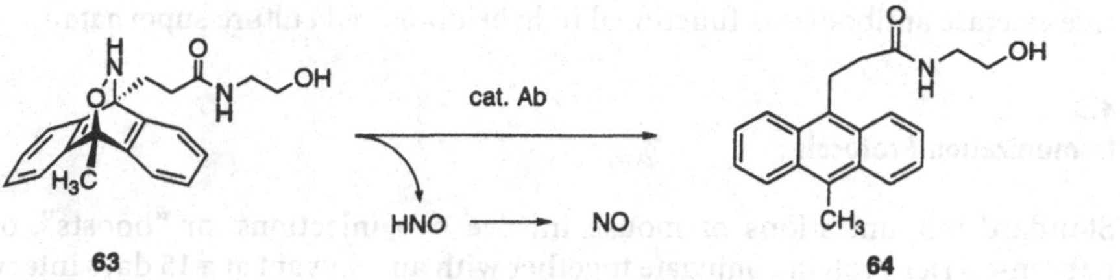

Scheme 22. Fluorogenic retro-Diels-Alder reaction

While only few reactions are fluorogenic, many interesting synthetic transformations can be arranged to become so. For example, compounds (R)-66 and (S)-66 are enantiomeric substrates for alcohol dehydrogenases (Scheme 23) [101]. Ketone 67 is formed upon oxidation, which then releases fluorescent umbelliferone by β-elimination. This second reaction is itself slow but can be acceler-

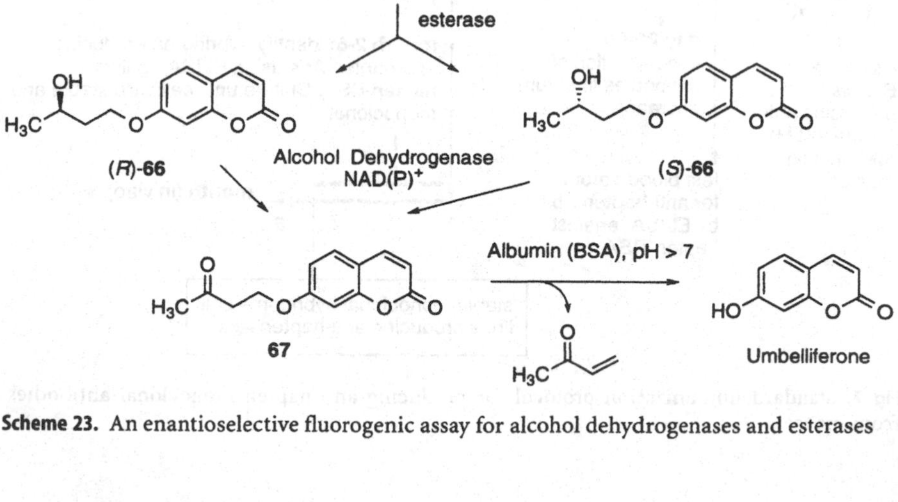

Scheme 23. An enantioselective fluorogenic assay for alcohol dehydrogenases and esterases

ated by addition of bovine serum albumin (BSA) to the medium. BSA has been shown to accelerate simple deprotonation reactions [102] and also catalyzes this β-elimination such that the half-life of ketone **67** is reduced from 10 h to 4 min with 2 mg/ml BSA. This indirect release strategy allows one to measure a typically non-fluorogenic oxidation process by fluorescence. It is readily applicable to other carbonyl releasing reactions such as the retro-aldol reaction or the hydrolysis of enol ethers. We have further expanded this technique to follow the hydrolysis of enantiomeric acetates (*R*)-**65** and (*S*)-**65** by fluorescence in the presence of alcohol dehydrogenase and BSA. This rapid screening assay for enantioselective esterase antibodies is functional in hybridoma cell culture supernatant.

4.3
Immunization Protocols

Standard immunizations of mouse involve two injections, or "boosts", of a hapten-carrier protein conjugate together with an adjuvant at a 15 days interval. At least eight weeks after the second boost, immune cells are stimulated by a final intravenous injection of hapten-conjugate, and the fusion carried out 4 days later to generate hybridoma from spleen cells (Fig. 7). Experiments with catalytic polyclonal antibodies [103] have shown that over-immunization re-

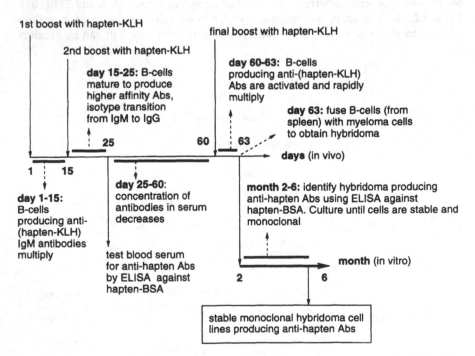

Fig. 7. Standard immunization protocol for producing anti-hapten monoclonal antibodies from mouse

duces the concentration of catalytic antibodies in serum, suggesting that excessive maturation towards binding is detrimental to catalysis. Nevertheless boosting twice with antigen is necessary to induce the transition from pentameric IgM to globular IgG antibodies that occur during maturation.

4.3.1
Heterologous and In Vitro Immunizations

Masamune et al. have reported that utilizing two structurally distinct haptens for the first and second boost may produce more efficient catalytic antibodies for hydrolytic reactions [104], a strategy which perhaps avoids affinity maturation while allowing isotype transition. Alternatively a procedure called "in vitro" immunization can be carried out, whereby spleen cells from unimmunized mice are cultured in vitro for five days in the presence of a non-conjugated soluble transition state analog and an immune-stimulatory peptide. This results in activation of cells producing hapten binding antibodies, which are then fused as normal to generate hybridoma. Although this protocol produces antibodies of the less convenient pentameric IgM form, it has been reported to give many efficient catalytic antibodies for an ester hydrolysis reaction [105].

4.3.2
Anti-Idiotypic Antibodies

A series of experiments suggests that design and synthesis of transition state analogs can be circumvented altogether if one wants simply to reproduce an existing enzymatic activity with catalytic antibodies [106]. Thus immunization against the enzyme acetyl-choline esterase (AChE) leads in part to active-site specific antibodies. These antibodies are used as antigen in a second immunization. Some "anti-idiotypic" antibodies resulting from this second immunization recognize the binding pocket of the first antibodies and therefore reproduce the original acetyl-choline esterase active site. These show catalytic activity for the hydrolysis of acetyl choline, but this activity is reduced by a factor of 10^4 compared to the original enzyme. Whether this technique works generally remains to be seen.

4.3.3
Reactive Immunization

In an effort to induce catalytic residues during immunization, Lerner et al. reported that efficient catalytic antibodies can be generated by a process called "reactive immunization". In this method a chemically reactive hapten is used to create a covalent bond with a catalytic residue in the antibody binding pocket, thereby allowing its selection and amplification in the course of immunization. Aldolase antibodies were obtained using a reactive 1,3-diketone, which formed a covalent bond with a lysine residue in the antibody binding pocket (see Sect. 2.8). A similar experiment yielded esterase antibodies by immunization against activated phosphodiester haptens [107]. These experiments go beyond the concept of transition state analogs and mark a new level of refinement in the chemical design of catalytic antibodies. Due to the critical role of catalytic amino-acid side chains in enzymes, there is no doubt that reactive immunization is to become a prevalent strategy for inducing catalytic antibodies.

4.4
Evolutionary Techniques

Benkovic and Smiley have reported an orotate decarboxylase antibody obtained by genetic selection starting from a library of antibody genes from an unimmunized animal introduced into a bacterial strain lacking orotate decarboxylase enzymes [108]. In essence, bacteria expressing a functional catalytic antibody for the reaction survived and multiplied, and were induced to produce mutations beneficial for catalytic activity. The isolated catalytic antibody is highly active, which brilliantly demonstrates the power of this approach.

A Scripps group has isolated a galactosidase antibody starting from a library of antibody genes from an animal immunized with a transition state analog, displayed on the surface of phages (see Sect. 2.4.). In the presence of surface-bound covalent inactivator **68**, which is known to react covalently with galactosidase enzymes, phages with galactosidase antibodies on their surface convert **68** to quinone methide **69**, which rapidly reacts with a nucleophile on the catalytic antibody (Scheme 24). As a result, phage displaying catalytically active antibodies become attached to the surface. Washing unbound phages and amplifying surface bound, catalytically active clones by error prone PCR allows one to select and evolve catalytic antibodies with improved properties. By contrast, simple mutation and selection of a phage displayed catalytic antibody for transition state analog binding has been found to be of limited use to improve catalytic activity [109].

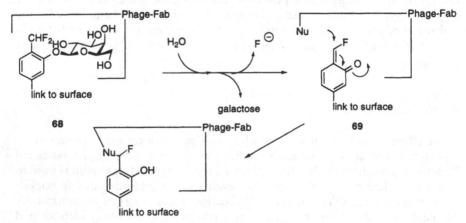

Scheme 24. Catalysis-dependent covalent attachement of phage-displayed Fab fragments on a solid surface allows selection of catalytic antibodies

These examples suggest that genetic manipulation of recombinant antibody libraries can be an efficient route to catalytic antibodies. These genetic strategies are limited to catalyzing key metabolic reactions in the first case, and to reactions for which covalent inactivators can be found in the second case. In any event, these encompass many synthetically relevant transformations. A major handicap there remains expression of recombinant antibodies in stable form and preparatively significant amounts, which is much more difficult to achieve than with hybridoma.

4.5
Preparative Applications

Antibodies can be utilized under a variety of non-physiological conditions and are excellently suited for preparative applications due to their intrinsic stability. We have found that antibodies catalyzing the retro-Diels-Alder reaction of 63 function equally well between pH 4 and pH 11. Aldolase antibody 72D4 operates in the presence of 10% acetone. Janda et al. have used an immobilized esterase antibody with up to 40% dimethylsulfoxide [110]. Esterase catalytic antibodies have been used in reverse micelles and in lipid-coated form to transform lipophilic substrates [111]. Catalytic antibodies can also be used in a biphasic alkane/water system [112]. The lipophilic substrate remains in the alkane phase where it does not undergo any reaction, which suppresses any uncatalyzed reaction. In case that the reaction product is still lipophilic and returns to the alkane phase, product inhibition is also suppressed under these conditions.

We reported the first gram scale synthesis of an optically active product using catalytic antibody 14D9 (anti-10) for the conversion of 55 to 56 (see Sect. 3.3). Preparative reactions were possible due to a high turnover number of the catalytic antibody ($k_{cat}=0.4$ s^{-1}), and allowed applications in natural product synthesis. This enantioselective reaction was of particular interest since it was not realizable otherwise, thereby illustrating the potential of catalytic antibody technology to provide completely new biocatalysts. Aldolase antibody 38C2 (anti-41) stemming from the work of Barbas et al. is now being used for several preparative reactions. For these two antibodies, reactions are carried out in homogeneous aqueous phase, from which the products can be recovered by dialysis and extraction. The catalytic antibody can be recovered quantitatively with little loss of activity. For antibody 14D9 we have carried out 5000 turnovers per catalytic site while losing only 20% of total sample activity, mainly due to manipulations during dialysis. Once a useful catalytic antibody has been identified, there is no principle obstacle for its utilization in large scale, particularly if the antibody is produced by a hybridoma cell line.

Acknowledgement. Financial support was provided by the Swiss National Science Foundation, the Koordinationsgruppe für Forschungsfragen der Basler Chemischen Industrie (KGF) and the Wander Stiftung. The author thanks Prof. U. Baumann of the Department of Chemistry & Biochemistry in Bern for preparing Fig. 1b and Fig. 5.

5
References

1. (a) Schultz PG, Lerner RA (1995) Science 269:1835; (b) Thomas NR (1996) Natural Product Reports 479
2. Litman GW (1996) Sci Am Nov 96:67
3. Tijssen P (1985) In: Burdon RH, Knippenberg PH (eds) Practice and theory of enzyme immunoassays. Elsevier, Amsterdam
4. Kohler G, Milstein C (1975) Nature 256:495
5. Pauling L (1946) Chem Eng News 24:1375
6. Jencks WP (1969) In: Catalysis in chemistry and enzymology. McGraw-Hill, New York
7. (a) Tramontano A, Janda KD, Lerner RA (1986) Science 234:1566; (b) Pollack SJ, Jacobs JW, Schultz PG (1986) Science 234:1570
8. (a) Mader MM, Bartlett PA (1997) Chem Rev 97:1281; (b) Stewart JD, Benkovic SJ (1995) Nature 375:388; (c) Kirby AJ (1996) Angew Chem Int Ed Engl 35:707; (d) Fujii I, Tanaka F, Miyashita H, Tanimura R, Kinoshita K (1995) J Am Chem Soc 117:6199
9. (a) Kurz JL (1963) J Am Chem Soc 85:987; (b) Kurz JL (1972) Acc Chem Res 5:1
10. For an analysis of cyclodextrin catalysis in terms of KTS, see: (a) Tee OS, Mazza C, Lozano-Hemmer R, Giorgi JB (1994) J Org Chem 59:7602; (b) Tee OS, Du X (1992) J Am Chem Soc 114:620
11. Tee OS (1994) Adv Phys Org Chem 29:1
12. (a) Wolfenden R (1972) Acc Chem Res 5:10; (b) Lienhard GE (1973) Science 180:149
13. Janda KD, Benkovic SJ, Lerner RA (1989) Science 244:437
14. Wirshing P, Ashely JA, Benkovic SJ, Janda KD, Lerner RA (1991) Science 252:680
15. Charbonnier J-B, Golinelli-Pimpaneau B, Gigant B, Tawfik DS, Chap R, Schindler DG, Kim S-H, Green BS, Eshhar Z, Knossow M (1997) Science 275:1140
16. (a) Guo J, Huang W, Scanlan TS (1994) J Am Chem Soc 116:6062; (b) Fox T, Scanlan TS, Kollman PA (1997) J Am Chem Soc 119:11,571
17. (a) Janda KD, Weinhouse MI, Schloeder DM, Lerner RA (1990) J Am Chem Soc 112:1274; (b) Suga H, Ersoy O, Tsumuraya T, Lee J, Sinskey AJ, Masamune S (1994) J Am Chem Soc 116:487
18. Nakatani T, Hiratake J, Shinzaki A, Umeshita R, Suzuki T, Nishioka T, Nakajima H, Oda J (1993) Tetrahedron Lett 34:4945
19. Iwabuchi Y, Miyashita H, Tanimura R, Kinoshita K, Kikuchi M, Fujii I (1994) J Am Chem Soc 116:771
20. (a) Janda KD, Schloeder D, Benkovic SJ, Lerner RA (1988) Science 241:1188; (b) Ersoy O, Fleck R, Sinskey A, Masamune S (1996) J Am Chem Soc 118:13,077; (c) Ersoy O, Fleck R, Sinskey A, Masamune S (1998) J Am Chem Soc 120:817
21. Martin MT, Angeles TS, Sugasawara R, Aman NI, Napper AD, Darsley MJ, Sanchez RI, Booth P, Titmas RC (1994) J Am Chem Soc 116:6508
22. (a) Hirschmann R, Smith AB III, Taylor CM, Benkovic PA, Taylor SD, Yager KM, Sprengeler PA, Benkovic SJ (1994) Science 265:234; (b) Smithrud DB, Benkovic PA, Benkovic SJ, Taylor CM, Yager KM, Witherington J, Philips BW, Sprengeler PA, Smith AB III, Hirschmann R (1997) J Am Chem Soc 119:278
23. Iverson BL, Iverson SA, Roberts VA, Getzoff ED, Tainer JA, Benkovic SJ, Lerner RA (1990) Science 249:659
24. Scanlan TS, Prudent JR, Schultz PG (1991) J Am Chem Soc 113:9397
25. Weiner DP, Wiemann T, Wolfe MM, Wentworth P, Janda KD (1997) J Am Chem Soc 119:4088
26. Lavey BJ, Janda KD (1996) J Org Chem 61:7633
27. (a) Sinnott ML (1990) Chem Rev 90:1171; (b) Kirby AJ (1987) CRC Crit Rev Biochem 22:283
28. Wang A, Withers SG (1995) J Am Chem Soc 117:10,137
29. Ganem B (1996) Acc Chem Res 29:340
30. Reymond J-L, Janda KD, Lerner RA (1991) Angew Chem Int Ed Engl 30:1711

31. Bols M (1998) Acc Chem Res 31:1
32. Yu J, Hsieh LC, Kochersperger L, Yonkovich S, Stephans JC, Gallop MA, Schultz PG (1994) Angew Chem Int Ed Engl 33:339
33. Suga H, Tanimoto N, Sinskey AJ, Masamune S (1994) J Am Chem Soc 116:11,197
34. Janda KD, Lo L-C, Lo C-HL, Sim M-M, Wang R, Wong C-H, Lerner RA (1997) Science 275:945
35. (a) Vic G, Tran CH, Scigelova M, Crout DHG (1997) Chem Commun 169; (b) Li J, Wang PG (1997) Tetrahedron Lett 38:7967
36. (a) Sinha SC, Keinan E, Reymond J-L (1993) Proc Natl Acad Sci USA 90:11910; (b) Shabat D, Sinha SC, Reymond J-L, Keinan E (1996) Angew Chem Int Ed Engl 35:2628
37. (a) Faber K, Mischitz M, Kroutil W (1996) Acta Chem Scand 50:249; (b) Archelas A, Furstoss R (1997) Annu Rev Microbiol 51:491
38. Janda KD, Shevlin CG, Lerner RA (1993) Science 259:490
39. Janda KD, Shevlin CG, Lerner RA (1995) J Am Chem Soc 117:2659
40. (a) Na J, Houk KN, Shevlin CG, Janda KD, Lerner RA (1993) J Am Chem Soc 115:8453; (b) Na J, Houk KN (1996) J Am Chem Soc 118:9204
41. Sinha SC, Keinan E, Reymond J-L (1993) J Am Chem Soc 115:4893
42. Shevlin CG, Hilton S, Janda KD (1994) Bioorg Med Chem Lett 4:297
43. Janjic N, Tramontano A (1989) J Am Chem Soc 111:9109
44. Cochran AG, Schultz PG (1990) J Am Chem Soc 112:9414
45. Hsieh LC, Yonkovich S, Kochersperger L, Schultz PG (1993) Science 260:337
46. Nakayama GR, Schultz PG (1992) J Am Chem Soc 114:780
47. Schröer J, Sanner M, Reymond J-L, Lerner RA (1997) J Org Chem 62:3220
48. Koch A, Reymond J-L, Lerner RA (1994) J Am Chem Soc 116:803
49. Chen Y, Reymond J-L (1995) Tetrahedron Lett 36:4015
50. Chen Y, Reymond J-L (unpublished)
51. Hsieh LC, Stephans JC, Schultz PG (1994) J Am Chem Soc 116:2167
52. (a) Shokat KM, Leumann CJ, Sugasawara R, Schultz PG (1989) Nature 338:269; (b) Uno T, Schultz PG (1992) J Am Chem Soc 114:6573; (c) Shokat K, Uno T, Schultz PG (1994) J Am Chem Soc 116:2261
53. Cravatt BF, Ashley JA, Janda KD, Boger DL, Lerner RA (1994) J Am Chem Soc 116:6013
54. Koch T, Reymond J-L, Lerner RA (1995) J Am Chem Soc 117:9383
55. Flaganan ME, Jacobsen JR, Sweet E, Schultz PG (1996) J Am Chem Soc 118:6078
56. (a) Hajos ZG, Parrish DR (1974) J Org Chem 39:1615; (b) Eder U, Sauer G, Wiechert R (1971) Angew Chem Int Ed Engl 10:496; (c) Agamic C (1988) Bull Soc Chim Fr 3:499
57. (a) Fessner WD (1992) Kontakte (Darmstadt) 3:3; (b) Fessner W-D, Watter C (1997) Top Curr Chem 184:97
58. (a) Reymond J-L, Chen Y (1995) Tetrahedron Lett 36:2575; (b) Reymond J-L, Chen Y (1995) J Org Chem 60:6970
59. Reymond J-L (1995) Angew Chem Int Ed Engl 34:2285
60. Wagner J, Lerner RA, Barbas CF (1995) Science 270:1797
61. Barbas CF, Heine A, Zhong G, Hoffmann T, Gramatikova S, Björnestedt R, List B, Anderson J, Stura EA, Wilson IA, Lerner RA (1997) Science 278:2085
62. Zhong G, Hoffmann T, Lerner RA, Danishefsky S, Barbas CF (1997) J Am Chem Soc 119:8131
63. List B, Shabat D, Barbas CF, Lerner RA (1998) Chem Eur J 4:881
64. (a) Wendt KU, Poralla K, Schulz GE (1997) Science 277:1811; (b) Starks CM, Back K, Chappell J, Noel JP (1997) Science 277:1815; (c) Lesburg CA, Zhai G, Cane DE, Christianson DW (1997) Science 277:1820
65. Fish PV, Sudhakar AR, Johnson WS (1993) Tetrahedron Lett 34:7849
66. (a) Li T, Janda KD, Ashley JA, Lerner RA (1994) Science 264:1289; (b) Li T, Lerner RA, Janda KD (1997) Acc Chem Res 30:115
67. (a) Li T, Hilton S, Janda KD (1995) J Am Chem Soc 117:3308; (b) Hasserodt J, Janda KD, Lerner RA (1996) J Am Chem Soc 118:11,654
68. Li T, Janda KD, Lerner RA (1996) Nature 379:326

69. (a) Hasserodt J, Janda KD, Lerner RA (1997) J Am Chem Soc 119:5993; (b) Hasserodt J, Janda KD (1997) Tetrahedron 53:11,237
70. Hilvert D, Hill KW, Nared KD, Auditor M-TM (1989) J Am Chem Soc 111:9261
71. (a) Braisted AC, Schultz PG (1990) J Am Chem Soc 112:7430; (b) Meekel AAP, Resmini M, Pandit UK (1996) Bioorg Med Chem 4:1051
72. Suckling CJ, Tedford MC, Bence LM, Irvine JI, Stimson WH (1993) J Chem Soc Perkin Trans I 1925
73. Gouverneur VE, Houk KN, de Pascual-Teresa B, Beno B, Janda KD, Lerner RA (1993) Science 262:204
74. Yli-Kauhaluoma JT, Ashley JA, Lo C-H, Tucker L, Wolfe MM, Janda KD (1995) J Am Chem Soc 117:7041
75. (a) Romesberg FE, Spiller B, Schultz PG, Stevens RC (1998) Science 279:1929; (b) Heine A, Stura EA, Yli-Kauhaluoma JT, Gao C, Deng Q, Beno BR, Houk KN, Janda KD, Wilson IA (1998) Science 279:1934
76. (a) Jackson DY, Jacobs JW, Sugasawara R, Reich SH, Bartlett PA, Schultz PG (1988) J Am Chem Soc 110:4841; (b) Hilvert D, Nared KD (1988) J Am Chem Soc 110:5593; (c) Jackson DY, Liang MN, Bartlett PA, Schultz PG (1992) Angew Chem Int Ed Engl 31:182
77. (a) Braisted AC, Schultz PG (1994) J Am Chem Soc 116:2211; (b) Driggers EM, Cho HS, Liu CW, Katzka CP, Braisted AC, Ulrich HD, Wemmer DE, Schultz PG (1998) J Am Chem Soc 120:1945
78. (a) Yoon SS, Oei Y, Sweet E, Schultz PG (1996) J Am Chem Soc 118:11,686; (b) Zhou ZS, Jiang N, Hilvert D (1997) J Am Chem Soc 119:3623
79. Chen Y, Reymond J-L, Lerner RA (1994) Angew Chem Int Ed Engl 33:1607
80. (a) Vedejs E, Lee N (1995) J Am Chem Soc 117:891; (b) Fehr C (1991) Chimia 45:253
81. Reymond J-L, Janda KD, Lerner RA (1992) J Am Chem Soc 114:2257
82. Shabat D, Itzhaky H, Reymond J-L, Keinan E (1995) Nature 374:143
83. Reymond J-L, Jahangiri GD, Stoudt C, Lerner RA (1993) J Am Chem Soc 115:3909
84. Reymond J-L, Chen Y (1996) Isr J Chem 36:199
85. Kresge AJ, Chen HL, Chiang Y, Murrill CE, Payne MA, Sagatys DS (1971) J Am Chem Soc 93:413
86. Reymond J-L, Reber J-L, Lerner RA (1994) Angew Chem Int Ed Engl 33:475
87. Sinha SC, Keinan E (1995) J Am Chem Soc 117:3653
88. Wade H, Scanlan TS (1997) Annu Rev Biophys Biomol Struct 26:461
89. Patten PA, Gray NS, Yang PL, Marks CB, Wedemeyer GJ, Boniface JJ, Stevens RC, Schultz PG (1996) Science 271:1086
90 Zhou GW, Guo J, Huang W, Fletterick RJ, Scanlan TS (1994) Science 265:1059
91. Wedemayer GJ, Patten PA, Wang LH, Schultz PG, Stevens RC (1997) Science 276:1665
92. Jahangiri GK, Reymond J-L (1994) J Am Chem Soc 116:11,264
93. Ulrich HD, Mundorff E, Santarsiero BD, Driggers EM, Stevens RC, Schultz PG (1997) Nature 389:271
94. (a) Paul S, Volle DJ, Beach CM, Johnson DR, Powell MJ, Massey RJ (1989) Science 244:1158; (b) Paul S, Li L, Kalaga R, Wilins-Stevens P, Stevens FJ, Solomon A (1995) J Biol Chem 270:15,257; (c) Shuster AM, Gololobov GV, Kvashuk OA, Bogomolova AE, Smirnov IV, Gabibov AG (1992) Science 256:665
95. (a) Tawfik DS, Green BS, Chap R, Sela M, Eshhar Z (1993) Proc. Natl. Acad. Sci. USA 90:373; (b) MacBeath G, Hilvert D (1994) J Am Chem Soc 116:6101
96. (a) Lane JW, Hong X, Schwabacher AW (1993) J Am Chem Soc 115:2078; (b) Fenniri H, Janda KD, Lerner RA (1995) Proc Natl Acad Sci USA 92:2278
97. (a) Taran F (1996) Ph D thesis, Paris-Orsay IX, France; (b) Caruelle D, Grassi J, Courty J, Croux-Muscatelli B, Pradelles P, Barritault D, Caruelle JP (1988) Anal Biochem 173:328; (c) Benedetti F, Berti F, Massimiliano F, Resmini M, Bastiani E (1998) Anal Biochem 256:67
98. Reymond J-L, Koch T, Schröer J, Tierney E (1996) Proc Natl Acad Sci USA 93:4251; (a) Haugland RP (1995) Handbook of Fluorescent Probes and Research Chemicals, 6th Edition, Molecular Probes, Inc., pp. 201–244

99. Bahr N, Güller R, Reymond J-L, Lerner RA (1996) J Am Chem Soc 118:3550
100. Bensel N, Bahr N, Reymond MT, Reymond JL (unpublished)
101. Klein G, Reymond J-L (1998) Bioorg Med Chem Lett 8:1113
102. (a) Kikuchi K, Thorn SN, Hilvert D (1996) J Am Chem Soc 118:8184; (b) Hollfelder F, Kirby AJ, Tawfik DS (1996) Nature 383:60
103. Wallace MB, Iverson BL (1996) J Am Chem Soc 118:251 and references cited therein
104. (a) Suga H, Ersoy O, Williams SF, Tsumuraya T, Margolies MN, Sinskey AJ, Masamune S (1994) J Am Chem Soc 116:6025; (b) Tsumuraya T, Suga H, Meguro S, Tsunakawa A, Masamune S (1995) J Am Chem Soc 117:11,390
105. Stahl M, Goldie B, Bourne S, Thomas NR (1995) J Am Chem Soc 117:5164
106. Izadyar L, Friboulet A, Remy M-H, Roseto A, Thomas D (1993) Proc Natl Acad Sci USA 90:8876
107. Wirsching P, Ashley JA, Lo C-HL, Janda KD, Lerner RA (1995) Science 270:1775
108. Smiley JA, Benkovic SJ (1995) J Am Chem Soc 117:3877
109. (a) Miller GP, Posner BA, Benkovic SJ (1997) Bioorg Med Chem 5:581; (b) Baca M, Scanlan TS, Stephenson RC, Wells JA (1997) Proc Natl Acad Sci USA 94:10,063
110. Janda KD, Ashley JA, Jones TM, McLeod DA, Schloeder DM, Weinhouse MI (1990) J Am Chem Soc 112:8886
111. (a) Durfor CN, Bolin RJ, Sugasawara RJ, Massey RJ, Jacobs JW, Schultz PG (1988) J Am Chem Soc 110:8713; (b) Okahata Y, Yamaguchi M, Tanaka F, Fujii I (1995) Tetrahedron 51:7673
112. Ashley JA, Janda KD (1992) J Org Chem 6691

Immobilized Enzymes: Methods and Applications

Wilhelm Tischer[1] · Frank Wedekind[2]

[1] Boehringer Mannheim GmbH, Nonnenwald 2, D-82372 Penzberg, Germany.
  E-mail: wilhelm.tischer@roche.com
[2] Boehringer Mannheim GmbH, Nonnenwald 2, D-82372 Penzberg, Germany.
  E-mail: frank.wedekind@roche.com

Immobilized enzymes are used in organic syntheses to fully exploit the technical and economical advantages of biocatalysts based on isolated enzymes. Immobilization enables the
separation of the enzyme catalyst easily from the reaction mixture, and can lower the costs of
enzymes dramatically. This is true for immobilized enzyme preparations that provide a well-
balanced overall performance, based on reasonable immobilization yields, low mass transfer
limitations, and high operational stability. There are many methods available for immobilization which span from binding on prefabricated carrier materials to incorporation into in situ
prepared carriers. Operative binding forces vary between weak multiple adsorptive interactions and single attachments through strong covalent binding. Which of the methods is the
most appropriate is usually a matter of the desired applications. It is therefore the intention of
this paper to outline the common immobilization methods and reaction technologies to
facilitate proper applications of immobilized enzymes.

Keywords: Enzyme immobilization, Mass transfer effects, Operational stability, Immobilization
methods.

Topics in Current Chemistry, Vol. 200
© Springer Verlag Berlin Heidelberg 1999

1
Introduction

Man-made usage of binding enzymes onto solid materials goes back to the 1950s, when immobilized enzymes, that is enzymes with restricted mobility, were first prepared intentionally [1,2]. Immobilization was achieved by inclusion into polymeric matrices or binding onto carrier materials. Considerable effort was also put into the cross-linking of enzymes, either by cross-linking of protein alone or with the addition of inert materials [3].

In the course of the last decades numerous methods of immobilization on a variety of different materials have been developed. Binding to pre-fabricated carrier materials appears to have been the preferred method so far. Recently, cross-linking of enzyme crystals has also been reported to be an interesting alternative [4].

Immobilized enzymes are currently the object of considerable interest. This is due to the expected benefits over soluble enzymes or alternative technologies. The number of applications of immobilized enzymes is increasing steadily [5]. Occasionally, however, experimental investigations have produced unexpected results such as a significant reduction or even an increase in activity compared with soluble enzymes. Thus, cross-linked crystals of subtilisin showed 27 times less activity in the aqueous hydrolysis of an amino acid ester compared to equal amounts of soluble enzyme [6]. On the other hand, in the application of lipoprotein lipase in the solvent-mediated synthesis of esters there was a 40-fold increase in activity using immobilized or otherwise modified enzyme preparations as compared to enzyme powder [7].

This is why it is mandatory to have some basic knowledge of the essential contributions of the chemical forces of binding and of the physicochemical interactions during an enzyme reaction which generally is a matter of heterogeneous catalysis.

2
Why Immobilize Enzymes?

There are several reasons for the preparation and use of immobilized enzymes. In addition to a more convenient handling of enzyme preparations, the two

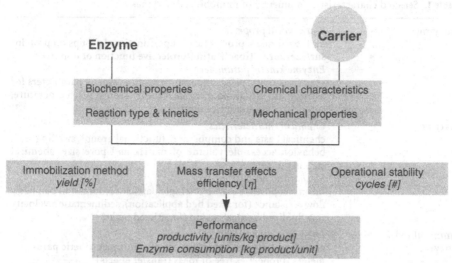

Fig. 1. Characteristics of immobilized enzymes

main targeted benefits are (1) easy separation of the enzyme from the product, and (2) reuse of the enzyme.

Easy separation of the enzyme from the product simplifies enzyme applications and supports a reliable and efficient reaction technology. On the other hand, reuse of enzymes provides cost advantages which are often an essential prerequisite for establishing an enzyme-catalyzed process in the first place.

The properties of immobilized enzyme preparations are governed by the properties of both the enzyme and the carrier material. The specific interaction between the latter provides an immobilized enzyme with distinct chemical, biochemical, mechanical and kinetic properties (Fig. 1).

Of the numerous parameters [8–10] which have to be taken into account, the most important are outlined in Table 1.

As far as manufacturing costs are concerned the yield of immobilized enzyme activity is mostly determined by the immobilization method and the amount of soluble enzyme used. Under process conditions, the resulting activity may be further reduced by mass transfer effects. More precisely, the yield of enzyme activity after immobilization depends not only on losses caused by the binding procedure but may be further reduced as a result of diminished availability of enzyme molecules within pores or from slowly diffusing substrate molecules. Such limitations, summarized as mass transfer effects, lead to lowered efficiency. On the other hand, improved stability under working conditions may compensate for such drawbacks, resulting in an overall benefit. Altogether, these interactions are a measure of productivity or of enzyme consumption, for example, expressed as enzyme units per kg of product. If we replace "enzyme units" by "enzyme costs" we obtain the essential product related costs, for example, in US$ per kg of product.

In order to estimate the cost advantages of immobilized enzymes, it is necessary to consider the individual manufacturing steps and their contribution to

Table 1. Selected characteristic parameters of immobilized enzymes

Enzyme	*Biochemical properties* molecular mass, prosthetic groups, functional groups on protein-surface, purity (inactivating/protective function of impurities) *Enzyme kinetic parameters* specific activity, pH-, temperature profiles, kinetic parameters for activity and inhibition, enzyme stability against pH, temperature, solvents, contaminants, impurities
Carrier	*Chemical characteristics* chemical basis and composition, functional groups, swelling behavior, accessible volume of matrix and pore size, chemical stability of carrier *Mechanical properties* mean wet particle diameter, single particle compression behavior, flow resistance (for fixed bed application), sedimentation velocity (for fluidized bed), abrasion (for stirred tanks)
Immobilized enzyme	*Immobilization method* bound protein, yield of active enzyme, intrinsic kinetic parameters (properties free of mass transfer effects) *Mass transfer effects* consisting of partitioning (different concentrations of solutes inside and outside the catalyst particles), external and internal (porous) diffusion; this gives the effectiveness in relation to free enzyme *determined under appropriate reaction conditions,* *Stability* operational stability (expressed as activity decay under working conditions), storage stability *"Performance"* productivity (amount of formed product per unit or mass of enzyme) enzyme consumption (e.g. units kg^{-1} product, until half-life)

the overall costs. Firstly, these comprise the costs for biomass from plant and animal sources, or from microbial fermentations. In the latter case, the costs are determined mainly by the fermentation scale and the expression rate of the enzymes. Secondly, downstreaming is needed to achieve the required purity but is accompanied by loss in activity. The use of larger fermentation scales may be necessary in order to compensate for loss in activity and also for some increase in cost (Fig. 2). Thirdly, costs for the immobilization procedure further increase the manufacturing expense. Thus, aside from the potential advantage of easier removal of the enzyme from the product formed, immobilized enzymes so far provide no cost benefit. However, cost savings will be achieved by multiple reuse of an immobilized enzyme. On the other hand, prolonged use also means downscaling of the unit operation and thus increased costs for manufacture of the enzyme itself which must also be taken into account. In conclusion, only multiple use will lead to dramatic cost reduction. In practice, this can be monitored by consideration of the amount of enzyme required per kg of formed product.

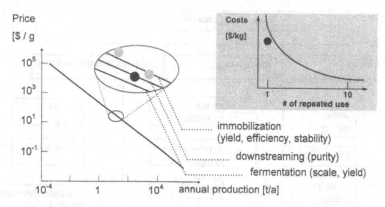

Fig. 2. Cost and productivity

3
Immobilization Methods

Many reviews and books on the immobilization of enzymes have been published during the last two decades [5, 11–20]. The intent of this part of the review is to explain the basic principles and to show recent developments of enzyme immobilization for the purpose of preparative biotransformation. The attachment of enzymes onto prefabricated artificial or natural carriers will be given special emphasis (see Sect. 3.2).

Immobilization of macromolecules can be generally defined as a procedure leading to their restricted mobility. A classification of immobilization methods according to different chemical and physical principles is shown in Fig. 3.

Enzymes are proteins that have been optimized by evolution for more or less specific metabolic reactions in living cells. Reaction conditions therein are often far away from those in industrial bioreactors. By immobilization novel characteristics can be induced to make enzymes tolerant to even harsh reaction environments. Some attempts to intentionally modify the catalytic behavior have been reviewed [15]. The immobilization procedure should be conducted in a way that allows the enzyme to maintain its active conformation and necessary catalytic flexibility. Furthermore, catalytically essential residues of the enzyme should be undisturbed and conserved.

Although the development of a suitable immobilization procedure often follows empirical guidelines, knowledge of the structural characteristics of the enzyme is helpful for achieving a high performance biocatalyst.

Basically, there are four ways to immobilize enzymes onto surfaces:

1. A suitable reaction to activate the enzyme for the immobilization process is performed prior to binding. This approach often suffers from significant loss of activity, because the protein is modified by highly reactive chemical compounds that are often not strictly group specific and may alter catalytically or

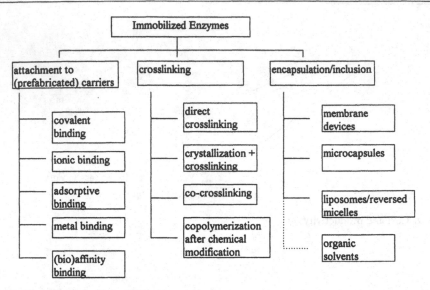

Fig. 3. Classification of immobilization methods

structurally essential residues. Also intra- and intermolecular cross-linking has to be considered.

2. The support is modified and activated. The native enzyme is bound in a subsequent step under well-defined conditions using the natural reactivity of the molecule. This is the most prominent technique to covalently bind enzymes to carrier surfaces.
3. A bi- or multifunctional coupling agent is used to mediate between carrier and enzyme functional groups. This can also lead to intra- and intermolecular enzyme cross-linking.
4. The enzyme is modified by recombinant DNA techniques to generate a protein with "(bio)specific" groups, so that it can adsorb onto special carriers using (bio)affinity binding.

3.1
Enzyme Functional Groups

3.1.1
Native Functional Groups

Enzymes are composed of amino acid chains linked via peptide bonds and can be considered as polyfunctional and (multi-)charged macromolecules with a defined, more or less rigid three-dimensional structure. A typical enzyme molecule with a molecular weight of 30,000 Da resembles a compact spherical particle of approx. 4 nm in diameter. Many enzymes, especially those which are regulated by various effectors, consist of more than one amino acid chain (subunit), that are covalently (via disulfide linkages) or non-covalently associated. In addition, the amino acid chains of enzymes can be posttranslationally modified in vivo by the cellular machinery. Only carbohydrate residues linked to serine/threonine or asparagine will be considered here, because they can be a target for immobilization reactions.

3.1.1.1
Amino Acid Side Chains

The chemical reactivity of these complex molecules depends on the reactivity of the amino acid side chains in their special microenvironment. Studies on chemical modification have revealed that only a few amino acid side chains are really reactive [21]. Of the 20 proteinogenic amino acids, the alkyl side chains of the hydrophobic residues are chemically inert for all purposes. The aliphatic hydroxyl groups of serine and threonine can be considered as water derivatives and, therefore, have a low reactivity in competition with a high concentration of water molecules (55 M). Thus, only nine side chains are chemically active (see Table 2). These are the guanidinyl group of arginine, the γ- and β-carboxyl groups of glutamic and aspartic acid, respectively, the sulfhydryl group of cysteine, the imidazolyl group of histidine, the ε-amino group of lysine, the thioether moiety of methionine, the indolyl group of tryptophan, and the phenolic hydroxyl group of tyrosine.

Most of the enzyme modification reactions, and hence of the coupling reactions, are nucleophilic reactions, in particular bimolecular nucleophilic substitution reactions following an S_N2-type mechanism. Therefore, the chemical reactivity is basically a function of nucleophilicity of the amino acid side chain. Following the overall nucleophilic order of Edwards and Pearson [22], the sulfhydryl group of cysteine is the most potent nucleophile in the protein, especially in its thiolate form.

The nitrogen in the amino group is a considerably weaker nucleophile but due to its abundance and omnipresence in enzymes is the most important target. This is particularly true as most reaction products from thiol nucleophilic attack are more or less unstable.

The reactivity of functional residues is strongly influenced by the pH. Generally reaction rates are higher at more alkaline pH. Lowering the pH below the pK_a of accessible amino groups will decrease the reaction rate as protonated amino groups can be regarded as unreactive. As the pK_a is a function of temperature, ionic strength and microenvironment, the values given in Table 2 are approximations. The ratio of protonated to deprotonated species at a certain pH can be estimated by the Henderson–Hasselbalch equation:

$$pH = pK_a + \log([A-]/[AH])$$ (1)

Accessibility of functional groups for an immobilization reaction is further influenced by microenvironmental effects. Steric hindrance by neighbouring groups, interaction with neighbouring groups and solvent molecules, or alterations of pK_a-values of dissociable groups can lead to a shifted reactivity.

Table 2. Reactive amino acid side chains[a]

Amino Acid	3/1 Letter Code	Functional group	pKa free A.A.	pKa M.P.	Susceptible to following reagents/reactions	Comments
C-terminus	–	$-\!\!\!\!\overset{\displaystyle O}{\underset{\alpha}{C}}\!\!-OH$	1.8–2.6	3.1–3.7	acylation carbodiimide	activation necessary via carbodiimide mediated coupling, in most examples dramatic loss of enzyme activity
N-terminus	–	$-\!\!\underset{\alpha}{N}H_2$	8.8–10.8	7.6–8.0	acylation, alkylation, arylation, diazotation	
cysteine	Cys/C	$-SH$	8.3	8.5–8.8	acylation, alkylation, arylation,	
lysine	Lys/K	$-NH_2$	10.8	10.0–10.2	acylation, alkylation, arylation,	most important site for immobilization reactions
methione	Met/M	$-S-CH_3$			alkylation, arylation, oxidation, cyanogen bromide	susceptible to oxicyanogen dation, cleavage site
histidine	His/H	(imidazole)	6.0	6.5–7.0	acylation, alkylation, arylation, aryl halides	
tyrosine	Tyr/Y	(phenol, –OH)	10.9	9.6–10.0	acylation, alkylation, arylation, iodination, nitration, diazotation	cross-linking reactions
tryptophan	Trp/W	(indole)			alkylation, arylation	

Table 2 (continued)

Amino Acid	3/1 Letter Code	Functional group	pK_a free A.A.	pK_a M.P.	Susceptible to following reagents/reactions	Comments
aspartic acid	Asp/D	β‑COOH	3.9	4.4–4.6	acylation carbodiimides	
glutamic acid	Glu/E	γ‑COOH	4.3	4.4–4.6	acylation carbodiimides	
arginine	Arg/R	guanidino (NH–C(=NH)–NH$_2$)	12.5	>12		

a Modified according to [38].

3.1.1.2
Enzyme-Linked Carbohydrates

Carbohydrates are attached to many secreted enzymes of higher organisms. Two basic ways exist to utilize carbohydrates for immobilization purposes.

Firstly, immobilized lections may be used. Lectins are proteins that display high binding selectivity for defined carbohydrate structures and sugar residues [23, 24]. The interaction with the glycosylated enzyme is non-covalent, and binding strength depends on the specific association constants. This approach is cost intensive because it affords the production of a further immobilized protein.

Secondly, selective chemical modification may be performed because the carbohydrate residues have a distinct low reactivity. This can be done by periodate oxidation which cleaves C–C bonds bearing adjacent hydroxyl groups and converts them to aldehydes [25,26]. The generated dialdehyde can react with a variety of nucleophiles – usually primary amino groups on the surface of carrier materials. The resulting Schiff bases can be further stabilized by sodium borohydride, sodium cyanoborohydride or pyridine–borane reduction [27].

3.1.2
Synthetic Functional Groups

Additional reactive groups can be introduced into enzyme molecules by special chemical modification. Of particular interest is the conversion of carboxylic acids to amines because it converts the original residue into a more potent nucleophile and changes the surface charge. This can be done by soluble carbodiimide chemistry [28]. Activation of the carboxylic acid of enzymes by condensing reagents and subsequent reaction with nucleophiles on the surface of the carrier material often leads to a significant loss of enzyme activity. The amount of condensing reagent, usually water-soluble carbodiimides [e. g. 1-ethyl-3-(3-dimethylaminopropyl)carbodiimide, EDC], has to be balanced carefully for those carboxyl groups that are not necessary for the catalytic integrity of the enzyme. However, mild conditions for enzyme immobilization on special amino functionalized agarose has recently been worked out [29].

The surface charge of the molecule can be also altered, for instance, by transforming thiols or amines into carboxylic acids. Thereby, stronger interactions with anion exchange matrices can be attained. Thiols can be modified by α-haloacetates [30]. Amino groups are usually modified using dicarboxylic acid anhydrides (e.g. succinic anhydride). Reaction with tyrosine, histidine, cysteine, serine and threonine side chains produces unstable products, especially at alkaline pH [31], so the reaction is fairly specific for primary amino groups. In a similar way, amyloglucosidase has been modified by an ethylene/maleic anhydride copolymer resulting in a polyanionic conjugate that was bound to polycationic carriers. The carrier-fixed enzyme was used repeatedly in starch hydrolysis. However, operational stability was not very high, allowing 10 cycles for the best preparation (30% substrate concentration, 50°C, 0.5 g L^{-1} active enzyme) [32].

Recombinant DNA technology offers new ways to design immobilized enzymes by constructing fusion proteins with specific binding tags. However, it should kept in mind that the expression and structure of an altered enzyme may also be influenced in a negative way. Furthermore, matrix-modified enzyme interactions are non-covalent. Only some examples of binding or affinity tags should be mentioned here, such as IgG binding domains [33], histidin tags [34], arginine tags [35], cellulose binding domains [36], streptavidin [37], binding domain of streptavidin. These tags have been fused to enzymes which then were bound onto special carriers.

Manipulations regarding cross-linking of enzymes to yield macromolecular aggregates should also be mentioned here. A huge variety of homo- or heterofunctional cross-linking agents have been developed [38], some of which are group specific. In biotechnology most often bis-epoxides, bis- or trisfunctional aziridines, or dialdehydes (e.g. glutardialdehyde) are used. Direct cross-linking of enzymes produces biocatalysts with poor characteristics concerning mechanical rigidity, compressibility, and hydrodynamic behaviour. Improvements are achieved by co-cross-linking with other inert materials like polymines [18] or by combination with other methods (cross-linking on solid supports or in preformed gels, see below, [39]). Cross-linking of enzyme crystals or amorphous enzyme precipitates has already been mentioned [109].

A different strategy for preparing enzymes for immobilization is to introduce vinyl groups by alkylating or acylating enzymes with activated vinyl monomers [40]. The modified enzymes are then polymerized with mono- and bifunctional acrylamide derivatives to yield elastic particles of irregular shape after crushing of the formed polymer blocks. Such copolymerization processes have yielded stable industrial biocatalysts for pharmaceutical application which are especially suited for stirred tank applications [41].

Recently, an efficient way to encapsulate lipases in hydrophobic sol-gel materials has been published [42]. By introducing methyltrimethoxysilane/polydimethylsilane in water/lipase/polyvinyl alcohol mixtures, gel formation occurs as catalyzed by sodium fluoride. The gels were dried and washed with organic solvents. For *Pseudomonas* lipase the esterification activity in organic solvents was dependent on the methyltrimethoxysilane content with activity yields up to 93%.

Modification of the surface properties by introducing additional hydrophobic residues via mixed carboxylic acid anhydrides of fatty acids and oxa derivatives [43] or amphipathic molecules like polyethylene glycol [44] has been generally used to effect the stability or reaction rates of enzymes in non-polar organic media. In some cases the enzyme became insoluble in aqueous systems [43] or soluble and active in organic solvents [45, 46].

3.2
Carrier Materials and Functional Groups

The characteristics of the carrier have a strong influence on the performance of an immobilized enzyme. The following properties should be well selected and balanced for a specific biotransformation:

- Functional groups: The type of activation, presence, distribution and density of functional groups determines the activity yields of an immobilization reaction, and stability and operational stability of the carrier-fixed enzyme. As shown above most immobilization procedures proceed via a nucleophilic attack of amino groups on activated carrier functional groups. This is in many cases the appropriate way to couple an enzyme on a carrier surface, because it avoids the enzyme getting into contact with highly reactive organic compounds of low molecular weight. The type of activation determines the coupling conditions, which of course should be in most cases as mild as possible. Exceptions are, for instance, the immobilization of enzymes from crude or semipurified mixtures, when the coupling conditions inactivate or lower undesirable enzymatic side reactions. Stability and operational stability may be explained by thermodynamic destabilization of the denatured state as a result of a decrease of the conformational space available for the denatured form through multi-point attachment. Furthermore, immobilization sometimes increases the operational stability by prevention of aggregation and/or proteolysis.

- Permeability and surface area: In most cases a large surface area (>100 m^2 g^{-1}) and high porosity are desirable, so that enzyme and substrate can easily penetrate. A pore size of >30 nm seems to make the internal surface accessible for immobilization of most enzymes.
- Hydrophilicity/hydrophobicity of the carrier matrix, which influences type and strength of non-covalent protein–matrix interaction. In addition, it can influence the adsorption, distribution and availability of the substrate and product.
- Insolubility: This is essential, not only for prevention of enzyme loss, but also to prevent contamination of the product by dissolved matrix and enzyme.
- Mechanical stability/rigidity: These properties are dependent on the type of reactor. If used in a stirred tank reactor, the support should be stable against sheer forces to minimize abrasion. Production of fines (particles below $100-50$ µm) can lead to plugging of sieve plates and filters.
- Form and size of support: The particle size will have an influence on filtration times from stirred tank reactors in repeated batch mode. Furthermore, this factor is important for the performance in column reactors regarding back pressure and flow rates, which of course are correlated. For this purpose a size of spherical particles in the range of $150-300$ µm is preferred.
- Resistance to microbial attack: During long term usage the support has to be stable against microbial degradation.
- Regenerability: This property is of interest in case of expensive carrier materials.

A detailed classification of carrier materials is available [47]. Basically, carrier materials can be divided into those of inorganic and organic origin.

3.2.1
Inorganic Carriers

Inorganic carriers (e.g. glass, silica gel, alumina, bentonite, hydroxyapatite, nickel/nickel oxide, titania, zirconia) often show good mechanical properties, thermal stability, and resistance against microbial attack and organic solvents. On the other hand, non-porous materials like metal and metal oxides only have small binding surfaces. Minerals usually display a broad distribution of pore size.

Silica gels are available under the trade names Promaxon (Promat), Spherosil (Rhone-Poulenc) or Aerosil (Degussa). Silica compounds can be prepared with defined pore sizes and binding surfaces (controlled pore glass, CPG), but they suffer from high production costs and show limited stability under alkaline conditions. Furthermore, silica carriers are chemically inert and need activation and modification. Usually, they are treated with aminoalkyl triethoxysilanes to introduce amino groups, which can subsequently be activated for enzyme coupling reactions by a variety of different methods as outlined in [48]. The most common procedure is the reaction with dialdehydes to transform amino groups into reactive unsaturated aldehyde residues which will be discussed below.

In this way, penicillin G amidase was first coupled to dextran (see below) with the modified enzyme showing significantly increased thermostability, and this preparation was then immobilized on amino-activated silica gels [49].

Organopolysiloxanes constitute macroporous inorganic synthetic materials carrying organic functional groups and are commercially available (Deloxan®, Degussa AG). They combine chemical resistance of the matrix and high loading capacity. Amino functional carriers are most interesting because they allow covalent coupling via glutardialdehyde chemistry. The straightforward reaction of the aldehyde groups with the primary amino groups of the carrier and the enzyme has been questioned. Simple Schiff bases are not that stable and the reactivity of proteins with freshly distilled glutaraldehyde is remarkably low [50]. So it was suggested that the reaction involves addition of amino groups to ethylenic double bonds of α,β-unsaturated oligomers present in commercial aqueous glutardialdehyde solution. Furthermore, the operational and storage stability of immobilized enzymes did not increase upon treatment with borohydride (own unpublished observation). Based on amino-functionalized organopolysiloxane, industrial biocatalysts have been developed. Immobilized glutaryl-7-ACA-acylase is used to split glutaryl-7-aminocephalosporanic acid to 7-aminocephalosporanic acid, a key intermediate for semi-synthetic cephalosporin antibiotics [51]. Immobilization times were found to be dependent on parameters like protein concentration, pH, and ionic strength. For instance, increasing the ionic strength led to remarkably reduced reaction times (own laboratory results).

3.2.2
Organic Carriers

3.2.2.1
Naturally Occurring Organic Carriers

Natural organic polymers – such as structural proteins (ceratin, collagen), globular proteins (albumin) or carbohydrates – are cheap starting materials for the production of support materials and are available in large quantities. From this group, carbohydrates are of special interest, because they do not suffer from biological safety aspects like protein matrices isolated from animal sources and because they are highly hydrophilic which provides a desirable microenvironment for many enzymes. Alginate [52], carrageenan [53], chitin or chitosan (prepared from chitin by deacetylation) are particularly useful for encapsulating microorganisms by ionotropic, respectively, acidic gelation [52,54]. Enzymes have been linked to carbohydrates simply by adsorption followed by cross-linking [39].

Chitosan is of importance because of its primary amino groups that are susceptible for coupling reactions. Furthermore, porous spherical chitosan particles are commercially available (Chitopearl, Fuji Spinning) allowing noncovalent or covalent attachment of enzymes [55]. This support matrix can be easily prepared [56] and activation methods have been summarized [57]. Treatment with polyethyleneimine or with hexamethylenediamine and glutardialdehyde can improve the mechanical characteristics [53,58] of the biocatalyst, which is poor otherwise. However, this is often accompanied by some activity loss or increase of diffusional limitations.

Dextran and agarose have to be cross-linked, for instance, with epichlorohydrin, to improve their mechanical characteristics and compressibility. Covalent immobilization to commercially available beaded forms (Sephadex, Sepharose) is summarized in [59]. The most widespread activation method is the cyanogen bromide method [60], yielding isourea and imidocarbonate functions which react with amine groups of enzymes to form N-substituted carbamates. Cyanogen bromide was more recently substituted by 1-cyano-4-dimethylaminopyridinium tetrafluoroborate (CDAP) [61]. However, it should be kept in mind that this linkage is unstable at pH < 5 and >10. A different way to activate carbohydrates for covalent coupling is the periodate cleavage that generates aldehyde functions. Reaction of the aldehyde moieties with amine groups of enzymes to Schiff bases is fast. A stable linkage is achieved by reduction with borohydrides (see Sect. 3.1.1.2). In a modification of this method, Sepharose was first transformed to glyoxyl-Sepharose by etherification with 2,3 epoxypropanol, then further oxidized by periodate to yield an agarose-aldehyde gel, which allows fast coupling reactions with amino groups [62].

Commercially available preactivated matrices [N-hydroxysuccinimide ester or hydrazide, Affi-Gel (BioRad) or a variety of different activated Sepharose derivatives (Amersham Pharmacia Biotech)] have been primarily developed for affinity chromatography and are generally too expensive for use as a biocatalyst matrix.

Cellulose is also an acceptable support and can be activated in a similar way. The binding capacity for enzymes is generally lower as compared to agarose but it is inexpensive and commercially available in fibrous and granular forms. Some drawbacks are the low particle sizes, which impairs the use in rapid high-pressure applications, and its susceptibility to microbial cellulases. Some immobilization and engineering aspects have been reviewed [63].

3.2.2.2
Synthetic Organic Carriers

Synthetic organic polymers display the greatest variability with regard to physical and chemical characteristics. In principle, they can be adapted to the requirements of nearly any enzymatic process. Furthermore, they are inert to microbial attack. They are commercially available as purely adsorptive resins, as ion exchangers with a variety of different basic and acidic groups, or as preactivated supports carrying for instance epoxide (Eupergit®, Roehm Pharma) or azlactone groups (Emphaze™, 3M). The main synthetic polymers are polystyrene, polyacrylate, polyvinyls, polyamide, polypropylene and copolymers based on maleic anhydride and ethylene or styrene, polyaldehyde, and polypeptide structures.

Polystyrene was the first synthetic polymer used for enzyme immobilization. Binding occurred mainly by adsorptive forces. Usually, binding of the enzyme is favoured at low salt conditions. However, the hydrophobicity of the matrix often leads to partial enzyme denaturation during the binding process and therefore to low activity yields. By combining adsorption and cross-linking, penicillin G amidase was covalently coupled to a fairly hydrophilic polyacrylic ester (XAD-7), which had been pre-coated with glutardialdehyde at alkaline pH [64]. The same enzyme was adsorbed on macroporous polymethacrylate, which was optimized with regard to monomer-co-monomer composition, and cross-linked by glutardialdehyde treatment [65]. However, the biocatalyst obtained in the latter study was of limited operational stability and could not match the demands of an industrial process.

In a different approach penicillin G amidase was coupled to Nylon 6, which was partially hydrolyzed and activated by conversion of the liberated amino groups with N-hydroxysuccinimide and dicyclohexylcarbodiimide [66]. Immobilization yields were fairly high, and preliminary data showed a good operational stability in a column reactor.

Ionic binding on ion-exchange matrices of mixed monomers [e.g. weakly basic styrene–divinylbenzene resin (Duolite A 568)] often leads to higher immobilization yields as compared to adsorption on hydrophobic adsorber resins. Immobilization can indeed be improved by the above mentioned cross-linking reactions. This is of particular importance if the biocatalyst is used in aqueous systems with a high salt content or in the case of processes which liberate charged compounds, thereby increasing the ionic strength (e.g. hydrolysis of esters or amides). However, most non-covalent immobilizations on (organic) supports are of interest for non-aqueous applications. Lipases and, to a lesser extent, proteases are the enzymes mainly used in organic solvents. Their immobilization and application technologies have been recently reviewed [16,67,68]. The hydrophobic polypropylene matrix (Accurel EP 100, Akzo Nobel) is one of the most often used supports for lipases [69,70]. Immobilization is very easy and straightforward. The enzyme protein is bound from aqueous solutions in which the support has been suspended. The carrier-fixed lipase is filtered off, washed and dried. However, there are a lot of parameters which influence the performance in organic solvents, e.g. water content of carrier-fixed enzyme or additives added prior to drying of the biocatalyst [71,72]. Reaction conditions for the production of bioesters, structured triacylglycerides, optically active compounds, or sensitive organic esters are dependent on the lipase specificity and the special performance of the biocatalyst [73]. A comparison of lipases immobilized on different carriers can be found in [74,75].

Activation of synthetic polymers can be achieved by a variety of methods. Basically, reactive groups can be introduced during polymerization by selection of suitable monomers or can be generated through modification of the polymer backbone.

Epoxyactivated organic carriers – the most often used "ready to go" enzyme support – can be prepared by copolymerization of glycidylmethacrylate with ethylenedimethacrylate [76] or methacrylamide with N,N'-methylenebisacrylamide and allylglycidyl ether (Eupergit) [77]. A different method of synthesis uses vinyl acetate and divinylethylene urea to build up the polymer backbone which is then surface modified with oxirane groups after hydrolysis of the acetate groups (VA-Epoxy BIOSYNTH®, Riedel-de Haen). Particle size, porosity, and the density of the epoxy groups can be adjusted by appropriate monomer mixtures and polymerization conditions. Generally, polymethacrylates are characterized by high mechanical strength and chemical stability.

They can be used directly for enzyme coupling primarily via protein amino groups in buffered aqueous solution, or the epoxy groups can be modified by a variety of different reagents, for example, introducing spacers with hydrophilic or hydrophobic groups. By this method a suitable distance of enzyme to carrier backbone can be adjusted to allow optimal enzymatic flexibility. Furthermore, the microenvironment of a bound enzyme can be influenced selectively. Excess of epoxide groups can be blocked after enzyme coupling with low molecular weight amino compounds (e.g. glycine, ethanolamine) as a further way to modify the biocatalyst. Again, a lot of work in this direction has been done with penicillin G amidase [78–80]. This enzyme is of special interest as the world-wide production of 6-aminopenicillanic acid (6-APA) proceeds via the one-step hydrolysis of penicillin G (penicillin G amidase) or penicillin V (penicillin V amidase). Our own experiments have shown that operational stability is influenced by the density of the oxirane groups and the carrier backbone.

Inactivation under more stringent production conditions (37°C) revealed a significantly more stable biocatalyst when immobilization was performed on a high density oxirane carrier, although some influence of the different carrier backbone could not be completely ruled out (see Fig. 4). One possible explanation is an improved conformational stabilization by a higher degree of multipoint attachment.

Compared to glutardialdehyde mediated coupling, reactions with oxirane groups are slower. However, here again coupling times can have a considerable influence on stability. This is illustrated in Fig. 5 which shows the storage stability of penicillin G amidase immobilized on a polymethacrylate carrier in relation to the coupling time.

Similar results were obtained for the immobilization of glutaryl-7-ACA-acylase (own laboratory experiments). Figure 6 demonstrates the decrease of activity in the supernatant of the coupling reaction mixture and the concomitant increase in carrier-bound activity. Maximum activity was measured after only 6 h, leaving about 20% of the initial activity in solution. During the next 14 h the remaining soluble enzyme was immobilized. However, an increase in activity could not be measured under standard conditions due to diffusional limitation and internal pH-shifts in the biocatalyst particles. According to these data,

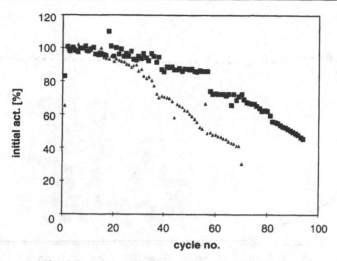

Fig. 4. Operational stability of penicillin G amidase immobilized on epoxy carrier. Batch cycles were run in a 1.5-l laboratory reactor at 10% penicillin G, 6 kU L^{-1}, 36°C. pH was maintained at 8.0 by the addition of 2 N NH$_3$. Initial splitting time was 65 min. The initial activity (relative base consumption per min at the beginning of each batch conversion) is plotted against the cycle number. PGA was immobilized on polymethacrylate with an epoxy density >2000 µmol g^{-1} (*squares*) or of 600 µmol g^{-1} (*triangles*)

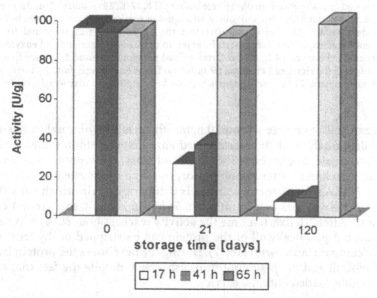

Fig. 5. Storage stability of penicillin G amidase immobilized on epoxy-activated polymethacrylate in relation to the immobilization time. The immobilization reaction was allowed to proceed for 17, 41, and 65 h, respectively, and was terminated by the addition of ethanolamine, the catalysts were washed and the activity (5 mM potassium phosphate pH 8, 28°C, 10% penicillin G, pH-stat 8.0) was determined immediately (t=0), and after storage in a sealed vessel at 25°C for 21 and 120 d

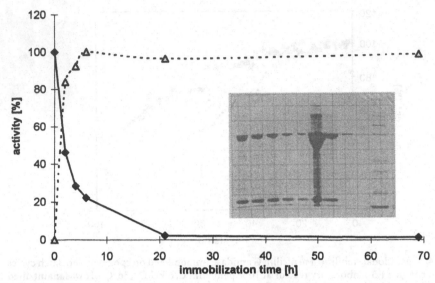

Fig. 6. Kinetics of immobilization of glutaryl-7-ACA-acylase on epoxy-activated polymethacrylate. The Gl-7-ACA-acylase was incubated with the epoxy-activated carrier. At definite times aliquots were taken from the reaction suspension. Supernatant and carrier-fixed enzyme were separated by centrifugation. The carrier-fixed enzyme was washed with water to remove non-covalently linked enzyme. The activities of the immobilized enzyme and supernatant were determined (5 mM potassium phosphate buffer pH 8, 37 °C, 2 % glutaryl-7-amino cephalosporanic acid, pH-stat 8.0). Simultaneously, an aliquot of carrier-fixed enzyme was boiled in sodium dodecylsulfate (SDS)/glycine buffer and the supernatant was subjected to SDS-polyacrylamide electrophoresis (see insert: from left to right: *lane 1* Carrier-fixed enzyme, 2 h; *lane 2* Carrier-fixed enzyme, 4 h; *lane 3* Carrier-fixed enzyme, 6 h; *lane 4* Carrier-fixed enzyme, 21 h; *lane 5* Carrier-fixed enzyme, 69 h; *lane 6* Dialyzed enzyme; *lane 7* Supernatant, 2 h; *lane 8* Supernatant, 21 h; *lane 9* Supernatant, 69 h; *lane 10* Molecular weight calibration markers)

immobilization is complete within 20 h, but the resulting biocatalyst has only poor stability. Analysis of the carrier-fixed enzyme via sodium dodecylsulfate polyacrylamide electrophoresis (SDS-PAGE) supports the hypothesis of further multipoint attachment after simple adsorption and monovalent linkage (see Fig. 6). Gl-7-ACA from *Acinetobacter* sp. is a dimeric protein which consists of an approximately 16 kDa α-subunit and a 54 kDa β-subunit, that are not covalently linked. After 2-h fixation time the activity reaches about 80 % of its maximum value, but practically all of the protein can be stripped off by treatment with SDS (compare lane 1 with lane 7). During the next hours the protein bands of the α-subunit and the β-subunit diminish slowly, despite the fact that additional protein is loaded onto the carrier.

4
Mass Transfer Effects

Immobilization of enzymes means a deliberate restriction of the mobility of the enzyme, which can also affect mobility of the solutes. The various phenomena, referred to as mass transfer effects, can lead to a reduced reaction rate, in other words to a reduced efficiency as compared to the soluble enzyme.

A reduced reaction rate may result from external diffusional restrictions on the surface of carrier materials. In stirred tanks external diffusion plays a minor role as long as the reaction liquid is stirred sufficiently. Further, partition effects can lead to different solubilities inside and outside the carriers. Partition has to be taken into account when ionic or adsorptive forces of low concentrated solutes interact with carrier materials [81 – 83]. The most crucial effects are observed in porous particles due to internal or porous diffusion as outlined below.

4.1
Porous Diffusion

Considerable importance is given in the literature to describe theoretical and practical aspects of mass transfer effects on immobilized enzymes.

All reactions of immobilized enzymes must obey the physicochemical laws of mass transfer and their interplay with enzyme catalysis. The question is, therefore, what are the reasons for restrictions caused by mass transfer, and how can they be avoided if necessary?

The mathematical description of diffusional limitations of enzyme kinetics in the combined action with mass transfer has for years been well established. The presentation of these interactions is very complex, in particular when terms for product inhibition, proton-generation or enzyme deactivation are incorporated in addition to comparably simple Michaelis–Menten kinetics.

For Michaelis–Menten based enzyme kinetics the extent of mass transfer control is usually expressed by the efficiency coefficient or effectiveness factor η, expressed as:

$$\eta = \frac{v' \text{ (substrate conversion rate of immobilized enzyme)}}{v \text{ (substrate conversion rate of free enzyme)}} \tag{2}$$

Numerically calculated values of η are available when only substrate diffusion in Michaelis–Menten type kinetics is considered. They can be presented in graphical form, expressing η as a function of the Thiele modulus Φ_R and of a dimensionless substrate concentration or occasionally in its reciprocal form as a dimensionless Michaelis constant β according to Lee and Tsao [84] (reviewed in [82,83]; see Table 3 for explanation of symbols), hence:

$$\eta = f(\Phi_R, \beta) \tag{3}$$

$$\Phi_R = R \times \sqrt{\frac{V_m}{D_{eff} \times K_M}} \tag{4}$$

$$\beta = \frac{S}{K_M} \tag{5}$$

Table 3. Effectiveness factors η for immobilized penicillin G amidases[a]

	S (mM)	Φ_R	η
PGA on Eupergit®C	268	68	1.0
	10	68	0.995
	0.013	68	0.023
PGA on Eupergit®250L	268	69	1.0
	10	69	0.995
	0.013	69	0.023

[a] Penicillin G amidase was immobilized on pre-fabricated carriers or insolubilized as cross-linked crystals. Eupergit-related value for R (mean particle radius of swelled carrier) was 80 µm [87]. V_m (assuming maximum intrinsic activity per accessible catalyst volume, based on active enzyme molecules; 1 unit = 1 µmol min^{-1} at 28°C) was 90 and 170 U cm^{-3} for Eupergit C and 250L, respectively [87]. D_{eff} (effective diffusion coefficient) was taken from literature [87] or calculated as shown in the text. K_M (intrinsic Michaelis constant) was uniformly taken as 13 mM [87] and S = 268 mM corresponds to the substrate concentration at catalyst surface of a 10% solution of penicillin G salt. η was calculated according to Atkinson et al. for spherical particles [85]. For simplification, surface and pore related indices have been omitted.

On the other hand, approximate calculations of η can also be used [85]. In this way, cumbersome solutions of the differential equations are not required if one merely wishes to obtain a general idea of substrate-related diffusional restrictions. A profound insight into effectiveness factors is given by Kasche [86] where a good correlation between calculated and experimental data is demonstrated.

To demonstrate limitations by porous diffusion, available data for penicillin G amidase on various carrier materials were collected to calculate and compare efficiency coefficients by insertion into Eqs. (3–5) (Table 3).

Data related to particle size and binding capacity are of major importance. The binding capacity of enzyme carriers for proteins is $\approx 0.1-10\%$ of the carrier weight. Prefabricated carriers are provided with 100–1000 µm in diameter. Spherical carriers such as Eupergit C particles have an average diameter of 160 µm [87].

Effective diffusion coefficients in carriers are dependent on their pore diameters (e.g. 25 and 250 nm in Eupergit® C and Eupergit® 250 L, respectively). The diffusion coefficient of penicillin G in free solution (D_0) was found to be 4.0×10^{-6} cm^2 s^{-1}. The effective diffusion coefficient (D_{eff}) in Eupergit® was 30 and 67% of this value when based on average pore diameters of 25 and 250 nm, respectively [87].

Assuming that the above values are representative for carrier-fixed enzymes, the effectiveness factors for penicillin G amidase from E. coli are presented in Table 3.

In conclusion, estimations of η indicate that there are no diffusional limitations as long as the substrate concentration is high. This is the case under the starting conditions of technical penicillin hydrolysis as long as only the substrate is taken into consideration. At lower substrate concentrations pore diffusion leads to low efficiency.

Low substrate concentrations occur in hydrolytic reactions when substrate conversion is close to completion. Under these conditions, however, other factors such as various kinds of product inhibition [87] will govern the reaction rate. Presumably, this will further increase the mass transfer limitations [88]. An additional effect, which can overshadow all other factors at any substrate concentration, is the formation of reaction-generated proton gradients (see below).

In practice it is desirable to be able to detect substrate-related porous diffusion reliably by simple means. A popular method is to assay the enzyme activity at varying substrate concentrations. At low substrate concentrations diffusional restrictions are likely to be predominant. They can be detected by graphical evaluation. Instead of straight lines somewhat curved lines are obtained in the case of v/S vs. v-plots according to Eadie-Hofstee [89]. However, the extent of the curvature is not necessarily as great as would be required to detect diffusional control beyond all doubt. Thus one cannot exclude diffusional effects if non-linearity is not observed.

To avoid diffusional limitations it is advisable to assay the enzyme activity under more drastic conditions. Amongst other things, this means increasing stirrer speed to exclude external diffusion, crushing the particles to reduce porous diffusion, increasing the substrate concentration to about $\geq$ 100-fold of K_M-value to avoid lack of substrate at the center of the particles or adding buffer to avoid pH-shifts. If the reaction rate is increased by any of these means it is likely that diffusional control is operative and can to some extent be reduced or even eliminated.

Appropriate means which can be used to increase efficiency are as follows:

- Decreasing the particle size of the carriers. In technical applications the lower limit is a diameter for spherical particles of $\geq$ 100 μm, which allows the convenient retention on a sieve plate even in large enzyme reactors. For smaller sized enzyme crystals other retention techniques have to be applied.
- Lowering of enzyme loading is recommended for enzymes with high specific activity which is easily achieved by common fixation methods. Enzyme activity in crystals was diluted by co-crystallization with inactivated enzyme [6]. On the other hand, in the case of enzymes with low specific activity, tight packing in crystals may be a useful fixation method when excessive inert carrier material would otherwise preclude reasonable reaction conditions.
- Preferential binding on the outer shell of carrier materials will enable increased efficiencies [90]. An increased efficiency of about a factor of two may be expected when only the outer shell at 10% of the radius is occupied by enzyme [91].

4.2
Reaction (Dynamic) and Support-Generated (Static) Proton Gradients

Cross-linked crystals from subtilisin exhibited 27 times less activity than soluble subtilisin in the hydrolysis of benzoyl-L-phenylalanine ethyl ester. Denaturation of the enzyme and restrictions from substrate-dependent internal diffusion were ruled out. A shift in the pH-dependence of the maximum activity to higher pH-values was observed which was explained by intermolecular electrostatic

interactions increasing the pKs of the catalytic residues of the active centers [6].
The observed pH-shifts in activity may indeed also be caused by:

1. Static proton or substrate gradients caused by partition of charged groups near surfaces with stationary charges [92,93].
2. Dynamic proton gradients due to hydrolytic reactions with liberation of protons as has frequently been observed for immobilized enzymes [10,94–97].

In the first case the observed pH-shift in activity can be reduced by using high ionic strengths which minimize the gradients caused by the stationary charges of the support for the immobilized enzyme. These static gradients are of little technical relevance in reactions where charged substrates with high ionic strengths are used.

Besides the static proton gradients, the dynamic gradient formed by reactions inside the particles must also be considered. In contrast to substrate and product related mass transfer limitations, even tiny amounts of protons may overshadow any other effects [94].

The reason for the latter is that in unbuffered proton forming reactions such as

$$R_1COOR_2 + H_2O \rightarrow R_1COO^- + H^+ + R_2OH \tag{6}$$

hydrolysis of only 1% of a 10^{-3} M substrate means formation of 10^{-5} M protons (corresponding to pH 5), which may be significantly different from the external pH-value and the optimum pH of the desired enzyme reaction.

This consideration is significant for hydrolytic reactions with hydrolases such as lipases, esterases, and amidases. These include penicillin amidases (synonymous with penicillin acylases) and cephalosporin acylases which are used for hydrolytic cleavage of penicillins and cephalosporins in thousands of tons per year [98]. These hydrolyses have to be performed at a pH-value of ~8 which is close to the optimum pH of the enzyme. Lower pH-values lead to lower reaction rates and reversibility of the reaction, and hence to a significant loss in product formation. Higher pH-values are not advisable owing to the instability of the reaction partners. Moreover, addition of buffers is not accepted because of the costly removal of the buffer components.

For a better understanding, several authors have created models for mathematical description. They have extended Michaelis–Menten-like kinetics by adding proton-generating terms [99] and have considered product inhibition and pH-values above the optimum pH of the soluble enzyme [100]. In addition, they included facilitated transport due to buffering capabilities of substrates and products [95] or of added buffers [96, 99, 101]. In view of the complexities of exact calculations only a basic outline can be given here.

Enzyme reactions are in many cases characterized by a bell-shaped pH/activity profile:

$$V_m = \frac{V_m'}{1 + \dfrac{I}{K_1} + \dfrac{K_2}{I}} \tag{7}$$

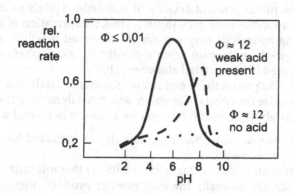

Fig. 7. Efficiency of proton-generating enzyme reactions. Curve *a* Soluble enzyme reaction with a bell-shaped activity/pH-profile $K_1 = 10^{-4}$ M, $K_2 = 10^{-8}$ M, $\Phi < 0.1$ (- - -). Curve *b* Severe diffusional control of an immobilized enzyme when $\Phi = 12$, S = 145 mM, $K_M = 14.5$ mM (- - -). Curve *c* Moderate internal diffusion control of an immobilized enzyme in the presence of a weak acid (conditions of curve b plus 10 mM weak acid) (......) (simplified graphic representation derived from [101])

where I is the actual proton concentration, K_1, K_2 are the pH-profile related constants, giving a profile such as that shown in Fig. 7, curve a.

Accumulated protons generated within the pores of a catalyst cause the formation of a proton gradient. The extent of such a gradient is predominantly a matter of the proton formation rate, which is dependent on the immobilized enzyme's activity and the mass transfer driven transport of protons to the outside of the catalyst particles. At steady state a mass balance occurs.

Ruckenstein [96] has calculated the resulting diffusion-restricted enzyme activities at high substrate concentrations. He teaches us in simple terms that pH-shifts and resulting reductions in activity occur even at substrate concentrations high enough to exclude any substrate-related diffusional restrictions, i.e. at effectiveness factors close to 1.

In his calculations, Ruckenstein introduced a modified and proton-related Thiele modulus. He replaced the substrate-related K_M by the pH-related K_1 [Eq. (7)] and used the effective diffusion coefficient of protons instead of substrates. It turned out that even at comparably low values of this modified Thiele modulus the reaction rates decrease and, instead of a bell-shaped profile, a curve which becomes flatter towards alkaline pH-values is obtained (Fig. 7, curve b).

In the presence of weak acids, facilitated transport of protons was assumed. In consequence, significantly increased transport of protons can be achieved, so that the activity/pH-profile is shifted to more alkaline pH-values (Fig. 7, curve c).

Experimental proof of diffusion-related pH-shifts was obtained in porous carriers for immobilized penicillin amidase. Direct fluorescence measurements of co-immobilized fluorochromes with pH-dependent fluorescence intensity revealed pH-gradients during hydrolysis [87, p 125]. These gradients decrease with increasing buffer capacity and are negligible for buffer concentrations of > 0.05 M.

For assaying the pH-dependent activity of subtilisin crystals as mentioned above, low buffer capacities were used [6,102]. Thus, the formation of a dynamic pH-gradient during hydrolysis may explain the observed pH-shift in activity. Also carrier-fixed α-chymotrypsin reveals exactly the same phenomena [86, 97], which is discussed in more detail elsewhere [103].

In order to optimize productivity and reduce the loss of catalyst or product it is advisable to minimize the effects of reaction-generated dynamic pH-gradients. Whenever practicable and useful this may be achieved in several ways:

- By reducing enzyme density and/or particle size as indicated for substrate-mediated diffusional control.
- By using buffers with sufficient capacity (> 0.05 M) that minimize the dynamic pH-gradients. Occasionally the substrates or products themselves provide such properties so that only the optimal external pH-value has to be adapted. It should not be lower than the pK-value of the weak acid and not lower than the optimum pH of the enzyme [95, 96].
- By operating at an external pH higher than the optimum pH of the enzyme.
- By co-immobilization of a proton-consuming enzyme. This was successfully done with urease, where *in situ* formed ammonia neutralized the generated protons in an extraordinarily effective manner [99].

Proton-consuming reactions such as the urease reaction are likely to show all the diffusion-related pH-shifts in the opposite direction. Obviously, they may be treated in an analogous manner.

4.3
Temperature Dependence

In a diffusion-free enzyme reaction the reaction rate increases up to a certain critical value exponentially and is described by the Arrhenius equation [82]. In diffusion-controlled reactions the reaction rate is a matter of the efficiency factor η [see Eqs. (3–5)]. In more detail, the maximum reaction rate is expressed within the root of Eq. (4). Conclusively, the temperature dependence is a function of the square root of the enzyme activity. In practice, immobilized enzymes are much less temperature dependent when their reaction rate is diffusion controlled.

4.4
Stability Assessment

It is useful to differentiate between storage stability and operational stability. Storage stability is provided by an appropriate formulation. Mostly, several additives protect the enzyme from denaturation under recommended storage conditions. By these means, shipment and distribution are supported. The operational stability is a matter of working conditions and of detrimental effects from the selected reaction conditions, such as pH, temperature, solvents, impurities, and other factors that contribute to protein denaturation or modification of functional groups and thus enhance inactivation. The operational stability determines the enzyme's performance and determines cost benefits as discussed above. For diffusional-free reactions time-dependent inactivation can proceed along various inactivation rates, such as exponential activity decay according to a first-order reaction. In this case, the rate of inactivation can be described in a similar manner to that given for temperature dependence by the Arrhenius equation (see above).

In strictly diffusion-controlled reactions, i.e. at low efficiencies, simply spoken, only a small fraction of the total immobilized enzyme quantity is working. If in the course of operation actively working enzyme is destroyed, some other previously "resting" fraction may substitute to some extent. This is one reason why immobilized enzymes may erroneously appear to be more stable than free enzymes. To avoid such misinterpretation requires the assay of the enzyme activity in the absence of diffusional limitations. For industrial applications it is more useful to consider the enzyme's performance as indicated in Fig. 1, instead of just assessing its half-life.

4.5
Other Contributions

Unfortunately, most enzymes do not obey simple Michaelis–Menten kinetics. Substrate and product inhibition, presence of more than one substrate and product, or coupled enzyme reactions in multi-enzyme systems require much more complicated rate equations. Gaseous or solid substrates or enzymes bound in immobilized cells need additional transport barriers to be taken into consideration. Instead of porous spherical particles, other geometries of catalyst particles can be applied in stirred tanks, plug-flow reactors and others which need some modified treatment of diffusional restrictions and reaction technology.

5
Performance of Immobilized Enzymes

5.1
Enzyme Formulation and Activity

For reactions in water-immiscible organic solvents the simplest method of immobilization is to use dried enzyme powders and to suspend them in the solvent [104]. They can be removed by filtration or centrifugation and reused afterwards. However, even this simple method can cause trouble.

A powder of lipoprotein lipase (LPL) esterified an organic substrate in toluene at a rather poor reaction rate (Table 4), which was to some extent explained by adhesion of the sticky enzyme powder to the surface of the reaction vessel [7]. When polyethylene glycol (PEG) was bound covalently to LPL and this modified enzyme was dissolved in toluene, approximately 3.5 U mg^{-1} of enzyme protein were assayed. After simple addition of PEG to the reaction mixture together with LPL powder, the same poor reaction rate of the enzyme powder alone was observed. On the other hand, when LPL powder was lyophilized together with PEG the resultant preparation had an activity of 1.8 U mg^{-1}. In this case, the enzyme

Table 4. Acetylation activities of dispersed lipases

Enzyme	Formulation	Activity[a]
Acetylation of (±)-sulcatol in toluene		
LPL	powder	<0.06
	powder + PEG	<0.06
	lyophilized together with PEG	1.8
	with covalently bound PEG	3.5
	carrier-fixed	2.3
Acetylation of sec-phenylethyl alcohol		
PSL	powder[b]	1.5 (15)
	carrier-fixed enzyme[b]	0.08 (8)
CAL-B	powder[b]	1.7 (17)
	carrier-fixed enzyme[b]	0.4 (40)

[a] Units/mg dried preparation; figures in parentheses per enzyme protein.
[b] In n-hexane, water saturated, at 2°C in units mg^{-1}; either with lyophilized preparations (powders) containing roughly 10% protein or with lyophilized carrier-fixed preparations (roughly 1% protein by weight) [120].

powder was highly dispersed in the organic medium and could easily be separated by centrifugation. Adsorption of LPL on a carrier material (Celite®) yielded a reaction rate of 2.3 U mg^{-1} [7].

These experimental findings indicate the importance of proper formulation, even if these results are not necessarily merely a matter of distribution, since water activity, enzyme conformation and stability also affect the assayed activity [6, 105–108]. Improper storage conditions and formulations of powders not specially developed for applications in solvents are likely to be responsible for low activities, for example, when air humidity or reaction-generated water make hygroscopic lyophilisates sticky. In conclusion, proper formulation of lyophilisates or of carrier-fixed preparations will result in appreciable activities.

In contrast to the poor reaction rates reported elsewhere [109], crude powder of lipase from *Pseudomonas cepacia* (PSL) acylates secondary phenylethyl alcohol at reasonable reaction rates [120] (Table 4). Based on active enzyme protein, the highest reaction rates were observed for carrier-fixed enzymes. This is also true for lipase from *Candida antarctica* (CAL-B). PSL crystals catalyzed with comparable activity when they had been pretreated with surfactants [109].

With regard to distribution there are several means of enhancing the reaction rates for immobilized enzymes:

- In the case of enzyme powders, proper formulation and storage conditions will ensure reasonable activities. This is achieved by the addition of various compounds prior to the drying process to improve distribution. Such compounds can also be quite useful as stabilizers and as protective agents. Suitable preparations are usually provided by enzyme manufacturers for dedicated enzymes.
- Solubilization of the enzyme in an organic solvent by covalent coupling of lipophilic compounds. Immobilization must then be achieved by inclusion

into membrane devices or by separation in two- or multi-phase reactions [110–112].

- Immobilization of the enzyme. Even simple adsorption onto porous carriers may significantly increase the availability of single catalytic centers, and also ensures easy separation from the product. Cross-linked enzyme crystals appear to require surfactants to compensate for their low activity in water-immiscible organic solvents [109].

Immobilization entails the introduction of inert carrier materials. With the exception of enzyme crystals or enzymes within enzyme membrane reactors, the inert carrier material is usually present in excess of the active enzyme protein. The carrier for crystals is the enzyme protein itself, whose specific activity strictly determines the weight-related activity of the crystal. This is different in the case of dedicated carrier materials. The range of active enzyme can be quite broad because enzyme loading can be adjusted according to the binding capacity of the carrier material. It is therefore possible to establish a well-balanced relationship between reaction volume and carrier by adjusting the amount of bound enzyme on the carrier.

This was investigated for the hydrolysis of cephalosporins with carrier-fixed glutaryl-7-ACA acylase, for example [113]. The application suffers from a rather low specific enzyme activity of about 4 U mg^{-1} [assayed with glutaryl-7-ACA (7-β-(4-carboxybutanamido)aminocephalosporanic acid at 37°C, pH 8]. About 60 g of dried carrier-fixed enzyme (5 kU L^{-1}) hydrolyzed more than 95% of 75 mM substrate within 50 min at 20°C. The enzyme occupies about 6% of the reactor volume of a stirred tank. Significantly greater amounts of carrier would retain more product within the pores and would necessitate additional washing steps and hence lead to dilution. On the other hand, higher enzyme loading on the carrier would further reduce its volume in the reactor and would probably lead to losses of enzyme activity during the frequently repeated recycling steps.

For poorly active enzymes a high catalyst density as in cross-linked crystals is of advantage. Limitations due to low activity are less a question of technology than of cost targets because enzymes are usually expensive when their specific activity is low. This is not necessarily true in the case of some proteases and lipases produced on a large scale as ingredients of detergents. Thus cost targets can be met more easily not by extreme stabilization but by low-priced enzyme. Unfortunately, some commercially available crude enzymes are a mixture of different enzyme species [114] which is the reason for some of the drawbacks described here. In order to obtain reproducible results independent of lot variations, it is advisable to use only enzymes which are at least provided with specifications ensuring adequate quality.

5.2
Stability

Increased operational stability of immobilized enzymes is essential in order to achieve the cost benefits already mentioned. Enzyme stability can be controlled by assaying the activity decay over time until half-life activity is reached. This is a useful means of control when thermal inactivation takes place according to first-order kinetics.

Complications occur when more or less stable mixtures of different enzyme species are present, either as a result of more or less tight binding or differing intrinsic stabilities.

The role of mass transfer effects, whether occurring accidentally or by design, is ambivalent, causing Trevan to ask the question "Diffusion limitation – friend or foe?" [115]. Lower activity as a result of low efficiency indicates that only a minor portion of enzyme is active during operation. The other unused portion may, in simple terms, replace the enzyme as it is inactivated step by step. In other words, mass transfer controlled reactions appear to be much less sensitive to decay of enzyme activity, thus falsely creating an impression of stabilization. Under harsh reaction conditions it may be advantageous to operate under these conditions to keep the reaction rate constant until the diffusion limitation disappears [82, 115, 116].

With regard to such effects it is advisable not only to determine the operational stability by tracing the time course of activity but to follow its productivity, or vice versa its consumption, related to the formed product.

Carrier-fixed penicillin G amidase in the multi-ton hydrolysis of penicillin G is a useful example to illustrate enzyme consumption. The enzyme is applied in stirred tanks with sieve plates at the bottom to retain the enzyme particles when the product solution is drained off. The pH-value is kept constant by controlled feed of ammonia solution. Fresh substrate solution is refilled about a thousand times or more. The consumption of enzyme in such a process is below 10 mg kg^{-1} (0.2 kU kg^{-1}) of isolated 6-APA when the enzyme activity is determined with penicillin G solutions at 28° C and pH 8.0. Under identical conditions the consumption of soluble enzyme for each tank filling would be beyond all reasonable cost.

Cross-linked crystals of lipase from *Candida rugosa* (CRL) were applied in the resolution of racemic ketoprofen chloroethyl ester. In batch-wise operation, the half-life of the catalyst was reached after about 18 cycles or, in terms of enzyme consumption, about 5.6 g of enzyme protein were consumed to prepare 1 kg of (S)-ketoprofen. CRL suffers from a low specific activity towards this poorly water-soluble substrate which may explain the high enzyme input [117].

When cross-linked crystals of thermolysin were applied in peptide synthesis in ethyl acetate, they were stable for several hundred hours at amazingly low enzyme consumption, whereas a soluble enzyme preparation became inactive within a short period of time. Again it is worthwhile to consider the quality of the soluble enzyme preparation. When soluble thermolysin was stored in mixed aqueous–organic solutions, it lost about 50% of its activity within the first day of incubation only to be then quite stable for the next 15 days. It is possible that the initial inactivation was caused by an unstable fraction of thermolysin and that crystals of thermolysin no longer contained this unstable fraction [118]. Productivity comparable to that of crystals was achieved with thermolysin adsorbed on Amberlite® XAD-7 resin which was employed in continuous plug flow reactors with *tert*-amyl alcohol as solvent [119].

With regard to purity, quality, and formulation, and hence to cost considerations, it can be useful to define "productivity" as the fermentation volume required to prepare the immobilized enzyme activity needed to synthesize a certain amount of product.

This is useful when the overall performance of an immobilized enzyme catalyzed process has to be competitive with other technologies, such as preparation of the desired product by fermentation or by whole cell biotransformations.

6
Conclusions

Enzymes have achieved acceptance as catalysts in the synthesis of chemical compounds, particularly in the fine chemicals industry for the manufacture of enantiopure compounds. Immobilization of enzymes is a useful tool to meet cost targets and to achieve technological advantages. Immobilization enables repetitive use of enzymes and hence significant cost savings. From the technological point of view, immobilized enzymes can easily be separated from the reaction liquid and make laborious separation steps unnecessary. Additional benefits arise from stabilization against harsh reaction conditions which are deleterious to soluble enzyme preparations. Due to the wide variation in the properties of the individual enzyme species and due to the varying requirements of reaction technology for the target compounds it is advisable to exploit fully the wealth of methods and techniques of immobilization. Which of the available methods is the best in the end will be decided by both the specific technical requirements and the overall business framework.

7
References

1. Hartmeier W (1986) Immobilisierte Biokatalysatoren, Springer, Berlin Heidelberg New York, pp 18–20
2. Buchholz K, Kasche V (1997) Biokatalysatoren und Enzymtechnologie. VCH, Weinheim, pp 7–11
3. Silman ICH, Katchalski E (1966) Ann Rev Biochem 35:873
4. Zelinski T, Waldmann H (1997) Angew Chem 109:746
5. Katchalski-Katzir E (1993) TIBTECH 11:471
6. Schmidtke JL, Wescott CR, Klibanov AM (1996) J Am Chem Soc 118:3360
7. Ottolina G, Carrea G, Riva S, Sartore L, Veronese FM (1992) Biotechnol Lett 14:947
8. Buchholz K (ed) (1979) Characterization of immobilized biocatalysts. VCH, Weinheim
9. Multiple authors (1983) The Working Party on Immobilized Biocatalysts Enzyme Microb Technol 5:304
10. Buchholz K, Kasche V (1997) Biokatalysatoren und Enzymtechnologie, VCH, Weinheim, pp 248–256
11. Mosbach K (ed) (1976) Methods Enzymol 44
12. Mosbach K (ed) (1987) Methods Enzymol 135
13. Mosbach K (ed) (1987) Methods Enzymol 136
14. Kennedy JF, Cabral JMS (1983) Chemical analysis. In: Scouten WH (ed) Solid phase biochemistry, analytical and synthetic aspects, Vol 66. Wiley, New York, pp 253–392
15. Clark DS (1994) Trends Biotechnol 12:439
16. Balcao VM, Paiva AL, Malcata FX (1996) Enzyme Microb Technol 18:392

17. Abdul Mazid M (1993) Bio/Technology 11:690
18. Bahulekar R, Ayyangar NR, Ponranthnam S (1993) Enzyme Microb Technol. 13:858
19. Rao ZP, Raju DR, Baradarajan A, Satyanarayana M (1984) Indian Chem Eng 26:11
20. List D, Knechtel W (1980) Industrielle Obst- und Gemueseverwertung 65:415
21. Means G, Feeney R (1990) Bioconjugate Chem 1:2
22. Edwards JO, Pearson RG (1962) J Chem Soc 84:26
23. Sharon N (1993) TIBS 18:221
24. Elgavish S, Shaanan B (1997) TIBS 22:462
25. Bobbitt JM (1956) Adv Carbohydr Chem 11:1
26. Roxer GP (1987) Methods Enzymol 135:141
27. Cabacungan JC, Ahmed A, Feeney RE (1982) Anal Biochem 124:272
28. Kurzer F, Douraghi-Zadeh K (1967) Chem Rev 67:107
29. Fernandezlafuente, R Rosell CM, Rodriguez V, Santana C, Soler G, Bastida A, Guisan JM (1993) Enzyme Microb Technol 15:546
30. Gurd FRN (1967) Methods Enzymol 11:532
31. Gounaris AD, Perlman GEJ (1967) Biol Chem 242:2739
32. Solomon B, Levin Y (1974) Biotechnol Bioeng 16:1161
33. Kondo A, Teshima T (1995) Biotechnol Bioeng 46:421
34. Ljungquist C, Breitholz A, Brink-Nilsson H, Moks T, Uhlen M, Nilsson B (1989) Eur J Biochem 186:563
35. Stempfer G, Hoell-Neugebauer B, Kopetzki E., Rudolph R (1996) Nature Biotechnol 14:481–484
36. Ong E, Gilkes NR, Antony R, Warren J, Miller RC Jr, Kilburn DG (1989) Bio/Technology 7:604
37. Huang X, Wals MK, Swaisgood HE (1996) Enzyme Microb Technol 19:378
38. Wong SS (1991) Chemistry of protein conjugation and crosslinking. CRC Press Inc
39. Stanley WL, Watters GG, Kelly SH, Chan BG, Garibaldi JA, Schade JE (1976) Biotech Bioeng 18:439
40. Jaworek DJ, Botsch H, Maier J (1976) Methods Enzymol 44:195
41. Product Profile Enzygel PGA 150, 300, Boehringer Mannheim GmbH
42. Reetz MT, Zonta A, Simpelkamp J (1996) Biotechnol Bioeng 49:527
43. Urabe I, Yamamoto M, Yamada Y, Okada H (1978) Biochem Biophys Acta 524:435
44. Inada Y, Furukawa M, Sasaki H, Kodera Y, Hiroto M, Nishimura H, Matsushima A (1995) Trends Biotechnol 13:86
45. Ito Y, Fujii H, Imanashi Y (1992) Biotechnol Lett 14:1149
46. Takahashi K, Ajima A, Yoshimoto T, Inada Y (1984) Biochem Biophys Res Commun 125:761
47. Buchholz K (1979) DECHEMA Monographs Vol. 84, Verlag Chemie
48. Haller W (1983) Solid phase biochemistry. In: Scouten, W.H. (ed.) Analytical and synthetic aspects. Wiley, New York, pp 535–599
49. Burteau N, Burton S, Crichton RR (1989) FEBS Lett 258:185–189
50. Richards M, Knowles JR (1968) J Mol Biol 37:232
51. Giesecke U, Wedekind F, Tischer W (1992) DECHEMA Biotechnology Conferences 5:609
52. Guiseley KB (1989) Enzyme Microb Technol 11:706
53. Tosa T, Sato T, Mori T, Yamamoto K, Takata I, Nishida Y, Chibata I (1979) Biotechnol Bioeng 21:1697
54. Gerbsch N, Buchholzm R (1995) FEMS Microbiology Rev 16:259–269
55. Itoyama K, Tokura S, Hayashi T (1994) Biotechnol Progress 10:225
56. Vorlop KD, Klein J (1987) Methods Enzymol 135:259
57. Scouten WH (1987) Methods Enzymol 135B:30
58. Takata I, Tosa T, Chibata I (1978) J Solid-Phase Biochem 2:225
59. Porath J, Axen R (1976) Methods Enzymol 44:19
60. Kohn J, Wilchek M (1984) Appl Biochem Biotechnol 9:285
61. Kohn J, Wilchek M (1983) FEBS Lett 154:209
62. Guisan JM (1988) Enzyme Microbiol Technol 10:375

63. Gemeiner P, Stefuca V, Bales V (1993) Enzyme Microb Technol 15:551
64. Carleysmith SW, Dunnill P, Lilly MD (1980) Biotechnol Bioeng 22:735
65. Koilpillai L, Gadre RA, Bhatnagar S, Raman RC, Ponrathnam S, Kumar KK, Ambekar GR, Shewale JG (1990) J Chem Tech Biotechnol 49:173
66. Boccu E, Gianferrara T, Gardossi L, Veronese FM (1990) IL Farmaco 45:203
67. Malcata FX, Reyes HR, Garcia HS, Hill CG, Amundson CH (1992) Enzyme Microb Technol 14:426
68. Malcata FX, Reyes HR, Garcia HS, Hill CG, Amundson CH (1990) J Amer Oil Chem Soc 67:890
69. Bosley J (1997) Biochem Soc Trans 25:174
70. Chirazyme Lipases Product Information, Boehringer Mannheim GmbH
71. Mattiasson B, Adlercreutz P (1991) Tibtech 9:394
72. Adlercreutz P (1992) Eur J Biochem 199:609
73. Mustranta A, Forssell P, Poutanen K (1993) Enzyme Microb Technol 15:133
74. Valivety RH, Halling PJ, Peilow AD, Macrae AR (1992) Biochim Biophys Acta 1122:143
75. Shaw J-F, Chang R-C, Wang FF, Wang YJ (1990) Biotechnol Bioeng 35:132
76. Svec F, Hradil J, Coupek J, Kalal J (1975) Angew Makromol Chem 48:135
77. Eur Patent 058767
78. Kolarz BN, Trochimczuk A, Bryjak J, Wojaczynska M, Dziegielewski K, Noworyta A (1990) Angew Makromol Chem 179:173
79. Erarslan A, Güray AJ (1991) Chem Tech Biotechnol 51:181
80. Dbrobnik J, Saudek V, Svec F, Kalal J, Vojtisek V, Barta M (1979) Biotechnol Bioeng 21:1317
81. Goldstein L (1976) Methods Enzymol 44:397
82. Gloger M, Tischer W (1983) In: Bergmeyer HU, Bergmeyer J, Graßl M (eds) Methods of enzymatic analysis, 3rd edn. VCH, Weinheim, vol 1, pp 142–163
83. Tischer W (1995) In: Drauz K, Waldmann H (eds) Enzyme catalysis in organic synthesis. VCH, Weinheim, pp 73–87
84. Lee YY, Tsao GT (1974) J Food Sci 39:667
85. Yamane T (1981) J Ferment Technol 59:375
86. Kasche V (1983) Enzyme Microb Technol 5:2
87. Schlothauer RC (1996) Thesis TU Hamburg-Harburg
88. Engasser J-M, Horvath C (1974) Biochemistry 13:3845
89. Horvath C, Engasser J-M (1974) Biotechnol Bioeng 16:909
90. Carleysmith SW, Dunnill P, Lilly MD (1980) Biotechnol Bioeng 22:735
91. Horvath C, Engasser J-M (1973) Ind Eng Chem Fundam 12:229
92. McLAren AD, Packer L (1970) Adv Enzymol 33:245
93. Goldstein L, Levin Y, Katchalski E (1964) Biochemistry 3:1913
94. Trevan MD (1980) Immobilized enzymes – an introduction and applications in biotechnology. Wiley, New York, pp 11–55
95. Halwachs W, Wandrey C, Schügerl K (1978) Biotechnol Bioeng 20:541
96. Ruckenstein E, Sasidhar V (1984) Chem Eng Sci 39:1185
97. Kasche V, Bergwall M (1977) In: Salmona M, Saranio M, Garattini S (eds) Insolubilized enzymes. Raven Press, New York, pp 77–86
98. Rolinson GN (1988) J Antimicrob Chemother 22:5
99. Liou JK, Rousseau I (1986) Biotechnol Bioeng 28:1582
100. Bailey JE, Chow MTC (1974) Biotechnol Bioeng 16:1345
101. Ruckenstein E, Rajora P (1985) Biotechnol Bioeng 27:807
102. ChiroCLEC-BL Information booklet (1997), Altus Biologics Inc.
103. Tischer W, Kasche V, Tibtech, submitted
104. Dickinson M, Fletcher PDI (1988) Enzyme Microb Technol 11:55
105. Klibanov AM (1997) Trends Biotechnol 15:97
106. Carrea G, Ottolina G, Riva S (1995) Tibtech 13:63
107. Tsai S-W, Dordick JS (1996) Biotechnol Bioeng 52:296
108. Zacharis E, Moore B, Halling PJ (1997) J Am Chem Soc 119:12396

109. Khalaf N, Govardhan CP, Lalonde JJ, Persichetti RA, Wang Y-F, Margolin AL (1996) J Am Chem Soc 118:5494
110. Lilly MD, Woodley JM (1985) In: Tramper J, van der Plas HC, Linko P (eds) Biocatalysts in organic synthesis. Elsevier, Amsterdam, pp 179–191
111. Scheper T (1990) Adv Drug Delivery Rev 4:209
112. Chang HN, Furusaki S (1997) Adv Biochem Eng 44:27
113. Tischer W, Giesecke U, Lang G, Röder A, Wedekind F (1992) Ann N Y Acad Sci 672:502
114. Schmid RD, Verger R (1998) Angew Chem 110:1694
115. Trevan M (1987) Tibtech 5:7
116. Naik SS, Karanth NG (1978) J Appl Chem Biotechnol 28:569
117. Lalonde JJ, Govardhan CP, Khalaf N, Martinez AG, Visuri KL, Margolin A (1995) J Am Chem Soc 117:6845
118. Persichetti RA, Clair NLS, Griffith JP, Navia AL, Margolin AL (1995) J Am Chem Soc 117:2732
119. Nagayasu T, Miyanaga M, Tanaka T, Sakiyama T, Nakanishi K (1997) Biotechnol Bioeng 43:1118
120. Biocatalysts for industrial application – Product catalogue (1997) Boehringer Mannheim GmbH; powder (PSL: Chirazyme L-1, Cat no 1827642; CAL-B: Chirazyme L-2, Cat no 1836021); carrier-fixed preparations (PSL: Chirazyme L-1, Cat no 1831429; CAL-B: Chirazyme L-2, Cat no 1835807)

Phospholipases as Synthetic Catalysts

Stefano Servi

Dipartimento di Chimica, Politecnico di Milano, Via Mancinelli 7, I-20131 Milano, Italy.
E-mail: *stefano.servi@polimi.it*

Natural phospholipids can be completely modified with a group of hydrolytic enzymes found in living organisms. Phospholipase A_1, phospholipase A_2, and 1,3-specific lipases are hydrolases which can selectively cleave the ester bonds at position *sn*1 and *sn*2. The compounds of partial hydrolysis can be reacylated chemically. Each of the compounds obtained can then be modified with the very efficient phospholipase D which can effect polar head substitution in the presence of an alcohol as a phosphoryl acceptor. The enzymes from bacterial sources are readily available from culture broth and are highly selective. Phospholipase C can be used to obtain diacylglycerol and organic phosphates as hydrolysis products. The sequential use of the latter enzymes allows the preparation of organic diphosphates. The factors affecting the specificity and selectivity of these enzymes is discussed.

Keywords: Phospholipids, Phospholipases, Transphosphatidylation, Organic phosphates, Diacylglycerol.

Topics in Current Chemistry, Vol. 200
© Springer Verlag Berlin Heidelberg 1999

1
Phospholipids

Glycerophospholipids (PL) are abundant lipid components found in Nature [1]. Most vegetable oils, fish oil and egg yolk are particularly rich in mixtures of phospholipids. They are characterized by the presence of a polar *head* and two fatty acid chains in the apolar part of the molecule. The two acyl chains mainly consist of saturated fatty acid residues in the *sn*1 position and mainly (poly)unsaturated fatty acid chains in the *sn*2 position. Mixtures of phospholipids at low cost are obtained from the *degumming* process of vegetable oils. Lecithin, the main component of the mixture, has the polar head characterized by the choline residue. It is usually defined as phosphatidyl choline (PC) and it is understood that the composition of the apolar part is composed of mixtures of fatty acid residues dependent to a large extent on the source of the raw material (fatty acid chains composition of PC from soy beans: palmitic 11.6%, stearic 3.4%, oleic 4.6%, linoleic 66.4%, linolenic 8.7%). Scheme 1 shows a PC with two defined acyl chains at the glycerol backbone: 1-palmitoyl-2-linoleoyl-*sn*-glycero-3-phosphocholine (PLPC).

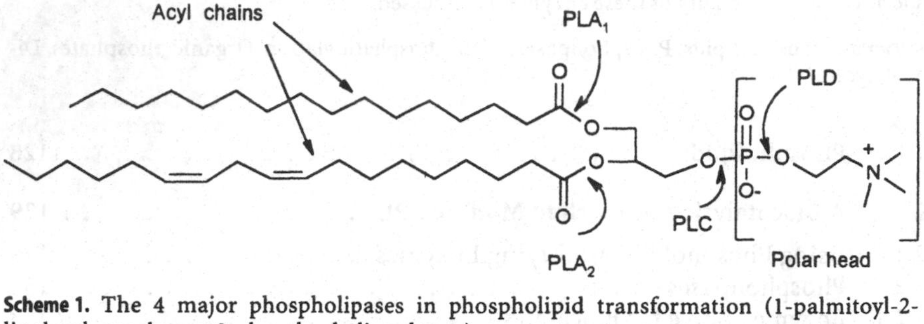

Scheme 1. The 4 major phospholipases in phospholipid transformation (1-palmitoyl-2-linoleoyl-*sn* -glycero-3-phosphocholine shown)

PC is commercially available with different degrees of purity with respect to phospholipids characterized by different polar heads. Other minor natural phospholipids can also be obtained by extraction and crude purification of raw materials. Recently, phospholipids are finding increasing applications in the most diverse fields of application as is witnessed by a vast technical and scientific literature [2–4]. Due to their amphiphilic properties they find applications in the food industry as surfactants and emulsifiers [5]. PC is employed for this property as a lung surfactant in the treatment of the neonatal respiratory distress syndrome [6]. Phosphatidyl serine has useful pharmaceutical properties and is used as a dietary supplement while other phospholipids are employed in the cosmetics field [7, 8]. Liposome technology makes considerable use of chemically defined phospholipids of (semi)synthetic origin in the field of drug delivery [9–15] while special phospholipids are designed as drug targeting molecules or pro-drugs [16–19]. Labeled phospholipids are required for fundamental biochemical studies [20–23]. An entire class of phospholipid analogues is in use for antitumor and antiviral applications as inhibitors of mitogenic signal transduction [24–28]. Some natural or modified phospholipids are proposed for their antioxidant properties [29–32] while other new PL are required for diverse applications [33–35].

Natural PL having a defined polar head other than choline are relatively more expensive than PC itself. PL with identical, defined acyl chains at the two positions are less accessible, while PL defined in their polar head with two different acyl chains can be considered as rather sophisticated compounds for complex synthesis [36]. Their preparation requires the use of enantiomerically pure glycerol equivalent synthons and the extensive use of protective groups and activating agents.

2
A Biocatalytic Approach to Modified PL

There are several reasons why a biocatalytic approach to structurally defined PL is desirable: the first concerns the aim to reduce the amount of chemical reagents to be used in the synthetic steps. PL are non-crystalline materials difficult to purify and of limited stability. Their purification often requires chromatography. Due to their ability to form aggregates with a wide variety of compounds of different classes their separation from unreacted material is not always an easy task. This fact is of particular concern since the destination of PL is often in the food, cosmetics, and pharmaceutical field. Enzymatic catalysis is expected to simplify the purification procedure. Moreover, due to their intrinsic selectivity and specificity it is expected that fewer by-products will be formed. Another consideration associated with PL purification is that, in case of incomplete reaction, possible impurities in the final product will constitute a much more acceptable problem if the starting materials are compounds which are already GRAS (generally regarded as safe) as natural phospholipids are, than in the case of residual chemical reagents. The potential of phospholipid modifying enzymes appears therefore of great interest in synthesis, particularly on an industrial scale. There is also the possibility to envisage a third approach to the preparation of new or less abundant phospholipids which is the combined chemo-enzymatic synthesis of PL. Such an approach for the large scale preparation of PL makes use of glycerol as starting material, which is stereoselectively phosphorylated using glycerol kinase and a phosphate donor [37]. The glycerol phosphate (GP) constitutes the crude chiral backbone which needs to be acylated and esterified at the phosphoric acid functionality.

In this article we will discuss the approach to the practical synthesis of modified phospholipids using natural PLs as starting materials, and phospholipases as biocatalysts.

2.1
Using Phospholipid Modifying Enzymes

In Scheme 1 are shown the more common enzymes named phospholipases (PLases) which selectively recognize each of the four individual ester bonds in the PL molecule. With the aid of these biocatalysts it is possible to *remove* (hydrolysis) or *substitute* (hydrolysis+reacylation or transesterification) each of the four residues, finally ending with an entirely modified phospholipid. This basic concept was proposed several years ago [38]. Recently the availability of new bacterial enzymes with improved phosphoryltransfer ability [39–51] has made it possible to prepare a large number of compounds in an efficient manner. In this operation the key step consists of the polar head substitution via a phospholipase D (PLD) catalyzed transfer. The modification of the chain can be effected in different phases of the synthesis and with different methodology. Some possible combinations are outlined in Scheme 2.

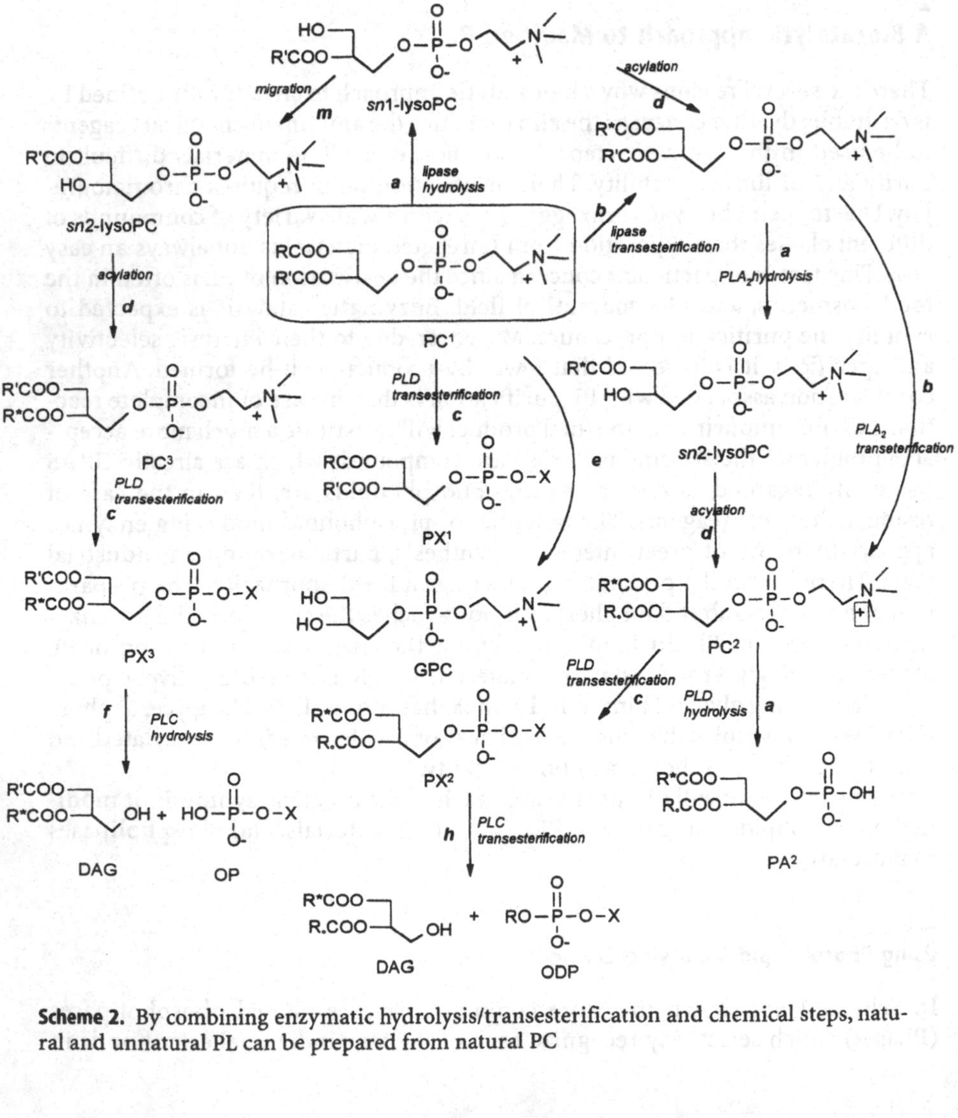

Scheme 2. By combining enzymatic hydrolysis/transesterification and chemical steps, natural and unnatural PL can be prepared from natural PC

Starting material is PC of natural origin, i.e., with undefined acyl chains. Target of the transformation are PL differing in structure and grade:

(1) polar head modified PL with the natural acyl composition (PX^1);
(2) PC with defined (and different) acyl chains at both positions (PC^2);
(3) polar head modified PL with defined (and different) acyl chains at both positions (PX^2);
(4) PC and PX with the same defined acyl chains at both positions (obtained from glycerophosphorylcholine, GPC);
(5) PA of any acyl chain composition (from any PC or PX);
(6) DAG and organic mono- and diphosphates.

The general strategy for reaching the goals in points 1–5 will be briefly commented on. A more detailed discussion of the various reaction pathways will be given using specific examples.

(1) Polar head exchanges rely on the wide substrate specificity of PLD from bacterial sources. From low molecular weight primary alcohols to large secondary ones, many structurally diverse compounds have been shown to be substrates for PLD with different yields and selectivity (path c). The hydrolysis compound phosphatidic acid (PA) will be present as a by-product.
(2) Insertion of defined acyl chains requires selective recognition of the two sn1 and sn2 positions. In the hydrolysis direction this is easily achieved although it can be laborious in practice. Hydrolysis with an enzyme having phospholipase A$_1$ (PLA$_1$) activity gives the $lyso$-compound. Chemical acylation gives a compound with defined acyl chains at the sn1 position. Further phospholipase A$_2$ (PLA$_2$) hydrolysis and chemical reacylation gives a PC^2 which has been modified in both acyl components. The direct transesterification reaction with the same enzyme would constitute a considerable advantage in the transformation. Acyl chain transfer with both types of enzymes is not efficient enough. A second and simpler way is to use as a starting material PC or PX with the same and defined acyl chains prepared by acylation of GPC (obtained by path e).
(3) From PC^2 transphosphatidylation gives the PX^2 modified in its polar head.
(4) PC and PX of this type are best prepared from GPC obtained by chemical hydrolysis of natural PC (e) and subsequent chemical acylation. From this kind of PC, PX are obtained by PLD catalyzed transesterification. PLA$_1$ or PLA$_2$ hydrolysis of these compounds and chemical reacylation allows one to obtain PL with different acyl chains. Since in $lyso$-PL the acyl chain in sn2 position tends to migrate to the sn1 position, PLA$_2$ hydrolysis gives a more suitable substrate for further acylation.
(5) From each of the previously obtained PC and PX, the corresponding PA can be obtained in practically quantitative yields.
(6) Phosphatidylcholine specific phospholipase C (PLC) can be used for the hydrolysis of PC (and PX) affording natural diacyl glycerol (DAG) and choline phosphate (CP) (or corresponding organic phosphate, OP). Transesterification has not been observed with this enzyme. The substrate specificity is less broad than desirable for the synthesis of OP that is otherwise

difficult to access. Phosphatidylinositol specific phospholipaseC (PLC_{PI}) can afford inositol phosphate (IP) by hydrolysis, and inositol alkyl phosphates (IAP) by transesterification.

2.2
Phospholipases

Phospholipases form a large class of lypolytic enzymes that are well distributed in most living organisms [52]. Since phospholipids are components of the cell membranes, the role of phospholipases in vivo is to hydrolyze membrane phospholipids, thus generating smaller molecules which are considered as second messengers or membrane signaling agents. Their function is of paramount importance [53,54]. The literature on the mechanism, function, structure activity, relationship etc. for these enzymes is so rich and complex that a simplification cannot even be attempted. Only aspects concerning application in biocatalysis will be accounted for.

Four major enzymes are involved in the transformation of PL. Each of them is specific for catalyzing the hydrolysis of the ester bond indicated. PLA_1 and PLA_2 recognize the carboxyl esters in the molecule and they are therefore hydrolases that are in some respect similar to lipases. PLC and PLD hydrolyze the phosphate ester bonds being therefore more similar to phosphatases and phosphodiesterases. Overall, the substrate specificity of these enzymes is rather broad. Numerous other phospholipases more specific for a particular type of PL (*lyso* compounds), a specific polar head (phosphatidyl inositol is hydrolyzed by a specific PLC, PLC_{PI}) are of less general availability and utility for biocatalysis. On the other hand, less specific hydrolytic enzymes (1,3-specific microbial lipases) can be used for phospholipid modification displaying the function attributed to PLA_1. PLases have the common feature of being catalytically active towards glycerophospholipids as a common class of substrates. They are therefore able to recognize and hydrolyze water insoluble substrates like lipases do with triglycerides. The mechanism and the substrate aggregation required for their action are diverse and still the subject of debate and research [52]. The uncertainty regarding the complete substrate specificity of these enzymes is increased by the fact that phospholipids exist in many different aggregated forms in water. Synthetic short-chain phospholipids and *lyso*-phospholipids either exist as monomers or as micelles. At concentrations below their critical micellar concentration (CMC) they are monomeric, while above that concentration they form micelles. The CMC of a particular phospholipid decreases with increasing fatty acid chain length. It can vary from 10 mmol l^{-1} for dihexanoyl PC to 10^{-7} mmol l^{-1} for dipalmitoyl PC [1]. While these values have to be seriously taken into account when performing kinetic studies, they are less crucial in the context of biocatalytic applications in synthesis. Since one of the main goals of biocatalysis is to achieve transformation in as high as possible space-time yields, under practical conditions the phospholipid substrates will always be above their CMC. The kinetics of phospholipase catalyzed hydrolysis strongly depends on the aggregation state: some of these enzymes hydrolyze aggregated substrates faster than monomeric ones. The rates with monomeric substrates are usually quite low. This holds especially for phospholipase A_2 (PLA$_2$) [55] for which kinetic information is obtained with sophisticated investigation techniques [56]. In this case, the use of a surfactant for the formation of mixed micelles can improve the reaction rates. Again, this technique is of little interest from a preparative point of view for the problems associated with phase separation and product recovery in the presence of considerable amounts of surfactants. PLC and PLD behavior is apparently less influenced by interfacial phenomena, although they tend to be active at the interface. PLD activity depends on the interfacial area as proved by experimental data [57]. The behavior of PLA$_2$ and PLC in mixed micellar systems has been extensively investigated because knowledge of their specificity is important in interpreting the action of these enzymes on natural membranes [58]. However this information is difficult to transfer to a preparative scene. Most of the extremely vast literature concerning these enzymes is on fundamental biochemical studies.

2.2.1
Enzyme Source for Biocatalysis

The PLA$_1$ (phosphatidylcholine 1-acyl-hydrolase, EC 3.1.1.32) hydrolyzes glycerophospholipids at the *sn*1 position giving a fatty acid and a *lyso*PL as products. It has a rather broad substrate specificity. Enzymes with PLA$_1$ activity have been well characterized in guinea pig pancreatic lipases and from other sources [59–62]. The same activity has been found in *Aspergillus oryzae* [63] and in *Serratia* [64]. However, enzyme preparations or microorganisms also displaying PLA$_1$ activity are readily accessible in microorganisms like *Rhizomucor*, *Mucor javanicus*, *A. niger* and others [65–68]. Some of these enzymatic preparations are commercially available [65]. They can be used for hydrolysis in emulsion systems in a free or immobilized form. Their tendency to catalyze a transesterification can be synthetically useful in systems of controlled water activity with a conversion of up to 60% [69]. Phospholipase A_2 (PLA$_2$, phosphatidylcholine 2-acyl-hydrolase, EC 3.1.1.4), hydrolyses L-α-phosphatidylcholine (PC) to L-α-lysophosphatidylcholine (LPC) and fatty acid. It also accepts as substrates phosphatidylethanolamine (PE), PA and other PL. Pancreatic PLA$_2$, the most easily accessible enzyme source, requires surfactants for the hydrolysis of PC. For biocatalytic applications the enzyme from invertebrates (bee or snake venom (*Crotalus adamanteus*, *Naja naja*, *Crotalus atrox*)) are more often used. Other sources are mammals (bovine pancreas, porcine pancreas) [70] and microbes (*Streptomyces violaceoruber*, *S. cinnamomeus*, *S. griseus*) [71, 72]. Use of this enzyme allows the removal/replacement of the acyl chain in position *sn*2 either via hydrolysis and subsequent chemical reesterification or through direct interesterification with an acyl donor. Hydrolysis reactions are much more efficient than transesterifications. Quantitative removal of the acyl chain in position *sn*2 can be effected in mixed micelles but biphasic systems are preferred because of the easier isolation of the product [9, 10, 35, 73–75]. Reactions other than hydrolysis occur with limited yield. The direct interesterification of PC with myristic acid in the presence of immobilized PLA$_2$ from different sources has been reported to give up to 43% incorporation of the myristoyl group in position sn2 [76]. Starting with sn2-lysoPC modest incorporation of fatty acids has been obtained either in a biphasic system or in an organic solvent with low water content [77–81]. PLA$_2$ is specific for PLs with the natural absolute configuration at position sn2 [82–84]. This property could be used with advantage in the resolution of racemic PLs or in assessing the absolute configuration of PLs of different origin. Interestingly, this enzyme has been shown to be susceptible to the chirality of the P atom [85]. Two important enzymes with phospholipase C activity are available for phospholipid modification: the so-called PC specific PLC [EC 3.1.4.3] in fact accepts quite a number of other PLs with a different polar head. It can be obtained from strains of *B. cereus* [86–90] as well as from *Clostridium welchii* and *C. perfringens* [91]. These enzymes have rather different substrate specificity and requirements. The PLC$_{PC}$ more often used is the enzyme from *B. cereus*. It is secreted outside the cell and the activity can be isolated with a rather simple procedure. PLC which hydrolyses phosphatidyl inositol (PLC$_{PI}$) [EC 3.1.4.10] can also be obtained from *B. cereus* and *B. thuringiensis* [92]. Application in the production of inositol phosphates has recently been reported [93].

PLD has a very broad distribution in living organisms. It was first isolated in various kind of cabbage and has since been recognized in a number of plants including *ricinus*, castor beans, spinach leaves, soy beans and others [94–102]. PLD found in yeast is a mitochondrial enzyme not used in biocatalysis [177, 178]. Numerous bacterial sources are rich in PLD. The enzyme can be obtained in the culture broth of various strains of *Streptomyces* [39–51] where it can be recognized directly with simple spectrophotometric methods on synthetic phospholipid analogues [103] or by measuring the choline formed from PLD catalyzed PC hydrolysis by combined enzymatic assays [104].

Table 1 collects the properties of some common PLD from different sources [43].

Table 1. Properties of PLD from different sources for biocatalytic applications [39–51]

Source	Molecular mass (KDa)	Isoelectric point	pH optimum	Specific activity (U/mg)	Activation factors	Ref
PMF	53.8	9.1	4.0–6	42	–	[43]
S. hachijoensis	16	8.6	7.5	631	+	[47]
S. lydicus	56	7.4	6	2390	–	[104]
S. chromofuscus	57	5.1	7–8.5	152	+	[50]
S. species	46	4.2	5.5	200	+	[176]
Cabbage	90	4.7	6–7	1492	+	[98]

3
Polar Head Modification

3.1
Phosphoryl Transfer

The synthetic utility of hydrolytic enzymes is greatly enhanced when they are able to catalyze effectively the reverse reaction of ester formation. With most hydrolytic enzymes the presence of water in the reaction medium practically prevents an ester synthesis since the equilibrium is completely shifted toward the formation of the hydrolysis products. Phospholipase D [EC 3.1.4.4] PLD is a unique enzyme in this respect since it is able to transfer the phosphatidyl moiety to an alcohol acceptor in the presence of more than stoichiometric amounts of water in the medium. The result of the transphosphatidylation reaction catalyzed by PLD is the formation of a new phospholipid modified in its polar head group (PX). This is usually accompanied by variable amounts of the hydrolysis product phosphatidic acid, PA (Scheme 3).

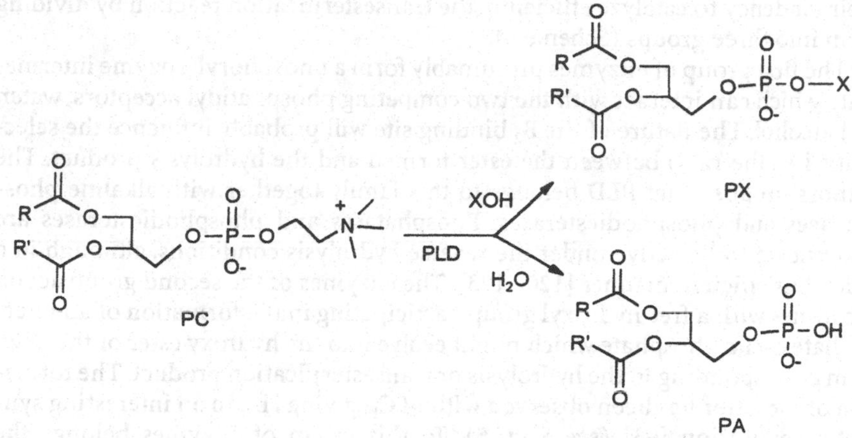

Scheme 3. PLD catalyzed polar head exchange gives a new phospholipid PX and PA as a by-product

This reaction constitutes by far the most synthetically useful phospholipid transformation catalyzed by a phospholipase [105–113]. Although this reaction has been known and exploited in PL modification for a long time [105, 113–115] the mechanism of PLD catalysis is still obscure. In analogy with other hydrolytic enzymes it is believed to occur with formation of an enzyme-substrate intermediate via a ping-pong mechanism, modified by a hydrolytic branch, from inhibition experiments showing the presence of an essential SH group in cabbage PLases [114] and from kinetic experiments in an emulsion system in bacterial PLases [57]. A common phosphatidyl-enzyme complex for the hydrolysis and transesterification reaction has been demonstrated by studying the intermediate partitioning between competing acceptors of different phospholipid substrates with the same nucleophile in a biphasic system [57]. Recently, the presence of an essential lysyl residue in the active site has been suggested [116]. Experiments with PC chiral at the phosphorus atom by dissymmetrical substitution with labelled oxygen atoms have proved that the reaction occurs with an overall retention of configuration at the P atom, in agreement with a double inversion (formation of a covalent intermediate with inversion and then a second inversion due to the action of the oxygen of the alcoholic group acting as a nucleophile) [85, 117]. It has been proposed that the two functions of hydrolysis and transesterification reside in two different enzymes [118]. However this possibility has been ruled out, mainly by the fact that homogeneous proteins from several different sources catalyze both reactions. It remains to be explained why, while Ca^{2+} is required by PLD in hydrolysis reaction, the transphosphatidylation appears not to be favored by the presence of the metal ions in some specific cases [119]. It is recognized in general that at least with some PLDs the influence of Ca^{2+} in the transesterification reaction is much less pronounced than in the hydrolysis of the same substrates. Interestingly, it has recently been observed that Ca^{2+} dependence in cabbage PLD is pH dependent [99]. In a recent work Bruzik et al. [93] compare the mechanism of different phosphoesterases and their tendency to catalyze efficiently the transesterification reaction by dividing them into three groups (Scheme 4).

The first group of enzymes presumably form a phosphoryl-enzyme intermediate which can interact with the two competing phosphatidyl acceptors, water, and alcohol. The nature of the B_2 binding site will probably influence the selectivity, i.e., the ratio between the ester formed and the hydrolysis product. The authors propose that PLD belongs to this family together with alkaline phosphatases and phosphodiesterases. Phosphatases and phosphodiesterases are also known to be active under the reverse hydrolysis conditions, although in a much less efficient manner [120–123]. The enzymes of the second group act on substrates with a free hydroxyl group participating in the formation of an intermediate cyclic phosphate which might evolve into the hydroxy ester of the open form corresponding to the hydrolysis or transesterification product. The formation of the latter has been observed with PLC_{PI} giving rise to an interesting synthetic application [93] (see Sect. 5). To this group of enzymes belongs the ribonuclease A [124, 125]. PLD and PC specific phospholipase C (PLC_{PC}) have been reported to give cyclic intermediates from substrates possessing free hydroxyl groups that are a few carbon atoms apart from the phosphate group [126, 127]. PLC_{PC} from *B. cereus*, however, has not been reported to give transfer reactions. The usual pathway by which hydrolysis by PLC_{PC} occurs is the one referring to the third group of enzymes. This mechanism is supported by the crystal structure of the enzyme showing the presence of one activated water molecule which directly performs the nucleophilic displacement [128]. For this last group of enzymes, to which the enzyme hydrolyzing inositol-1-phosphate (inositol monophosphatase) [129] also belongs, one single report describes a transfer reaction catalyzed by PLC from *Clostridium prefringens* [130]. In the reaction of PC with *N*-oleoylsphingosine the phosphorylcholine moiety is transferred to *N*-oleoyl sphingosine with formation of sphingomyelin. This reaction is potentially of great interest, but its generality has not been further investigated. PLC from *Bacillus cereus* was reported not to catalyze the same transformation. The transfer of the phosphorylcholine moiety to an acceptor alcohol would allow the preparation of phosphodiesters.

3.1.1
Role of Solvents

Natural phospholipids with long chain fatty acids are soluble in organic solvents and insoluble in water in the absence of a surfactant. The typical configuration

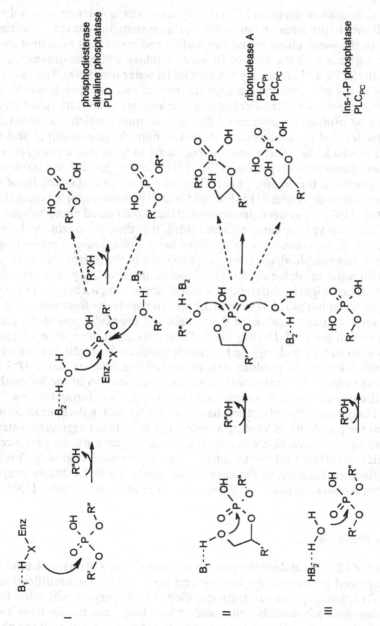

Scheme 4. Enzymes having phosphoric esters as substrates grouped by mechanism analogy (adapted from [93])

for the modification of the polar head is a biphasic system water-organic solvent which will eventually form an emulsion. In this system, the enzyme is initially dissolved in the water phase containing buffer and metal ions as activators (if required) together with the alcohol (if water soluble), while the organic phase will contain the PL and eventually the alcohol (if water insoluble). The rate and selectivity of the phosphatidyl-transfer reaction of course depends on the enzyme source, the nature of the alcohol acceptor and its concentration, but it also appears to be profoundly influenced by any parameter which can affect the phase behavior and the partition coefficient of both the phospholipid and the protein. It is well known that phospholipids tend to form stable complexes of varying stoichiometry with proteins [131 – 133]. These aggregates are stable and bring the protein in the organic phase. They are one of the first examples of the catalytically active derivatised [134] or lipid coated proteins reported recently in the literature [135]. It has been demonstrated that lipid coated phospholipase D is catalytically active in organic solvents [136]. It is then reasonable to believe that the catalytic reaction can readily occur both in the aqueous phase (especially with water soluble alcohols as acceptors) and in the organic phase (when dealing with water insoluble alcohols). Of course, the selectivity of the reaction in the two phases might be different. The nature of the complex between the protein and the phospholipids has been attributed in part to the formation of ionic pairs between the basic residues of the protein and the acidic part of the phospholipids (when present). During the PLD catalyzed transformation of phospholipids their nature is changing: hydrolysis produces an acidic species while transesterification might produce new phospholipids that are acidic (PG for instance) or neutral (PC analogues) in nature. The partition of the biocatalyst between the two phases is therefore likely to change even during the reaction progress. The nature of the solvent will have a prominent part in the biotransformation due to the possibility of forming a wide range of different aggregation states. Due to the strong interaction solvents-phospholipids-proteins, the presence of the organic solvent is considered to enhance the enzyme stability greatly. This has been verified, for instance, in the case of the highly purified cabbage enzyme [137]. Other authors consider the organic solvent to act as an activator [138].

3.1.2
Transfer to Primary Alcohols

The ability of PLD to catalyze the phosphoryl transfer to a primary alcohol has been recognized in the cabbage enzyme and exploited for the modification of natural PC (Scheme 3). The substrate specificity of this enzyme, which has been the only commercially available biocatalyst for a long time, has been reviewed [105]. Considerations on the enzyme specificity concern the nature of the phospholipid in its polar head and the acyl chains, and the nature of the acceptor alcohol. The best substrate of the cabbage enzyme is phosphatidyl choline. The nature of the acyl chains changes the kinetics of the reaction but this can be attributed mainly to the influence of the chains on the phase properties of the PL. This goes together with the medium and additives used in the reaction, and a rationale of the influence of the chain composition isolated from the medium engineering is not available. Small water soluble primary alcohols have been considered as co-substrates in the reaction catalyzed by the cabbage enzyme. Acceptors like ethanol are so efficient that the formation of phosphatidylethanol (PEt) provides the basis of an assay for the presence of PLD or ethanol [139,140]. Among secondary alcohols, only 2-propanol has been used successfully. Large hydrophobic primary alcohols are not good acceptors which means that the hydrolysis reaction becomes prevalent.

The reaction of phosphoryl transfer acquires a preparative value when high ratios between the formed PX and PA are obtained, taking into account the difficult purification of PL in general. The concentration of the alcohol in the water phase influences the selectivity. Phosphatidylglycerol (PG) 3 (Scheme 5) can be obtained with very high selectivity using 10 mol l^{-1} glycerol solution as the aqueous phase [141]. Both L- and D-serine are accepted for the preparation of the corresponding phosphatidylserine (PS) 4 and 5 with a rather low discrimination of the two enantiomers [142–145]. The reaction appears in general not to be stereoselective with respect to chiral or prochiral alcohol donors. This is related to the presence of free hydroxyl groups on stereogenic carbons in the acceptor molecules (see Sect. 3.1.4). PLD from bacterial sources, in particular from *Streptomyces*, are superior to the enzyme from cabbage from a preparative point of view. They can be obtained in the culture broth of easily grown non-pathogenic bacteria, and they can be used without further purification as biocatalysts (Table 1) [39–51]. They have a very wide substrate specificity as indicated in Scheme 5, e.g., a series of PE analogues 10–13 can be obtained in very high yields and selectivity [146]. Phosphatidyl genipin 15 and phosphatidyl arbutin 16 incorporate large bioactive molecules from natural sources [147, 148]. The idea of preparing PL bearing a pharmacologically active molecule in the polar head has produced a number of new molecules including phosphatidyl nucleotides 17 proposed as antivirals [18], phosphatidyl ascorbic acid 19 for its antioxidant activity [149], or a phospholipid 20 containing a modified sialic acid moiety as an enzyme inhibitor [17]. Compound 18 has also been prepared using PLD from *Streptomyces* as a biocatalyst in good yield [150]. The compound is used in the preparation of sterically stabilized liposomes (stealth liposomes) [13].

All these compounds and many others cited in the technical and patent literature [7–23] can be obtained in variable yields and purity. What makes the method attractive is that many structurally diverse compounds can be prepared with a single strategy and one biocatalyst. The method is used for the large scale preparation of the most common commercial PL like PS, PG, and PE with mixed acyl chains. One of the general strategies to industrial compounds with defined acyl chains, as already mentioned, is to prepare glycerophosphorylcholine (GPC) by hydrolysis of natural PC (path *e*, Scheme 2), chemical reacylation and exchange of the polar head via PLD catalyzed transesterification.

3.1.3
Transfer to Secondary Alcohols

Recently, the preparation of new PL has been reported, obtained by the transfer of a phosphoryl moiety to a secondary alcohol (Scheme 6). With PLD from

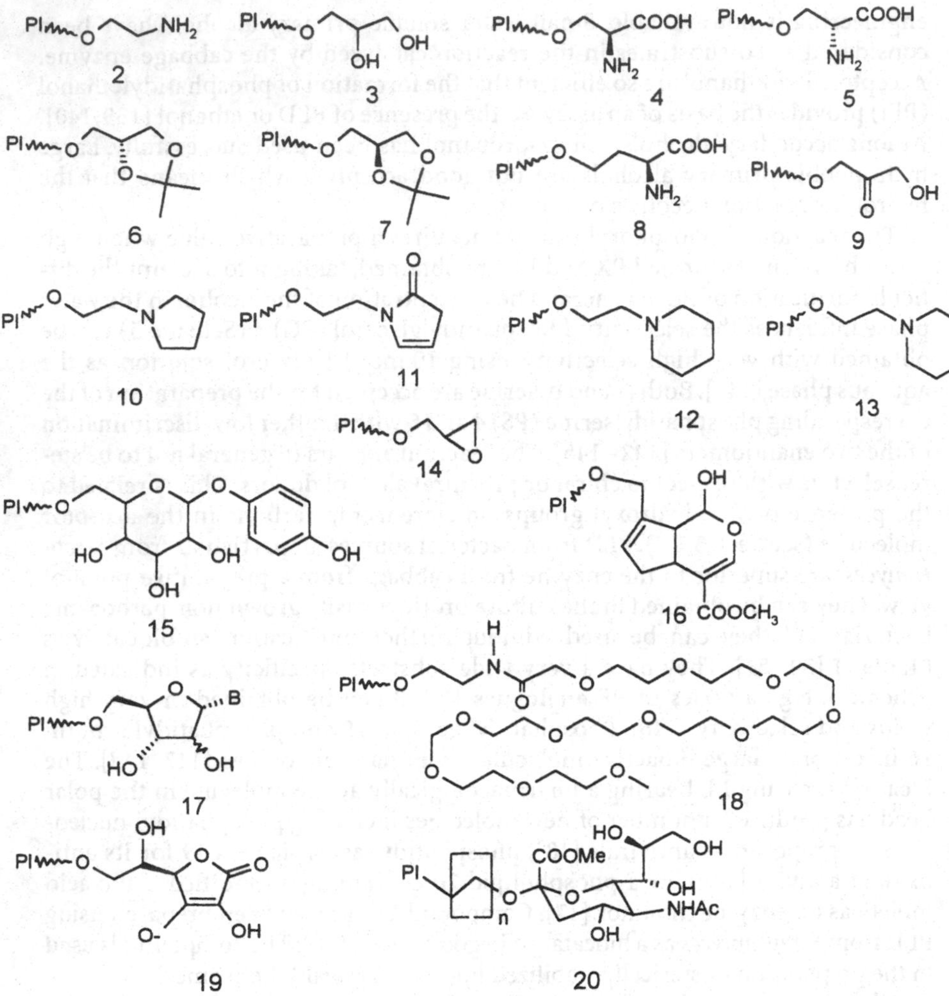

Scheme 5. Phospholipids with polar heads of various types obtained by PLD catalysed trans-phosphatidylation from PC and primary alcohols (PI = phosphatidyl-)

Streptomyces the reaction occurs in a synthetically useful manner [151]. The PL obtained from PC and cyclopentanol 21, cyclohexanol 22, cyclohexanediols 23, 24, 27, and related structures like 24 and 25 are obtained in a biphasic system. Interestingly, all the alcohols used are water insoluble compounds.

The attempt to obtain structural analogues to phosphatidyl inositols failed. Trihydroxy-cyclohexanes proved not to be substrates of such enzymes, like *myo*-inositol itself and derivatives [151]. The PLD isolated from spinach leaves was reported to be able to give PI as a transphosphatidylation product from PC and inositol [94]. To our knowledge no further exploitation of this method has followed the original article. Other authors have reported the formation of

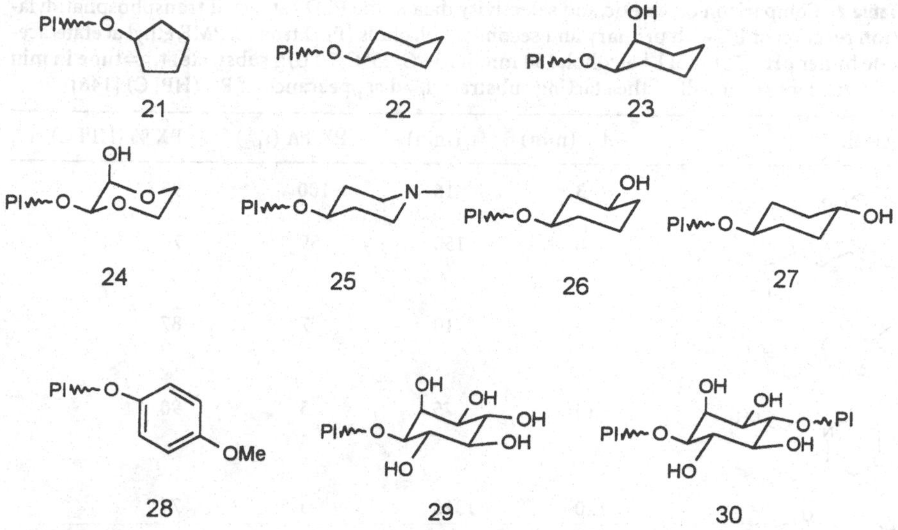

Scheme 6. Possible transphosphatidylation products from PC and secondary alcohols

bis-phosphatidyl inositol starting from PI and a PLD preparation obtained from cauliflower florets [152]. This has to be considered as a new enzymatic activity as compared with the cabbage phospholipase of common use. In fact, the report is quite in contrast to what is reported in the literature for that enzyme [105] which considers PI not to be a substrate for cabbage PLD. The authors describe the formation of compounds of type 30 as a mixture of regioisomers. From these two examples one can argue that probably some of the numerous PLD preparations from higher plants could be used for the synthesis of modified PI. The use of inositols as acceptor alcohols in the transphosphatidylation reaction is of interest both from a mechanistic and from a synthetic point of view: synthesis of phosphatidylinositols and their analogues is the target of intensive research due to the biological implication associated with these compounds. Secondary alcohols are inferior as acceptors to primary ones. Competition between the two kinds of functionalities results in the exclusive formation of one product as is proved by the synthesis of PG from glycerol and PC. Nevertheless, secondary alcohols can be profitably used for the formation of useful derivatives. Table 2 compares the results obtained from transphosphatidylation reactions using natural PC as a substrate and 1 mol l^{-1} alcohol solutions as acceptors in a biphasic system (ethyl acetate-acetate buffer pH 5.5, 0.1 mmol l^{-1} Ca^{2+}, PLD from *Streptomyces* PMF 30 U/g of substrate at 25 °C) [146]. The table reports the time in minutes at which half of the substrate has disappeared as measured by HPLC with UV detection at 205 nm (t$_{1/2}$), the time in minutes at which all the starting PC has disappeared (t$_f$) and the ratio for the selectivity of the reaction (i.e., the PX/PA ratio at t$_{1/2}$). In the fifth column of Table 2 the apparent purity of the PL obtained at the end of the reaction as measured by HPLC is indicated. Please note that the PX/PA ratio is not always congruent with the final purity of the pro-

Table 2. Comparison of kinetic and selectivity data in the PLD catalyzed transphosphatidylation reaction of PC with primary and secondary alcohols (PLD from S. PMF, ethyl acetate/acetate buffer pH 5.5, 1 mol l^{-1} alcohol, 150 mmol l^{-1} PC, 25 °C, 30 U/g substrate) $t_{1/2}$=time in min of consumption of half of the starting substrate. t_f=disappearance of PC (HPLC) [146]

Alcohol	$t_{1/2}$ (min)	t_f (min)	PX/PA ($t_{1/2}$)	PX % (HPLC) at t_f
(n-propanol, CH₃CH₂CH₂OH)	3	18	>100	95
(N,N-dimethyl propanolamine, HO–CH₂CH₂CH₂–N(CH₃)₂)	8	150	59	76
(solketal, 2,2-dimethyl-1,3-dioxolane-4-methanol)	5	10	7	87
(N-(2-hydroxyethyl)morpholine)	13	75	5	90
(HO–(CH₂CH₂O)$_n$–)	120	1310	1.5	75
(Pl∿∿–O–cyclopentyl)	53	430	1.1	42
(cyclopentanol)	7	90	8.5	87
(trans-cyclohexane-1,2-diol)	21	300	1	70

duct. In fact, it has been noted that the rate of PA formation is not the same during the entire reaction course. PA is formed relatively more rapidly at the beginning of the reaction. A simple comment to the data is that primary alcohols reacts faster than secondary ones, but that this is not solely dependent of the nature of the alcoholic group: cyclopentanol is more reactive and the reaction is more selective if compared, for instance, with *N,N*-dimethyl propanolamine. A rationale to this behavior is unclear. Some of the large new phospholipids formed are less prone to hydrolysis: which results in a higher selectivity.

3.1.4
Multiple Hydroxyl Groups

When more than one hydroxyl group is present in the acceptor, the newly formed phospholipid PX has a free hydroxyl group, thus potentially being also a phospholipid acceptor. Moreover, *intra*molecular displacement can occur. Thus polyhydroxy compounds constitute a special class of substrates (Scheme 7).

From the reaction of PC in the presence of glycerol, the initially formed PG presents one primary and one secondary hydroxy group. Thus, one molecule of

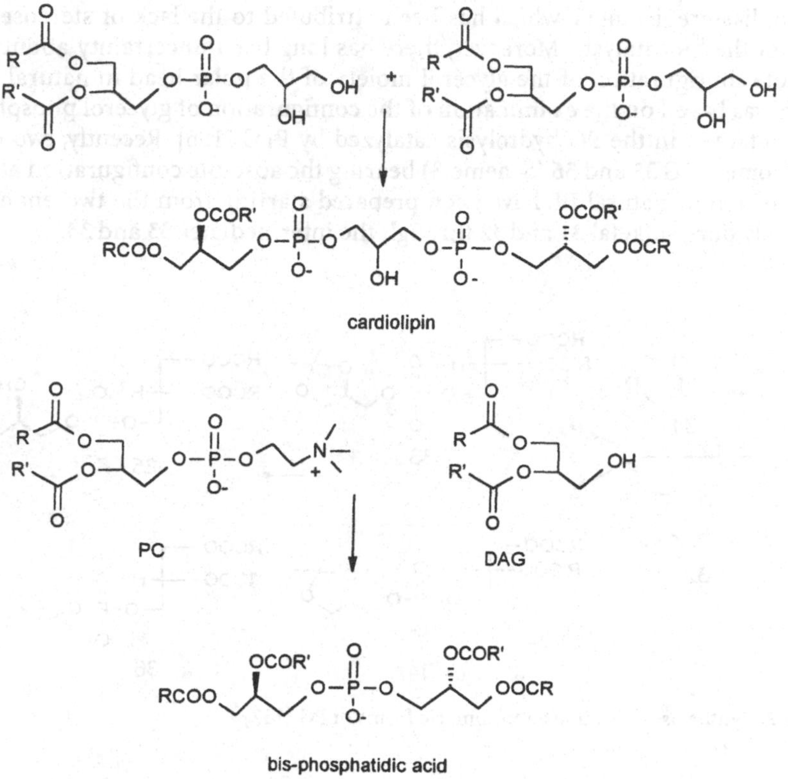

cardiolipin

PC

DAG

bis-phosphatidic acid

Scheme 7. Formation of dimeric compounds catalyzed by PLD [153, 154]

PG can act as a donor to a second molecule acting as an acceptor, thereby forming diphosphatidylglycerol (cardiolipin CDL) as a product. The formation of this common membrane PL has been observed in the presence of PLD from cabbage but only in trace amounts [119]. In more recent work it has been shown that, using PLD from *Streptomyces* as a catalyst, cardiolipin is formed in good yields and for longer reaction times the product can be obtained in substantial amounts [153]. This constitutes a drawback in the synthesis of PG from PC, but in this particular case the enzyme from cabbage should give a product with no CDL. In a similar reaction, one unit of DAG formed in vivo by the action of PLC on PC can act as a nucleophile on another PC molecule giving rise to the formation of bis-phosphatidic acid. This reaction has been observed in vivo and, although deprived of synthetic interest, it is remarkable in that the transphosphatidylation ability of PLD could play the role of signal attenuation due to the consumption of the second messenger DAG generated by PLC hydrolysis. The transesterification ability of PLD in vivo otherwise has no explanation [154]. Other authors have shown in bacteria that a remodeling of the membrane phospholipids can occur in the presence of exogenous alcohols by PLD catalysis [155]. The PG which is prepared from PC and glycerol is obtained as a mixture

of two diastereoisomers. which has been attributed to the lack of stereoselectivity of the biocatalysts. Moreover, there has long been uncertainty about the
absolute configuration of the glycerol moiety of the polar head in natural PG,
which was based on the examination of the configuration of glycerol phosphate
(GP) obtained in the PG hydrolysis catalyzed by PLC [156]. Recently, two dia-
steroisomers PG 35 and 36 (Scheme 8) bearing the absolute configuration at the
$sn2$ position of natural PL have been prepared starting from the two enantiomerically pure solketal 31 and 32 through the intermediates 33 and 34.

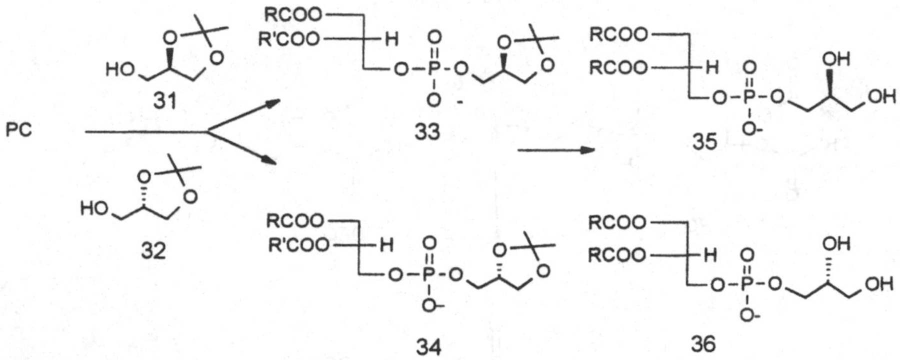

Scheme 8. Synthesis of two diastereoisomeric forms of PG [157]

PG 35 has the configuration at the polar head glycerol unit attributed to naturally occurring PG [157]. The preparation of diastereoisomerically pure 35 and
36 has not solved the uncertainty associated with the configuration of these
compounds obtained with PLase catalysis. However, a better understanding
comes from the consideration of the literature concerning the PLC_{PI}, PLC_{PC}, and
PLD catalyzed hydrolysis of some PL bearing free hydroxyl groups three to four
atoms away from the P atom (Scheme 9). Entry I in Scheme 9 shows the mechanism of hydrolysis of PI catalyzed by PLC_{PI}. This occurs [158] through the intermediate formation of the cyclic phosphodiester as indicated in Scheme 4. The
attack of a water molecule then gives the actual hydrolysis compound. The
second step of the reaction is believed to occur through catalysis by the same
enzyme as a slower step which allows the intermediate to be detected. Entry II
illustrates the mechanism proposed by Shinitzky et al. [127] for the PLC catalyzed hydrolysis of PG. Studies of the chirality of the glycerophosphate obtained
were used to establish the absolute configuration of the polar head. The formation of the cyclic phosphate was not observed before. The authors of this work
isolated the six-membered cyclic phosphate which was hydrolyzed chemically.
In other studies the same intermediate was hydrolyzed by some phosphodiesterase or phosphatase present in the enzymatic preparation. Similarly, phosphatidyldihydroxyacetone prepared by transphosphatidylation of PC with dihy-
droxyacetone upon hydrolysis gave the isolated cyclic phosphate [127]. Other

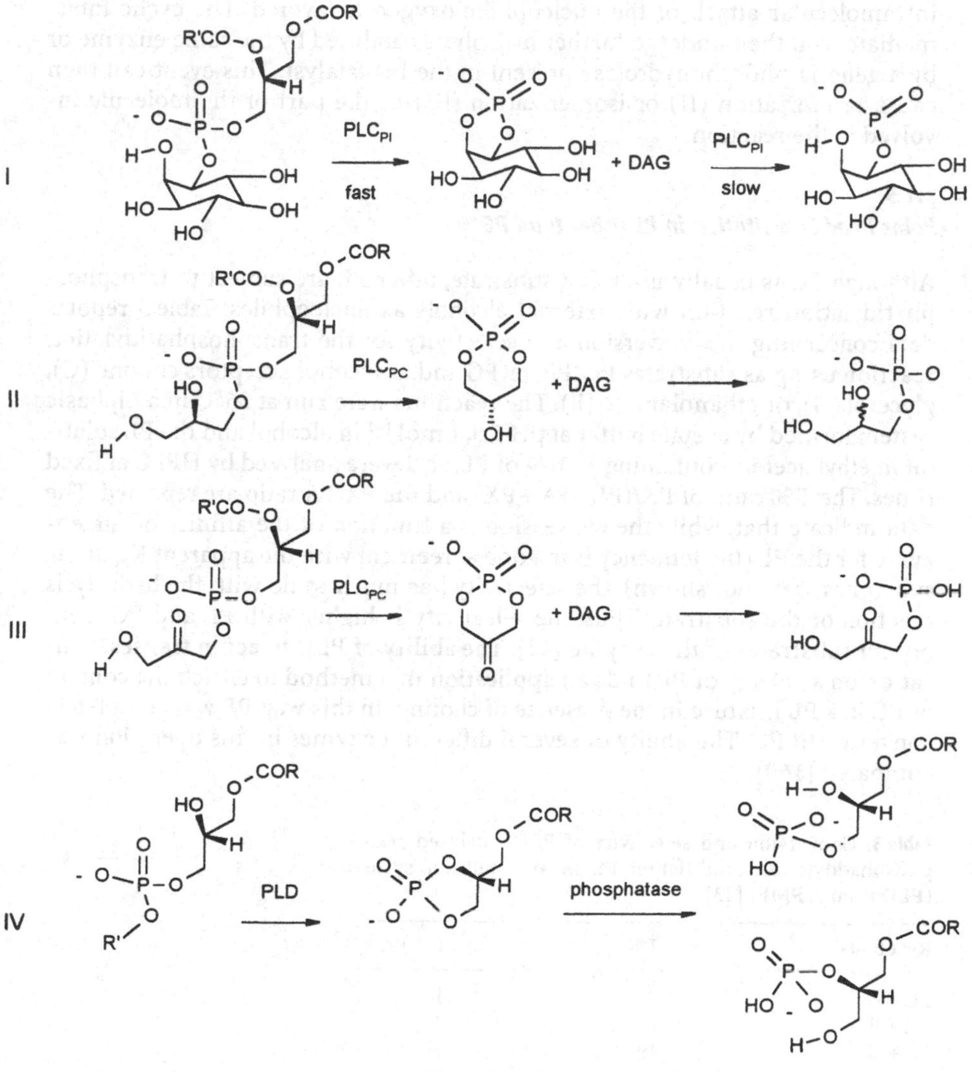

Scheme 9. Formation of cyclic phosphates in substrates bearing free hydroxyl groups [127, 158]

authors used the same reaction catalyzed by crude PLC preparations from *B. cereus* for the synthesis of glycerol phosphate and dihydroxyacetonephosphate [159]. Entry **IV** of Scheme 9 describes the PLD catalyzed hydrolysis of a *lyso-* phospholipid from which a five-membered ring phosphate is obtained prior to a phosphatase hydrolysis which can give the two isomeric phosphatidic acids. The reaction has been followed by ^{31}P NMR [126]. It appears from this work that in the phospholipase catalyzed hydrolysis of PL bearing free hydroxyl groups in a position suitable for the formation of a 5- or 6-membered cyclic phosphate the intramolecular attack of the nucleophilic oxygen is favored. The cyclic intermediates can then undergo further hydrolysis catalyzed by the same enzyme or by a generic phosphohydrolase present in the biocatalyst. This event can then cause racemization (II) or isomerization (IV) of the part of the molecule involved in the reaction.

3.1.5
Polar Head Substitution in PL Other than PC

Although PC is usually used as a substrate, other PL are subject to transphosphatidylation reaction with external alcohols as nucleophiles. Table 3 reports data concerning the conversion and selectivity for the transphosphatidylation reaction using as substrates PC, PE, or PG and as alcohol acceptors choline (C), glycerol (G), or ethanolamine (E). The reactions were run at 36 °C in a biphasic system formed by acetate buffer at pH 5.5, 1 mol l^{-1} in alcohol and the PL solution in ethyl acetate containing 30 U/g of PL, and were analyzed by HPLC at fixed times. The T% ratio of PX/(PL + PA + PX) and the PX/PA ratio are reported. The data indicate that, while the conversion is a function of the affinity of the enzyme for the PL (the tendency is in good agreement with the apparent K_m of the substrates, data not shown), the selectivity has much to do with the hydrolysis reaction of the substrate. Thus, the selectivity is higher with PE and PG, both poorer substrates of the enzyme [43]. The ability of PLD to act in transesterification on a variety of PL finds an application in a method to enrich the content of PC in a PL mixture in the presence of choline. In this way, PE was completely converted to PC. The ability of several different enzymes in this operation was compared [160].

Table 3. Conversion and selectivity of PLD catalyzed transphosphatidylation of different PL in an emulsion system (PLD from S. PMF) [43]

Reactants	T%	PX/PA
PC + G	77	19
PC + E	96	23
PE + G	49	∞
PE + C	26	∞
PG + C	7	∞
PG + E	18	∞

3.2
Reaction Configuration

Enzymatic conversion of PC to PX via transphosphatidylation reaction of PC has thus become an accepted industrial method for the modification of phospholipids at their polar head. The more frequent and simple application of the biocatalyst consists of the direct use of the culture broth in a biphasic system. The pH optimum and cofactor requirements for some of the most common PLD are reported in Table 1. One of the major problems associated with the scaling up process is inhibition at high substrate concentrations which results in decreased yields. The problem has been shown to be associated both with inhibition by the choline formed and the concentration effect displayed by this compound, which tends to reverse the transesterification equilibrium. Since choline is water soluble, the extent of the reaction also depends on the volume of the water phase. Removal of choline from the reaction medium is thus usually performed by operating in a biphasic system. When a water:organic solvent ratio of 2:1 is employed, efficient reaction can be obtained with up to 200 mmol l^{-1} PC concentration in the organic phase [146]. A complete removal however could be done by using specific choline-transforming enzymes like choline oxidase. This enzyme is usually used in combination with others in the enzymatic assay of PLD. Oxidation to the betaine releases hydrogen peroxide which needs to be further oxidized to dioxygen by catalase. Thus, a combination of the three enzymes can be applied to the complete removal of choline from the reaction medium [161]. Alternatives to the configuration of the emulsion system are represented by membrane reactors and the use of immobilized enzymes. PLD has been used in a flat membrane reactor for the production of PG [137]. The authors claim that a continuous production can be obtained by this method. In this case the enzyme is dissolved in the aqueous phase. An application where the enzyme is immobilized in a hollow fiber membrane bioreactor is described in Sect. 5. Enzyme immobilization on a solid support has also been reported by several groups [141, 162–165]. Advantages mainly concern the prolonged operativity of the biocatalyst. The solubilization of PLD in organic solvents has been proposed by modifying the enzyme according to a technique in which the enzyme is brought into the organic phase by coating with synthetic lipids. The reaction is then performed in a biphasic system. The water phase is necessary in order to remove the choline formed when PC is used as a substrate. The reaction outcome is reported to occur with high yields and selectivity. The method requires a purified enzyme as a starting material and a derivatization step [139].

With the same biphasic configuration as described above, PLD can be used in the simple hydrolysis step for the production of PA. This has some practical interest for the preparation of phosphatidic acids with a range of acyl chain patterns. It has been observed before that PI is not a substrate for most PLD from different sources. The extraction of PI from natural PL mixtures is complex and low yielding. When a crude PL extract containing PI is treated with PLD, all PL present are transformed into PA while PI is unchanged. Further hydrolysis with a phosphatase further degrades phosphatidic acid to DAG and inorganic phosphate leaving behind phosphatidylinositol as the only phospholipid component in the mixture from which it can be easily recovered by solvent partitioning [166].

3.3
Modification of PL Analogues

The broad substrate specificity of PLD from bacterial sources allows one to extend the polar head exchange technique to a wide range of structurally related

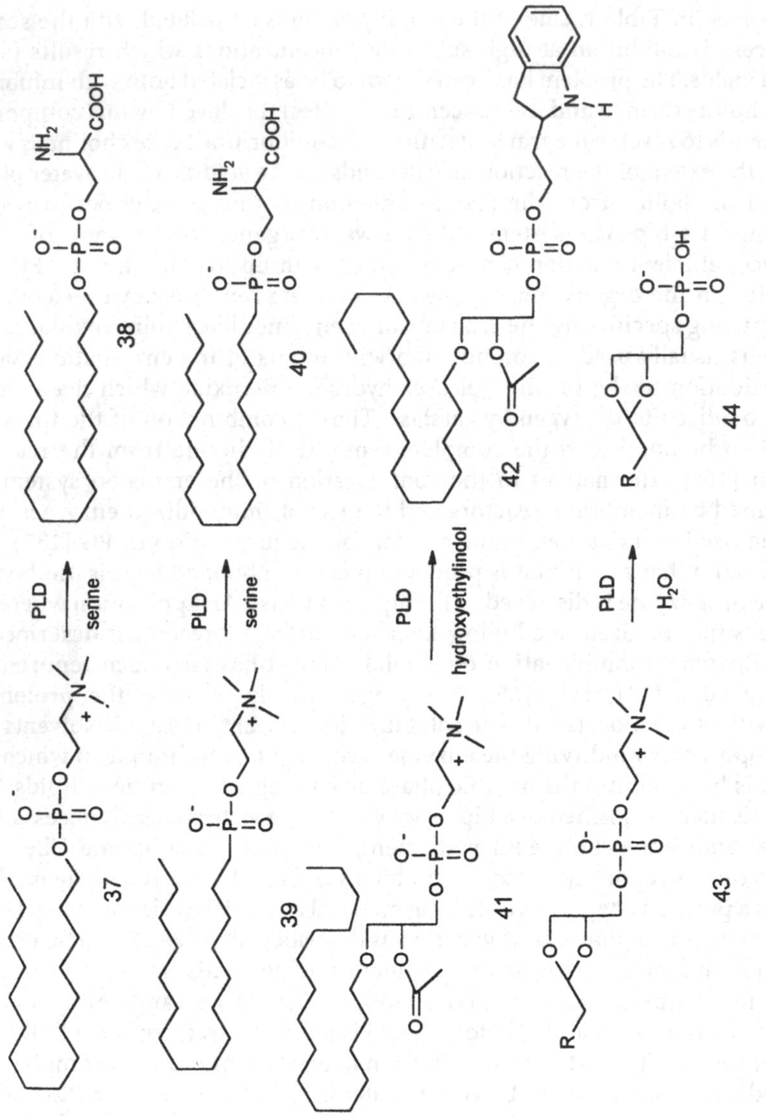

Scheme 10. Alkylphosphocholines as substrates of PLD [25–27, 171, 172]

phospholipid molecules containing phosphothioester bonds [167], plasmalogens [168, 169], or even phosphate diesters and phosphonates not characterized by the glycerol backbone (Scheme 10).

Alkyl phosphocholines of type **37** constitute a new class of antitumor agents with interesting properties in vivo [170]. They are substrates for the PLD catalyzed transesterification with serine to give the corresponding alkyl phosphoserine **38** [25, 26]. Surprisingly, the phosphono-analogue **39** is also a substrate for the transesterification allowing the preparation of a number of alkylphosphonates of type **40** [27]. PAF ethers **41** are also accepted as donors in the transphosphatidylation reaction and have been transformed into non-natural derivatives like the hydroxyethylindol **42** [171]. Finally, phosphocholines in which the acyl chains have been removed and the diol group transformed into an acetal of long chain aldehydes **43** are used as substrates for cabbage PLD as shown in the hydrolysis to **44** [172]. The data given in Scheme 10 are quite impressive: they show that apparently the left part of the molecule has little influence on the substrate recognition of the PL from the PLD. This fact broadens the synthetic utility and applications of this enzyme. It also raises the question about the actual structural requirements for catalysis. In fact, it is clear from the data shown in the preceding sections that, although the choline moiety is best recognized by the enzymes of different sources, it is not a strict requirement for activity. PLD then appears in this activity to be very near to a generic phosphodiesterase. The lack of structural information for these enzymes hampers further speculation on the catalytic mechanism and on similarity with other related enzymes.

4
Synthesis of Specific PL

Other phospholipases have properties ancillary to the ones displayed by PLD in synthetic applications. Their utility for the synthesis of specific compounds is limited by their modest transesterification capacities. However, there are some interesting applications besides the one cited in Sect. 2.2.1. Scheme 11 shows the

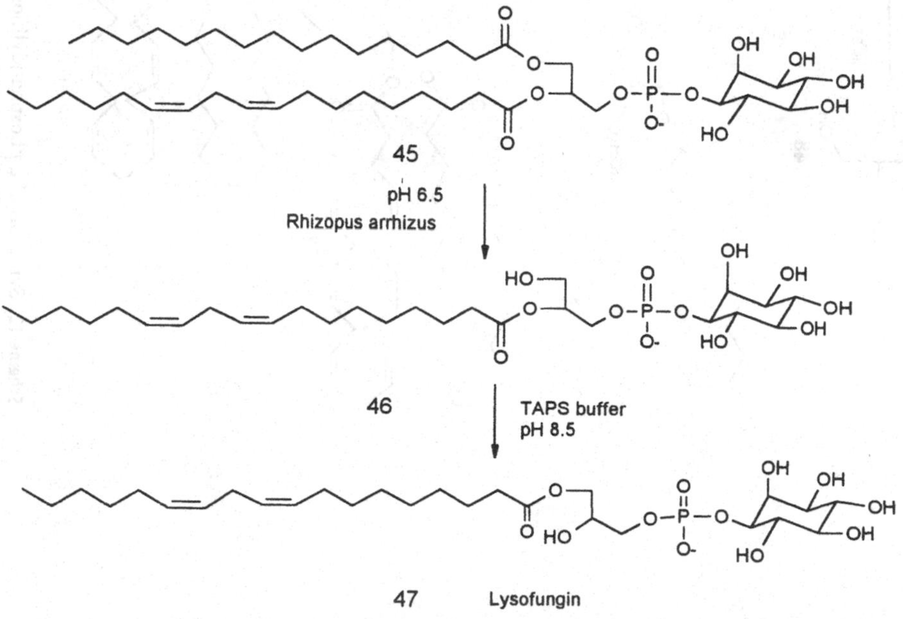

Scheme 11. Chemo-enzymatic synthesis of the anti-fungal agent lysofungin [173]

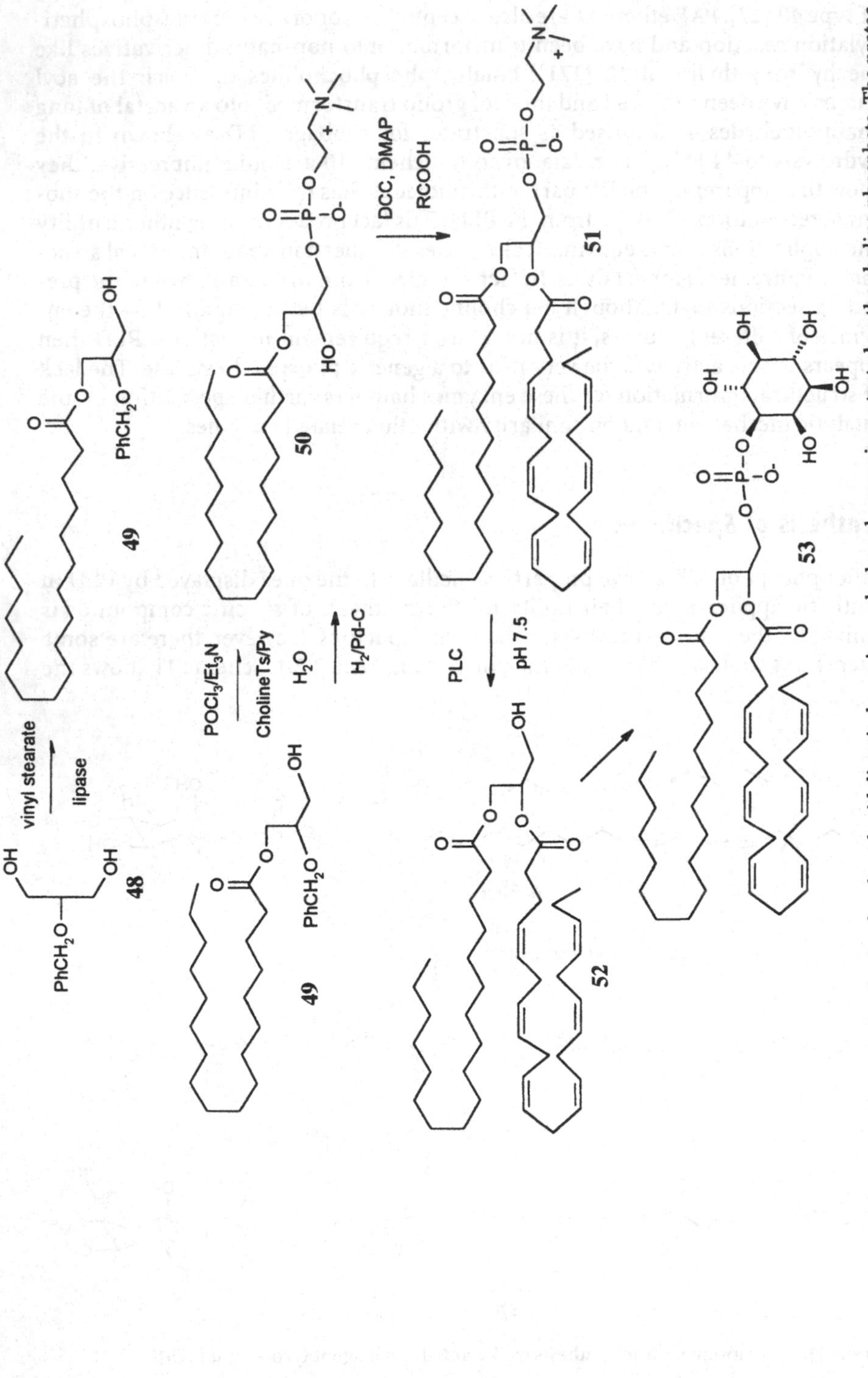

Scheme 12. Synthesis of a phosphatidylinositol: a primary hydroxyl group is protected as the choline phosphate. The protecting group is removed with PLC [174]

semisynthetic approach to the naturally occurring antifungal agent *lysofungin*. This can be taken as a general approach to modified phospholipids (path *a*, *m*, *d* of Scheme 2) [173]. The synthesis uses a natural product as a starting material. Hydrolysis catalyzed by the 1,3-specific lipase from *R. arrhizus* gives the *lyso* compound **46** which is isomerized to the thermodynamically more stable **47**, the desired compound. Scheme 12 reports an application of PLC for the preparation of a specific DAG which is then used for the chemical synthesis of a PI with defined polar chains at the two positions. Chirality is introduced with the dissymmetrization of **48** with a lipase. Compound **49** is protected as the phosphocholine **50** which is then debenzylated and acylated at the *sn*2 position to give the phosphatidylcholine **51**. Deprotection is then performed by PLC hydrolysis to the diacylglycerol **52**. Subsequent careful functionalization affords in several steps **53** [174]. Recently, PLC immobilized on various supports for the preparation of DAG from phospholipids has been prepared [175].

5
Organic Phosphates from PL

Organic phosphates (OP) can be considered as by-products in the PLC catalyzed hydrolysis of PL. Since chemical phosphorylation of multifunctional compounds makes use of very reactive reagents, it requires extensive use of protecting groups and product purification. This is reflected in the high costs of several phosphates of complex or multifunctional molecules. The PLC hydrolysis of a suitable PL precursor can be considered for the preparation of organic monophosphates. In order to obtain di-phosphates from phospholipids, transesterification is necessary. As mentioned in Sect. 3.1, PLC$_{PI}$ is able to carry out such a transfer as has been shown recently by using several different alcohols as phosphate acceptors (Scheme 13). The reaction occurs through the formation of a cyclic phosphate (IcP) according to the mechanism described in Scheme 4 [93]. IcP is the actual substrate for the transfer reaction. Twenty structurally diverse alcohols ranging in structural complexity from methanol to a serine containing tetrapeptide were transferred to give the corresponding phosphodiesters IXP in variable yields. The reaction was performed in water where the transesterification was competing with the hydrolysis. The rate was apparently insensitive

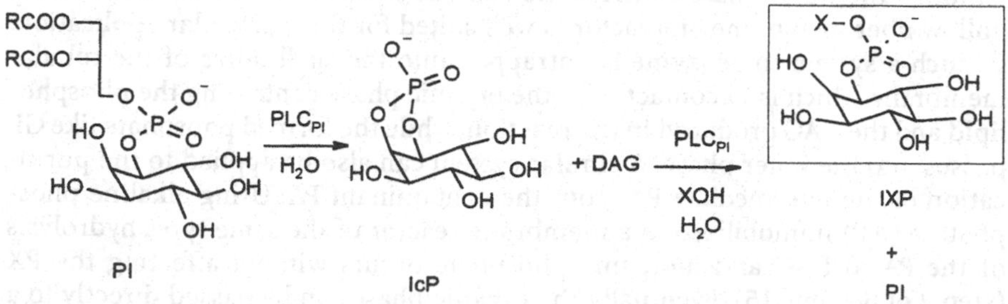

Scheme 13. Formation of inositolalkylphosphates via PLC$_{PI}$ catalysed transesterification [93]

from the alcohol structure. Primary hydroxyl groups were selective acceptors in the presence of secondary alcohols. The reaction showed a low stereoselectivity in the presence of chiral alcohols. A similar transesterification capacity has not been recognized in PLC$_{PC}$. However, *B. cereus* PLC$_{PC}$ has been used in combination with PLD for the indirect phosphorylation of primary alcohols [159].

Scheme 14 shows a proposed application in which PC is transformed into phosphatidylsolketal (PSK). This compound is hydrolyzed by PLC$_{PC}$ forming the OP corresponding to the alcohol acceptor in the first enzymatic reaction. The result of the sequential use of the two enzymes, one in transesterification and the second in hydrolysis, corresponds to the net phosphorylation reaction. The reaction can be of very wide applicability if the hypothesis is verified that both

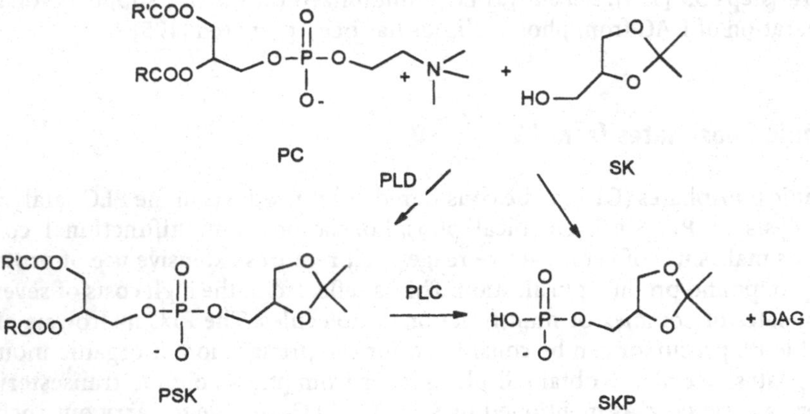

Scheme 14. Indirect enzymatic phosphorylation of primary alcohols [159]

enzymes have a wide substrate specificity. This is unfortunately only partly true in that PLC$_{PC}$ is of much less wide substrate specificity than desired. However, this property is very much dependent on the enzyme source and reaction conditions, especially in the presence of surfactants. Possible substrates for the reaction besides PSK have been found to be PG and phosphatidyldihydroxyacetone. In the PLC hydrolysis of PL, the organic phosphate is the only water soluble compound in the reaction mixture. Its isolation is therefore easy to accomplish, although the water phase contains the enzyme which needs to be recovered. A hollow fiber membrane bioreactor is well suited for this particular application. In such a system the enzyme is entrapped into the shell pores of the tubular membrane which is in contact with the organic phase containing the phospholipid and the DAG produced in the reaction, while the formed phosphate like GP passes into the water phase. A similar system can also be applied to the purification of the intermediate PX from the contaminant PA. Using alkaline phosphatase (AP) immobilized in a membrane reactor of the same type, hydrolysis of the PA to DAG and inorganic phosphate occurs without affecting the PX (step 2 of Scheme 15). Eventually the organic phase can be passed directly to a second membrane reactor for PLC hydrolysis (step 3 of Scheme 15). Since PLD

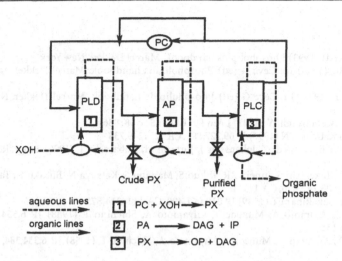

Scheme 15. The proposed continuous production of polar head modified phospholipid (1), its purification by PA removal (2) and hydrolysis to the corresponding OP (3). The three enzymes are immobilized in hollow fiber hydrophobic membrane bioreactors [179]

has proved to be extremely stable in the same kind of reactor, in principle three consecutive steps can allow a continuous production with possible withdrawal of the PX/PA mixture as the first product (step 1). The same mixture can act as the substrate for the second reactor in which the PA formed as a by-product is hydrolyzed, and purified PX results as the second product. The latter can be used as the substrate for the third reactor in which PLC operates, giving OP and DAG as final products [179].

6
Conclusions

Phospholipases are very versatile enzymes which allow the transformation of inexpensive natural products into highly valuable compounds like specific structurally defined phospholipids, organic monophosphates or diphosphates and DAG with the natural absolute configuration. Of particular synthetic utility is PLD from bacterial sources which is able to effect the phosphoryl transfer in a water-containing biphasic system. PLD shows a wide substrate specificity for both the polar head and the alcohol acceptors as well as for the lipophilic part of the molecule. The enzyme behaves like a generic phosphodiesterase with broad substrate specificity and high transphosphatidylation ability. The molecular basis of this behavior should become clear by inspection of the three-dimensional structure and comparison with other phosphoric acid ester hydrolytic enzymes. The crystal structure of this enzyme has not been elucidated. The potential of the many different PLD from plants which show peculiar substrate specificity should allow one to expand the synthetic utility to the hydrolysis-synthesis of natural and unnatural phosphatidylinositols.

7
References

1. Cevc G (ed) (1993) Phospholipids handbook. Marcel Dekker, New York
2. New RRC (1993) In: Cevc G (ed) Phospholipids handbook. Marcel Dekker, New York, p 855
3. Namba Y (1993) In: Cevc G (ed) Phospholipids handbook. Marcel Dekker, New York, p 879
4. Van Nieuwenhuyzen W (1981) J Am Oil Chem Soc 58:886
5. Noguchi K, Uchida N (1997) JP 09,227,895, CA 127:249,738
6. Robertson B, Curstedt T, Johansson J, Jornvall H, Kobayashi T (1990) Progr Respir Res 25:23
7. Shoko R, Taketoshi S, Juneja RR, Kikuo S, Manabu S, Katsuya N, Busaku K, Takehito Y (1993) JP 05,132,496, CA 119:158,396
8. Nanba Y, Sakakibara T (1989) JP 010,061,490, CA 111:214,875
9. Nagata Y, Akimoto A, Muneda Y, Miyamoto A, Shichino F (1988) JP 6,354,385, CA 109:170,133
10. Nagata Y, Akimoto A, Muneda Y, Miyamoto A, Shichino F (1988) JP 6,354,384, CA 109: 170,134
11. Kobayashi H, Hasegawa M, Kato S, Kondo M, Inui M, Otani Y (1989) JP 01,299,292, CA 112:235,768
12. Ladika M, Wu WSW, Jons SD (1994) J Am Chem Soc 116:12,093
13. Lasic DD (1994) Angewandte Chemie Int Ed Eng 33:1685
14. Markowitz M, Singh A (1991) Langmuir 7:16
15. Takano T, Nakagami H, Hashida R, Enomoto H (1986) JP 61,291,593, CA 106:176,800
16. Berkovic D, Goeckenjan M, Lueders S, Hiddemann W, Fleer EAM (1996) J Exp Ther Oncol 1:302
17. Wang P, Schuster M, Wang Y-F, Wong C-H (1993) J Am Chem Soc 115:10,487
18. Shuto S, Ueda S, Imamura S, Fukukawa K, Matsuda A, Ueda T (1987) Tetrahedron Lett 28:199
19. Shuto S, Imamura S, Fukukawa K, Sakakibara H, Murase J-I (1987) Chem Pharm Bull 35:447
20. Schmitt JD, Amidon B, Wykle RL, Waite M (1995) Chem Phys Lipids 77:131
21. Akoka S, Meir C, Tellier C, Belaud C, Poignant S (1985) Synth Commun 15:101
22. Perly B, Dufourc EJ, Jarrell HC (1984) J Labelled Compd Radiopharm 21:1
23. Sahai P, Viswakarma RA (1997) J Chem Soc Perkin Trans 1 1845
24. Grunicke HH, Maly K, Uberall F, Schubert C, Kindler E, Stekar J, Brachwitz H (1996) Adv Enzyme Regul 36:385
25. Brachwitz H, Langen P, Dube G, Schildt J, Paltauf F, Hermetter A (1990) Chem Phys Lipids 54:89
26. Brachwitz H, Schoenfeld R, Langen P, Paltauf F, Hermetter A (1989) EP 338,407, CA 112:199,123
27. Brachwitz H, Oelke M, Bergmann J, Langen P (1997) Bioorg Med Chem Lett 7:1739
28. Wells A, Chen P, Turner T (1997) WO 9,735,588, CA 127:303,328
29. Nagao A, Koga T, Terao J (1997) Kikan Kagaku Sosetsu 33:79 CA 127:345,351
30. Koga T JP 93-39,269, CA 126:276,530
31. Koga T, Terao J (1995) Kenkyu no Shinpo 5:99 CA 126:276,530
32. Nwosu CV, Boyd LC, Sheldon B (1997) J Am Oil Chem Soc 74:293
33. Vinggaard AM, Jensen T, Morgan CP, Cockcroft S, Hansen, HS (1996) Biochem J 319:861
34. Baba N, Kosugi T, Daido H, Umino H, Kishida N, Nakajima S, Shimizu S, (1996) Biosci Biotechnol Biochem 60:1916
35. Murakami S, Tokuyama S, Ozawa K, Kyota R, Nakachi O, (1988) JP 63,157,993, CA 111:134,685
36. Bittman R (1993) In: Cevc G (ed) Phospholipids handbook. Marcel Dekker, New York, p 141
37. Crans DC, Whitesides GM (1985) J Am Chem Soc 107:7019

38. Slotboom AJ, Verheij HM, De Haas GH (1973) Chem Phys Lipids 11:295
39. Shimbo K, Yano H, Miyamoto Y (1989) Agric Biol Chem 53:3083
40. Dinh T, McClure GD, Kennerly DA (1995) Int Arch Allergy Immunol 107:69
41. Iwasaki Y, Mishima N, Mizumoto K, Nakano H, Yamane T (1995) J Ferment Bioeng 79:417
42. Hasegawa M, Ota N, Aisaka K (1992) JP 04,088,981, CA 117:144,792
43. Carrea G, D'Arrigo P, Piergianni V, Roncaglio S, Secundo F, Servi S (1995) Biochim Biophys Acta 1255:273
44. Carrea G, D'Arrigo P, Secundo F, Servi S (1997) Biotechnol Letters 19:1083
45. Yoshioka I, Mizoguchi J, Takahara M, Imamura S, Beppu T, Horinouchi S (1991) EP 90-403,529 CA 115:176,720
46. Shimbo K, Yano H, Miyamoto Y (1990) Agric Biol Chem 54:1189
47. Okawa Y, Yamaguchi T (1975) J Biochem 78:363
48. Yamaguchi T, Okawa Y, Sakaguchi K, Muto N (1973) Agric Biol Chem 37:1667
49. Streikuvene IK, Nekrashaite GI, Kulene VV, Travkina VS, Peslyakas II, Glemzha AA (1982) Appl Biochem Microbiol 18:35
50. Imamura S, Horiuchi Y (1979) J Biochem (Tokyo) 85:75
51. Sawada H, Kudo S, Watanabe T, Kuroda A, Oki T (1988) JP 63,219,373, CA 111:113,768
52. Dennis EA (ed) (1991) Phospholipases. Methods in enzymology, vol 197. Academic Press, New York
53. Rhee SG, Dennis EA (1996) Modular Texts Mol Cell Biol 1:173
54. Liscovitch M, Cantley LC (1994) Cell 77:329
55. Dennis EA (1983) In: Boyer E (ed) The enzymes, 3rd edn, vol 16. Academic Press, New York, p 307
56. Ransac S, Moreau H, Rivière C, Verger R (1991) In: Dennis EA (ed) Phospholipases. Methods in enzymology, vol 197. Academic Press, New York, p 49
57. Carrea G, D'Arrigo P, Mazzotti M, Secundo F, Servi S (1997) Biocatal Biotrans 15:251
58. Roberts MF, Otnaess AB, Kensil CA, Dennis EA (1978) J Biol Chem 253:1252
59. Gassama-Diagne A, Fauvel J, Chap H (1991) In: Dennis EA (ed) Methods in enzymology, vol 197. Academic Press, New York, p 316
60. Slotboom AJ, de Haas GH, Bonsen PP, Burbach-Westerhuis G, van Deenen LLM (1970) Chem Phys Lipids 4:15
61. Kucera GL, Sisson PJ, Thomas MJ, Waite M (1988) J Biol Chem 263:1920
62. Paltauf F (1978) Lipids 13:165
63. Noguchi K, Uchida N (1997) JP 09,227,895, CA 127:249,738
64. Kim MK, Rhee JS (1996) J Microbiol Biotechnol 6:407
65. Haas MJ, Cichowicz DJ, Phillips J, Moreau R (1993) J Am Oil Chem Soc 70:111
66. Morimoto T, Murakami N, Nagatsu A, Sakakibara J (1993) Tetrahedron Lett 34:2487
67. Ikuta J, Tashiro S, Masano Y, Ando N, Asaoka S, Kobayashi H (1994) JP 06,327,486, CA 122:185,535
68. Mustranta A, Forssell P, Aura AM, Suortti T, Poutanen K (1994) Biocatalysis 9:181
69. Svensson I, Adlercreutz P, Mattiasson B (1992) J Am Oil Chem Soc 69:986
70. Verheji HM, Slotboom AJ, de Haas GH (1981) Rev Physiol Biochem Pharmacol 91:91
71. Okawa Y, Yamaguchi T (1976) Agric Biol Chem 40:437
72. Verma JN, Bansal VS, Khuller GK, Subrahmanyam D (1980) Indian J Med Res 72:487
73. Na A, Eriksson C, Eriksson SG, Oesterberg E, Holmberg K (1990) J Am Oil Chem Soc 67:766
74. Paltauf F (1976) Chem Phys Lipids 17:148
75. Aura AM, Forssell P, Mustranta A, Suortti T, Poutanen K (1995) J Am Oil Chem Soc 72 1375
76. Pedersen KB (1991) WO 91/05,056
77. Lilja-Hallberg M, Haerroed M (1994) Biocatalysis 9:195
78. Pernas P, Olivier JL, Legoy MD, Bereziat G (1990) Biochem Biophys Res Commun 168:644
79. Haerroed M, Elfman I (1995) J Am Oil Chem Soc 72:641
80. Wu FC, Lin G (1994) J Chin Chem Soc (Taipei) 41:97
81. Lin G, Wu FC, Liu SH (1993) Tetrahedron Lett 34:1959

82. Bonsen PPM, de Haas GH, Pieterson WA, van Deenen LLM (1972) Biochim Biophys Acta 270:364
83. Aarsman AJ, Roosenboom CFP, Van der Marel GA, Shadid B, Van Boom JH, Van der Bosch H (1985) Chem Phys Lipids 36:229
84. Orr GA, Brewer CF, Heney G (1982) Biochemistry 21:3202
85. Bruzik K, Tsai MD (1984) Biochemistry 23:1656
86. Little C (1981) Methods in Enzymology 71:725
87. Zwaal RF, Roelofsen B, Comfurius P, Van Deenen LLM (1971) Biochim Biophys Acta 233:474
88. Zwaal RF, Roelofsen B (1976) Methods in Enzymology 32:154
89. Otnaess AB, Little C, Sletten K, Wallin R, Johnsen S, Flensgrud R (1977) Eur J Biochem 79:459
90. Myrnes BJ, Little C (1980) Acta Chem Scand B 375
91. Titball RW, Hunter SEC, Martin KL, Morris BC, Shuttleworth AD, Rubidge T, Anderson DW, Kelly DC (1989) 57:367
92. Griffith OH, Volwerk JJ, Kuppe A (1991) In: Dennis EA (ed) Methods Enzymol 197:493
93. Bruzik K, Guan Z, Riddle S, Tsai MD (1996) J Am Chem Soc 118:7679
94. Mandal SB, Sen PC, Chakrabarti P (1980) Phytochemistry 19:1661
95. Sagatova FA; Rashidova DS, Yakubov IT, Rakhimov MM (1996) Prikl Biokhim Mikrobiol 32:500, CA 126:16,167
96. Madoery RR, Gonzalez C (1997) Inf Tecnol 8:23
97. Novotna Z, Valentova O, Daussant J, Kas J (1997) In: Williams JP, Khan MU, Lem NW (eds) Physiology, biochemistry and molecular biology of plant lipids. Proceedings of the 12th International Symposium, Toronto 7-12 July 1996. Kluwer, Dordrecht, p 404
98. Abousalham A, Riviere M, Teissere M, Verger R (1993) Biochim Biophys Acta 1158:1
99. Abousalham A, Teissere M, Gardies AM, Verger R, Noat G (1997) Plant Cell Physiol 36:989
100. Ueki J, Morioka S, Komari T, Kumashiro T (1995) Plant Cell Physiol 36:903
101. Wang X, Xu L, Zheng L (1994) J Biol Chem 269:20,312
102. Atwal AS, Henderson HM, Eskin NAM (1979) Lipids 14:913
103. D'Arrigo P, Piergianni V, Scarcelli D, Servi S (1995) Anal Chim Acta 304:249
104. Shimbo K, Yano H, Miyamoto Y (1990) Agric Biol Chem 54:1189
105. Heller, M (1978) Adv Lipid Res 16:267
106. Van Den Bosch H (1974) Annu Rev Biochem 43:243
107. D'Arrigo P, Servi S (1997) Trends Biotechnol 15:90
108. Shuto S, Matsuda A (1997) Yuki Gosei Kagaku Kyokaishi 55:207, CA 126:317,529
109. Yamane T (1997) Kagaku Kogyo 48:396
110. Yamane T, Iwasaki Y (1995) Yukagaku 44:875
111. Yamane T (1995) Yukagaku 44:623
112. Kovatchev S, Eibl H (1978) Adv Exp Med Biol 101:221
113. Eibl H, Kovatchev S (1981) Methods Enzymol 72:632
114. Yang SF, Freer S, Benson AA (1967) J Biol Chem 242:477
115. Dawson RMC, (1967) Biochem J 102:205
116. Secundo F, Carrea G, D'Arrigo P, Servi S (1996) Biochemistry 35:9631
117. Bruzik K, Tsai MD (1991) Methods Enzymol 197:258
118. Saito M, Kanfer J (1975) Arch Biochem Biophys 169:318
119. Stanacev NZ, Stuhne-Sekalec L, Domazet Z (1973) Can J Biochem 51:747
120. Garcia-Diaz M, Avalos M, Cameselle JC (1993) Eur J Biochem 213:1139
121. Han R, Coleman JE (1995) Biochemistry 34:4238
122. Pradines A, Klaebe A, Perie J, Paul F, Monsan P (1988) Tetrahedron 44:6373
123. Pradines A, Klaebe A, Perie J, Paul F, Monsan P (1991) Enzym Micob Technol 13:19
124. Breslow R (1991) Acc Chem Res 24:317
125. Hershlag D (1994) J Am Chem Soc 116:11,631
126. Friedman P, Haimovitz R, Markman O, Roberts MF, Shinitsky M (1996) J Biol Chem 271:953
127. Shinitzky M, Friedman P, Haimovitz R (1993) J Biol Chem 268:14,109

128. Hansen S, Hough E, Svensson LA, Wong YL, Martin S (1993) J Mol Biol 234:179
129. Atack JR, Broughton HB, Pollack SJ (1995) FEBS Lett 361:1
130. Kanfer JL, Spielvogel CH (1975) Lipids 10:391
131. Das ML, Daak ED, Crane FL (1965) Biochemistry 4:859
132. Das ML, Hiratsuka H, Machinist JM, Crane FL (1962) Biochim Biophys Acta 60:433
133. Green DE, Fleischer S (1963) Biochim Biophys Acta 70:554
134. Matzushima A, Kodera Y, Hiroto M, Nishinura H, Inada Y (1996) J Mol Cat B:Enzymatic 2:1
135. Okahata Y, Mori T (1997) Trends Biotechnol 15:50
136. Okahata Y, Niikura K-I, Ijiro K (1995) J Chem Soc Perkin I 919
137. Lee SY, Hibi N, Yamane T, Shimizu S (1985) J Ferment Technol 63:37
138. Kudo S, Sawada H, Watanabe T, Kuroda A (1988) EP 285,421, CA 110:230,192
139. Mueller GC, Fleming MF, LeMahieu MA, Lybrand GS, Barry KJ (1988) Proc Natl Acad Sci USA 85:9778
140. Metz SA, Dunlop M (1990) Arch Biochem Biophys 283:417
141. Juneja LR, Hibi N, Yamane T, Shimizu S (1987) Appl Microbiol Biotechnol 27:146
142. Comfurius O, Zwaal RFA (1977) Biochim Biophys Acta 488:36
143. Comfurius P, Bevers EM, Zwaal RFA (1990) J Lipid Res 31:1719
144. Juneja LR, Kazuoka T, Goto N, Yamane T, Shimizu S (1989) Biochim Biophys Acta:1003:277
145. de Ferra L, Massardo P, Piccolo O, Servi S (1997) EP 776,976, CA 127:64,623
146. D'Arrigo P, de Ferra L, Pedrocchi-Fantoni G, Scarcelli D, Servi S, Strini A (1996) J Chem Soc Perkin I 2651
147. Takami M, Hidaka N, Miki S, Suzuki Y (1994) Biosci Biotech Biochem 58:1716
148. Takami M, Suzuki Y (1994) Biosci Biotech Biochem 58:1897
149. Nagao A, Ishida N, Terao J (1991) Lipids 26:390
150. D'Arrigo P, Pedrocchi-Fantoni G, Servi S (unpublished results)
151. D'Arrigo P, de Ferra L, Piergianni V, Ricci A, Scarcelli D, Servi S (1994) J Chem Soc Chem Commun 1709
152. Clarke NG, Irvine RF, Dawson RMC (1981) Biochem. J 195:521
153. D'Arrigo P, de Ferra L, Pedrocchi-Fantoni G, Scarcelli D, Servi S, Strini A (1996) J Chem Soc Perkin I 2657
154. van Blitterswijk WJ, Hilkmann H (1993) EMBO J 12:2655
155. Walton PA, Goldfine H (1987) J Biol Chem 262:10,355
156. Joutti A, Renkonen O (1976) Chem Phys Lipids 17:264
157. Silvius JR (1993) In: Cevc G (ed) Phospholipids handbook. Marcel Dekker, New York, p 1
158. Griffith OH, Volwerk JJ, Kuppe A (1991) Methods Enzymol 197:493
159. D'Arrigo P, Piergianni V, Pedrocchi-Fantoni G, Servi S (1995) J Chem Soc Chem Commun 2505
160. Juneja LR, Yamane T, Shimizu S (1989) J Am Oil Chem Soc 66:714
161. Juneia LR, Taniguchi E, Shimizu S, Yamane T (1985) J Ferment Technol 73:357
162. Masoom-Yasinzai M, Yaqoob M (1995) Process Biochem 30:701
163. Takami M, Suzuki Y (1995) J Ferment Bioeng 79:313
164. Yamane T, Juneja LR, Li D, Shimizu S (1990) Ann NY Acad Sci 613:686
165. Tobback P, Weng Z, Fowue B (1988) Meded Fac Landbouwwet Rjiksuniv Gent 53:1785
166. Shimizu S, Yamane T, Li D, Juneja LR (1992) USP 5,100,787, CA 116:233,930
167. Cox JW, Snyder WR, Horrocks LA (1979) Chem Phys Lipids 25:369
168. Achterberg V, Fricke H, Gerken G (1986) Chem Phys Lipids 41:349
169. Wolf RA, Gross RW (1985) J Lipid Res 26:629
170. Houlihan WJ, Lohmeyer M, Workman P, Cheon SH (1995) Med Res Rev 15:157
171. Archaimbault B, Durand J, Rigaud M (1992) Biochim Biophys Acta 1123:347
172. Klykov VN, Ostapemko OV, Serebrennikova GA (1993) Bioorg. Khim 19:360
173. VanMiddlesworth F, Lopez M, Zweerink M, Edison AM, Wilson K (1992) J Org Chem 57:4753

174. Baba N, Kosugi T, Daido H, Umino H, Kishida N, Nakajima S, Shimizu S (1996) Biosci Biotechnol Biochem 60:1916
175. Anthonsen T, D'Arrigo P, Pedrocchi-Fantoni G, Secundo F, Servi S, Sundby E (1998) J Mol Cat B:Enzymatic (submitted)
176. Asahi Kasei, Tokyo (1992) Enzyme catalog, file T-39
177. Mayr JA, Kohlwein SD, Paltauf F (1996) FEBS Lett 393:236
178. Waksman M, Eli Y, Liscovitch M, Gerst JE (1996) J Biol Chem 271:236
179. Each step of the sequence in Scheme 15 gives good results. The three consecutive reactors configuration has not been realized. D'Arrigo P, Servi S (unpublished results)

Biocatalytic Approaches for the Synthesis of Enantiopure Epoxides

A. Archelas · R. Furstoss

Groupe Biocatalyse et Chimie Fine, ERS 157 associée au CNRS, Faculté des Sciences de Luminy, Case 901, 163 Avenue de Luminy, F-13288 Marseille Cedex 9 (France).
E-mail: furstoss@lac.gulliver.fr

Because of their high chemical versatility, enantiopure epoxides, as well as their corresponding vicinal diols, are recognized as being high value intermediates in fine organic chemistry, in particular for the synthesis of biologically active compounds in optically pure form. Therefore, research work aimed to set up efficient procedures allowing for the preparation of such target molecules has been intensively developed recently, leading to the emergence of various new methods based on either conventional chemistry or on biocatalytic reactions. In this review, we will focus on the use of such enzymatic approaches for the synthesis of enantiopure epoxides. Examination of the recent literature indicates that two general strategies have thus been developed, i.e. (a) the formation of the epoxide ring itself from an appropriate precursor (in general the corresponding olefin) and (b) the resolution of racemic substrates already bearing an oxirane ring. Several of these approaches have been shown to allow the synthesis of epoxides of different structures, which were thus obtained in enantiomerically enriched or even enantiopure form.

Keywords: Enantiopure epoxides, Oxidative enzymes, Cytochrome P-450, ω-hydroxylases, Methane monooxygenases, Lipases, Microbial oxidations, Epoxide hydrolases, Biotransformations.

1
Introduction

Among the numerous building blocks used for fine organic synthesis, epoxides are undoubtedly recognized as being highly valuable intermediates. As illustrated in Fig. 1, this is essentially due to the versatility of the oxirane function which can be chemically transformed into numerous, more elaborate intermediates en route to biologically active targets. The corresponding vicinal diols, which can itself be used as such or can be either transformed into the epoxide itself, or used as highly reactive cyclic sulfate or sulfite derivatives, are similarly very interesting "epoxide-like" compounds [1].

In recent years, considerable effort has been devoted to the synthesis of these intermediates in enantiomerically pure form, due to the fact that chiral molecules very often show different biological activity for each enantiomer [2, 3]. Both purely chemical and enzyme-catalyzed methodologies have been developed to achieve their preparation, and several reviews and monographs have been devoted to this topic [4–11]. The Sharpless method was the first conventional chemical breakthrough, which allowed the titanium-tartrate catalyzed asymmetric epoxidation of allylic alcohols [12–14]. More recently, several other methodologies have been explored toward this goal. Among these, the so-called Jacobsen-Katzuki catalysts allowing for the epoxidation of various olefins have been extensively developed. Interestingly, this method was shown to lead to high stereoselectivities when applied to the epoxidation of simple alkenes, essential-

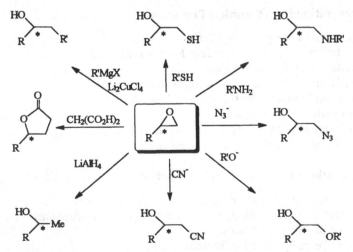

Fig. 1. Reaction of epoxides with various nucleophiles (Adapted from Faber et al. [172])

ly *cis*-disubstituted and trisubstituted olefins conjugated with aryl, alkenyl and alkynyl groups, whereas for simple (non-conjugated) *cis* and terminal olefins, as well as for *trans* olefins, the selectivity is less satisfactory [4, 7]. To these pioneering approaches, three determining achievements have been added over the very recent years, i.e. (a) the Sharpless dihydroxylation process based on the use of the so-called AD-*mix* osmium based catalysts, allowing for direct dihydroxylation of olefins, which is particularly satisfactory – for yield and optical purity – when applied to *trans* substituted olefins [8, 15], (b) the catalytic kinetic resolution of terminal epoxides, achieved for instance by hydrolysis using a Co(II) salen complex [16] and (c) the asymmetric opening of *meso* epoxides [17].

As an overall observation, it appears that, depending on the structure of the substrate implied, one out of these approaches should make possible the preparation of a particular epoxide in a satisfactory way, as has been recently demonstrated for 3-chlorostyrene oxide for example [18]. However, although very elegant and efficient in certain cases, these approaches suffer from various drawbacks which can become crucial for large scale industrial production. These include the fact that they require the use of heavy metal-based catalysts (possible sources of industrial pollution) and are patent pending and, thus, cost-generating processes. Complementary to these chemical approaches, a number of enzyme-catalyzed methods have been studied during the last decade and now provide new, efficient and environmentally frienly alternatives. Here again, several reviews and monographs have been focused on the possible use of enzymes for fine organic synthesis, including ways to synthesize optically enriched epoxides [9, 11, 19–21]. Therefore, this review does not aim to provide an exhaustive catalogue of the possible enzymatic reactions allowing the preparation of enantiopure epoxides, but rather to discuss more recent information about these biotechnological techniques.

2
Preparation from Chemical Precursors

2.1
From the Corresponding Olefins

Numerous examples of metabolic pathways, describing the catabolism of molecules that show very diverse structures, have been described [22]. Many imply the direct incorporation, on an olefinic double bond, of one single oxygen atom which can originate from either water or molecular oxygen. However only few of these enzymes have been identified, which indicated that at least three types of enzymes, i.e. heme monooxygenases (cytochrome P-450), ω-monooxygenases, and methane monooxygenases, can be involved in these bio-epoxidations. Therefore, due to the asymmetric nature of these biocatalysts, the use of such enzymes was examined by numerous authors in the context of obtaining enantiopure epoxides.

2.1.1
Oxidation by Heme-Dependent Monooxygenases

One of the most important classes of oxygenating enzymes is the cytochrome P-450 family. These hemoproteins have been detected in all types of living cells, including insects, plants, yeast, bacteria and fungi, but because of their involvement in detoxication processes, the mammalian enzymes have been by far the most thoroughly studied [23]. These enzymes oxygenate lipophilic xenobiotics as the first and essential step in the detoxication process, the second step being conjugation with various natural counterparts, such as glutathione, glycosides, and sulfates, thus leading to water-soluble – and thus excretable – metabolites.

The cytochrome superfamily of heme monooxygenases catalyzes many reactions involved in the oxidative degradation of endogenous or exogenous compounds. These enzymes all share the same overall catalytic mechanism, and their varied substrate specificities arise from differences in their active site topologies. The prosthetic moiety implied in these cytochrome P-450 enzymes is a ferroporphyrin entity called "heme" which can activate molecular oxygen and incorporate one of its two oxygen atoms into an organic molecule [24, 25]. Depending on the molecular characteristics of the substrates, these monooxygenases will allow, for instance, hydroxylation of non-activated carbon atoms or epoxidation of olefinic double bonds [26 – 28].

Epoxidation of various olefins by cytochrome P-450 enzymes has been studied using rat liver microsomes [29, 30] as well as using enzymes from microbial origin. For example, Ruettinger and Fulco [31] reported the epoxidation of fatty acids such as palmitoleic acid by a cytochrome P-450 from *Bacillus megaterium*. Their results indicate that both the epoxidation and the hydroxylation processes are catalyzed by the same NADPH-dependent monooxygenase. More recently, other researchers demonstrated that the cytochrome P-450$_{cam}$ from *Pseudomonas putida*, which is known to hydroxylate camphor at a non-activated carbon atom, is also responsible for stereoselective epoxidation of *cis-β*-methylstyrene [32]. The (1S,2R)-epoxide enantiomer obtained showed an enantiomeric purity (ee) of 78%. This result fits the predictions based on a theoretical approach (Fig. 2).

During recent years, several cytochrome P-450 enzymes have been cloned and overexpressed in various hosts [33] and studies aimed to modify the active site topology of these enzymes, i.e. of the P-450$_{cam}$ for example, have been

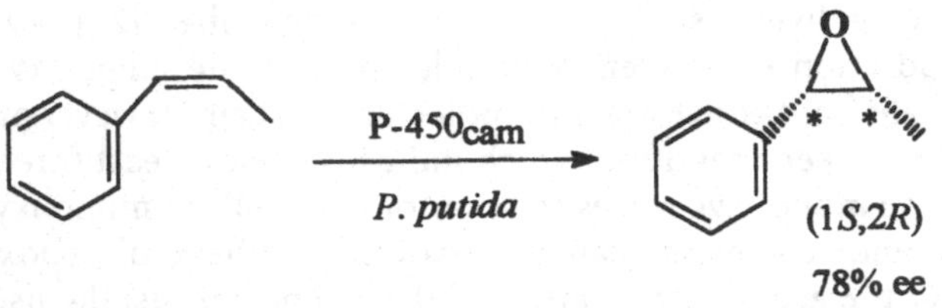

P-450$_{cam}$

P. putida

(1S,2R)

78% ee

Fig. 2. Stereoselective epoxidation of *cis-β*-methylstyrene using cytochrome P-450$_{cam}$ from *Pseudomonas putida* [32]

recently described [34]. Interestingly, the result of this work enabled the rational redesign of this enzyme, thus leading to an improvement in the activity and coupling efficiency towards unnatural substrates, and engineering of the regioselectivity of substrate oxidation. Furthermore, the construction of a "self-sufficient" fusion protein containing all three proteins involved in the multienzymatic system brings one step nearer the in vitro and in vivo biotechnological applications of this type of enzyme. Similarly, an elegant result has been obtained concerning another heme-dependent protein, human myoglobin [35]. Thus, the phenylalanine-43 in the heme pocket of this protein was replaced by tyrosine using site-directed mutagenesis: the tyrosine-43 mutant was shown to be approximately 25 times more active than the wild-type protein in mediating the oxidation of styrene by hydrogen peroxide. Moreover, it was shown to afford (R)-styrene oxide (60% yield), which had an ee as high as 98%, whereas the wild-type protein produced the racemic product. These examples are undoubtedly an early illustration of the potential that molecular biology approaches have for direct epoxidation of olefinic double bonds.

2.1.2
Oxidation by ω-Hydroxylases

The ability to activate and transfer molecular oxygen into organic molecules is not restricted to the cytochrome P-450 family, and some non-heme monooxygenases have been described as being involved in such processes as well. This is the case, for instance, for ω-hydroxylases which have been detected in several microorganisms. As early as 1973, Abbott et al. [36, 37] unequivocally established that the *Pseudomonas oleovorans* strain was able to perform ω-hydroxylation of aliphatic molecules as well as stereoselective epoxidation of a terminal double bond of the corresponding olefins. These authors observed that, for instance, 1,7-octadiene led exclusively to 7,8-epoxy-1-octene, which could be further re-epoxidized to the corresponding diepoxide [38]. Application of these oxidations to allyl alcohol derivatives have also been described recently, albeit the yields were low [39]. The enzymatic mechanism of these reactions has been studied by Coon et al. [40, 41], who isolated the so-called ω-hydroxylase enzyme. They showed that this enzyme was capable of performing either hydroxylation or epoxidation. Furthermore, they observed that both types of reactions can be in competition on an olefinic substrate, depending upon the structure of the substrate. However, cyclic olefins were neither hydroxylated nor epoxidized by this enzyme [42]. More recently, the gene encoding the ω-hydroxylase from *P. oleovorans* was expressed in *E. coli* and its structure has been explored using Mössbauer studies [43].

These results have led to an interesting industrial application for the synthesis of the β-blockers Metoprolol and Atenolol. Thus, epoxidation of the prochiral allyl ethers by several bacteria, including the *P. oleovorans* strain mentioned above, led to the corresponding (S)-epoxides which showed excellent enantiomeric purities (Fig. 3). Further on, these chirons (i.e. chiral building blocks) were transformed into the corresponding (S)-enantiomers of the drugs developed by the Shell and Gist-Brocades companies [44]. Refinement of this approach

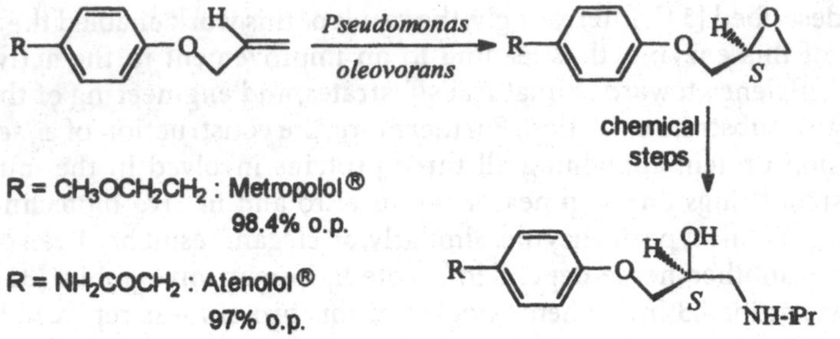

Fig. 3. Synthetic route to Metoprolol and Atenolol involving stereospecific microbial epoxidation [44]

has led to the development of interesting molecular biology strategies, which led to the construction of a *P. putida* strain equipped with an *alk* gene from *P. oleovorans*. Thus, this recombinant biocatalyst can accumulate the products that are not oxidized further. This *P. oleovorans* monooxygenase has been shown to epoxidize various types of olefins, including allylphenyl ethers and allylbenzene with high stereoselectivity [45].

Other bacteria have also achieved asymmetric epoxidation of substrates such as straight chain aliphatic terminal or subterminal alkenes or aromatic olefins [46–48]. In this context, one of the most thoroughly studied enzymatic system is xylene oxygenase (XO), which originates from a *P. putida* strain capable of utilizing toluene and xylenes as carbon sources for growth. This XO system essentially acts in epoxidation/hydroxylation reactions of aromatic substrates [45], and it has been introduced by genetic engineering into an *E. coli* host. Interestingly, it has been observed that whereas this enzyme generally does not epoxidize simple olefinic double bonds, it operates nicely on styrene to produce (*S*)-styrene oxide, which is obtained in 93% enantiomeric purity. Enantiopure styrene oxide is a useful chiral building block for the synthesis of optically active α-substituted benzyl alcohols [49] and has been used to synthesize various pharmaceuticals [50, 51]. Similarly, *m*-chlorostyrene was epoxidized with high stereoselectivity (ee >95%). However, *p*-chlorostyrene only led to a 37% ee whereas *p*-methylstyrene was preferentially oxidized at the methyl substituent. Biotechnological improvements of these monooxygenase-based transformations, using two-liquid phase fermentations, have been elaborated to enhance the practicability and yields of these biooxydations [52]. However, although elegant, the real applicability of these techniques to large-scale production is still to be demonstrated.

2.1.3
Oxidation by Methane Monooxygenases

Other interesting non-heme oxygenases are the methane monooxygenases (MMOs) [53]. These enzymes can transform methane into methanol, a reaction extremely difficult to carry out using conventional synthesis. They are also able to introduce oxygen into a large number of hydrocarbons like alkanes and alkenes, as well as on alicyclic and aromatic substrates [54–56]. For instance, Elliot et al. have recently described the epoxidation of various olefins with the particulate methane monooxygenase from *M. capsulatus* (bath) [57]. It has been shown that the same enzymatic system was responsible for the hydroxylation and the epoxidation process. Different MMO enzymes have been isolated from methanotrophic microorganisms such as *Methylosinus trichosporium*, *Methylococcus capsulatus*, *Methylococcus organophilum*, *Nocardia corallina* or *Xanthobacter* strain Py2 [58]. It has been found for example that the soluble enzymatic activity from *Xanthobacter* Py2 is a four component system, each of which has been purified to homogeneity [59]. Ohno and Okura [60] also observed epoxidation of short-chain alkenes, but the ee's they obtained were low (14 < ee < 28 %). More recently, Seki et al. studied the epoxidation of halogenated allyl derivatives by the bacteria *M. trichosporium*. Interesting results – i. e. inversion of stereoselectivities – were observed, depending on the substrate substitution. However, the observed ee of the products were low again [61].

2.1.4
Reaction Mediated by Haloperoxidases and Halohydrine Epoxidases

Another way to synthesize epoxides is the enzymatic addition of hypohalous acid (XOH) on an olefinic double bond, followed by cyclisation of the halohydrine intermediate. A class of enzymes named haloperoxidases, are known to catalyze the formation of α-halohydrines from alkenes in the presence of a halide ion and a hydroperoxide [62]. These enzymes are available from a variety of sources [63] and, according to the type of halide they can utilize, they can be separated into three groups: chloroperoxidases, bromoperoxidases (algae and bacteria), and iodoperoxidases (algae). Iodoperoxidases catalyze the formation of carbon-iodine bonds, whereas bromoperoxidases catalyze iodination and bromination reactions, chloroperoxidases being able to handle either chlorine, bromine or iodine ions. These enzymes can be either heme-containing enzymes [63] or non-heme enzymes [64–66]. They can react with various types of organic molecules and a recent review has been devoted to these different aspects [67]. However, the most commonly used haloperoxidase is the chloroperoxidase from *Caldariomyces fumago*, available as a commercial preparation. This hemoprotein has been crystallized and shown to exist in solution at neutral pH as a heavily glycosylated ($\approx$ 25–30% by weight) monomer [68, 69] and its exact structure has been studied by resonance Raman studies and by extended X-ray fine structure spectroscopy [70]. In the presence of hydrogen peroxide, this enzyme converts a large variety of olefins into the corresponding halohydrin [71,72]. Several types of olefins can thus be transformed, including alkenes [62], dienes [73], allenes [62], as well as more complex substrates [74]. Interestingly, the products formed from these reactions are those that could be predicted for chemical attack of hypohalous acid (XOH) on the olefine substrate. Moreover, it happens that the obtained products are racemic, giving a clue to the mechanism implied. In fact, it appears that these enzymes, in the presence of halide ions, act as an enzymatic source of hypohalous acid, which then adds non-enzymatically to the substrate [63, 75]. Therefore, the synthetic utility of this approach for the synthesis of enantiopure epoxides is linked to the further use of these intermediates which cyclization into epoxides can possibly be achieved by using halohydrine epoxidase enzymes. This type of enzyme has been detected in several bacteria such as *Flavobacterium*, *Pseudomonas* and *Arthrobacter* species [76]. This last one has been purified and characterized [77]. It allows the cyclization of 3-chloro-1,2-propanediol and of 1,3-dichloro-2-propanol, thereby affording either glycidol or epichlorhydrine, respectively.

However, the enantioselectivity of these reactions is not indicated, presumably because the obtained products were racemic.

Despite the lack of enantioselectivity, such reactions may still be interesting for an individual process. For example, an elegant and efficient industrial application has been elaborated for the production of – racemic – propylene oxide (Cetus procedure) by Neidleman [78]. However, to the best of our knowledge, severe competition with conventional chemical processes occurred and this procedure has not been used industrially.

It was recently described that such a two-step process was also occurring in the course of the microbial oxidation of the olefinic precursor of fosfomycin, a clinically important, broad spectrum antibiotic. Fifteen strains of aerobic bacteria as well as two fungal strains were shown to produce optically active fosfomycin [79].

Enantioselective degradation of halohydrines is one other way to prepare optically enriched epoxides. Kasai et al. [80] used an immobilized *Pseudomonas* strain to degrade racemic 2,3-dichloro-1-propanol. The residual (S)-substrate, which showed an ee as high as 99%, was then further transformed chemically into the corresponding (R)-epichlorohydrine (Fig. 4). Obviously, in such transformations, the yield is limited to 50%. On the other hand, enzymatic conversion of prochiral substrates can allow a theoretical 100% yield. Such an approach was illustrated by Nakamura et al. [81, 82] starting from the prochiral 1,3-dichloropropan-2-ol, which was converted into a 62% yield of (R)-epichlorohydrine (75% ee). Although this is not sufficient for further synthesis of biologically active compounds, it indicates that such transformations, leading to enantiomerically enriched epoxides, may be useful.

Interestingly, it has also been shown that the *Caldariomyces fumago* chloroperoxidase is able to yield epoxides directly from olefins in the absence of halide

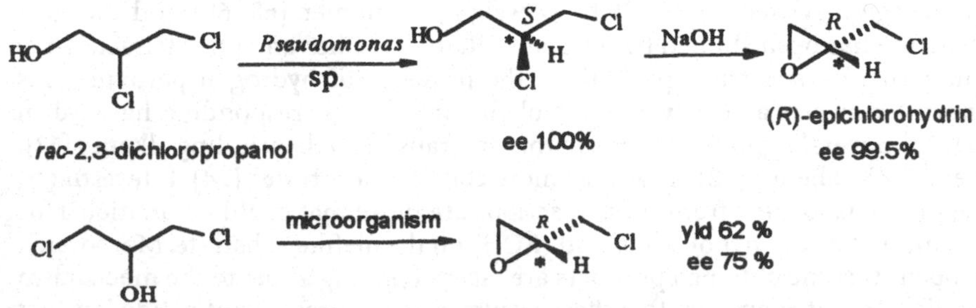

Fig. 4. Enantioselective degradation of halohydrines. Synthesis of (R)-epichlorohydrine [80–82]

ions. Extensive studies developed recently by Allain and Hager [83] and by Zaks and Dodds [84] showed that the best substrates were aliphatic cis-disubstituted olefins. In these cases, ee's as high as 95% could be reached but yields were generally moderate. Even lower yields and ee's were, in general, obtained from differently substituted olefins [85], but 2-methylalkenes appeared to lead to interesting enantiomeric purities in certain cases [86, 87] (Table 1).

Similarly, aromatic styrene, as well as some of its analogs, are oxidized in a stereoselective way in the presence of either t-butyl peroxide [85] or hydrogen peroxide [83, 88–90] leading to the corresponding epoxides. A first multistep synthesis featuring the use of chloroperoxidase for the synthesis of a biologically important target was described recently by Lakner and Hager [91]. (R)-(-)-mevalonolactone was synthesized through stereoselective epoxidation of methallylpropionate followed by chemical transformation. The lactone was obtained in 57% overall yield and 93% ee (Fig. 5). More recently, a potential α-methylamino acid synthon has been prepared the same way with 94% ee [92].

As a general feature, the use of such biocatalysts seems to be hampered by enzyme deactivation through, for instance, reaction with a number of aliphatic terminal olefins to give an inactive N-alkylated derivative [93] as well as by the sensitivity of the enzyme toward the peroxide which has to be used. However, interesting biotechnological approaches – i.e. continuous flow bioreactors – have been developed [94]. Also, very recently, the oxidation of indole to oxindole has been studied using various reactor types, resulting in a 20-fold increase of the total turnover number and space time yield [95, 96]. Industrial scale production has recently been claimed [97]. Chloroperoxidase from *C. fumago*,

Table 1. Asymmetric epoxidation reactions catalyzed by CPO

Substrates	ee epoxides (%)	Absolute configuration (epoxide)	Epoxide yield (%)	Ref.
cis-2-heptene	96	2R,3S	78	83
cis-2-octene	92	2R,3S	82	83
cis-2-methylstyrene	96	1S,2R	67	83
1,2-dihydronaphtalene	97	1R,2R	85	83
styrene	49	R	23	85
p-chlorostyrene	66	R	35	85
p-bromostyrene	68	R	30	85
p-nitrostyrene	28	R	5	85
α-methylstyrene	89	R	55	86
$PhOCH_2C(CH_3)=CH_2$	89	R	22	86
2-methyl-heptene	95	R	23	86
$EtCO_2CH_2C(CH_3)=CH_2$	94	R	34	86
3-bromo-2methyl-1-propene	62	S	61	87
4-bromo-2-methyl-1-butene	88	R	93	87
5-bromo-2-methyl-1-pentene	95	R	89	87
6-bromo-2-methyl-1-hexene	87	R	33	87
7-bromo-2-methyl-1-heptene	50	R	42	87

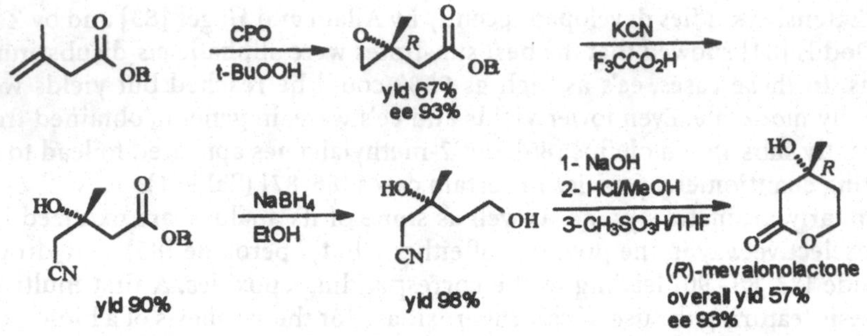

yld 67%
ee 93%

1- NaOH
2- HCl/MeOH
3-CH₃SO₃H/THF

yld 90%

yld 98%

(R)-mevalonolactone
overall yld 57%
ee 93%

Fig. 5. Synthesis of (R)-mevalonolactone through stereoselective bioepoxidation of methallyl-propionate using chloroperoxidase [91]

modified by genetic engineering of the cloned gene, was used in the presence of hydrogen peroxide and an organic solvent to epoxidize subterminal olefins.

2.1.5
Miscellaneous Microbial Epoxidations

Numerous other epoxidation reactions have been described throughout the literature. However, in many cases, the exact nature of the enzyme operative in these reactions has not been established. For instance, de Bont et al. [98, 99] isolated from soil several bacterial strains from, in particular, the *Mycobacterium*, *Nocardia* and *Xanthobacter* genera, which are able to grow on ethene, propene, 1-butene (or even butane) as sole carbon source. These bacteria were shown to be able to achieve the stereoselective epoxidation of the substrates. Detailed mechanistic studies using a *Mycobacterium* strain and $^{18}O_2$ have revealed that the oxygen incorporated indeed originated from molecular oxygen, and that the enzymes implied for hydroxylation reactions were different from those implied for epoxidation [100]. Furthermore, interesting ee's were obtained in some cases (ee ≈ 80%). In the same context, it was shown by Archelas et al. [101] that some of these strains are also able to epoxidize substituted alkenes, again leading to variations of the observed stereoselectivities. The *Xanthobacter* strains also allowed accumulation of 2,3-epoxybutane from the *cis* or *trans* olefin, although with low to moderate yields [102]. However, in spite of interesting technological improvements, all these biooxygenation reactions are of low preparative value because they show both low turnover values and severe inhibition of the monooxygenase as a result of product concentration [103].

Similar results were described by Mamouhdian and Michael, who isolated 18 bacterial strains able to produce optically enriched epoxides with excellent ee's (up to 98%) [104, 105]. However, in the case of *trans*-(2R,3R)-epoxybutane, it was shown that the enantiomeric enrichment is in fact due to a second-step enantioselective hydrolysis of the epoxide, which is first produced in racemic form. This, interestingly, is an unexpected example of the possible use of microbial epoxide hydrolases for the synthesis of enantiopure epoxides (see below).

This is not the case for a *Rhodococcus rhodochrous* strain that affords 1,2-epoxyalkanes from short-chain terminal olefins. Indeed, in the latter case, no product inhibition has been observed [106]. However, the ee's obtained were not determined, and it seems probable that they were quite low. On the other hand, a *Nocardia corallina* strain was described that afforded the corresponding 1,2-(*R*)-epoxy-2-methylalkane with ee's as high as 90% depending on the chain length [107]. These epoxides were used as chiral building blocks to prepare prostaglandin *ω*-chains.

Numerous other examples have been described indicating the ability of various different microbial strains [108, 109] and several patents have been filed in this context [see for example 110, 111]. However, although interesting, it is not clear whether some of these processes are used indeed for the production of enantiopure epoxides. (See Table 2 for a summary of bacterial epoxidations).

One of the drawbacks of the use of bacterial strains seems to be further degradation of the epoxide produced. However, an exception to this fact has been described recently using fungi instead of bacteria. Thus, Furstoss et al. have described the epoxidation of an (*S*)-sulcatol derivative by the fungus *Aspergillus niger*, which affords the corresponding (2*S*,5*S*) enantiomer in 100% d.e. and 50% yield. This can be further transformed into the natural enantiomer of the biologically active pheromone pityol [112] (Fig. 6). However, it is to be emphasized that such an accumulation of epoxide in the presence of a whole cell fungus appears to be an exception. Indeed, much more common is the observation that the epoxide formed is metabolized further on into the corresponding vicinal diol. Depending on reaction conditions, this can be achieved either spontaneously (acidic medium) or enzymatically (neutral pH). An interesting application of this approach is the possibility of a "tuned" preparation of either enan-

Table 2. Miscellaneous microbial epoxidations

Olefin	Bacterial strain	Epoxides		Ref.
		R	S	
3-chloro-propene	B3	4	96	98
	B4	3	97	98
	2 W	1	99	98
	Py2	20	80	98
1,7-octadiene	*Pseudomonas oleovorans*	92	8	38
1-hexadecene	*Corynebacterium equi* IFO 3730	100	0	108
2-methyl-1-hexene	*Nocardia corallina*	86	14	107
propene	*Methylosinus trichosporium* OB3b	57	43	60
1,3-butadiene	*Methylosinus trichosporium*	36	64	60
4-bromo-1-butene	*Mycobacterium* L1	82	18	101
	Mycobacterium E3	94	6	101
	Nocardia Ip1	90	10	101
3-buten-1-ol	*Mycobacterium* L1	70	30	101
	Mycobacterium E3	79	21	101
	Nocardia Ip1	92	8	101

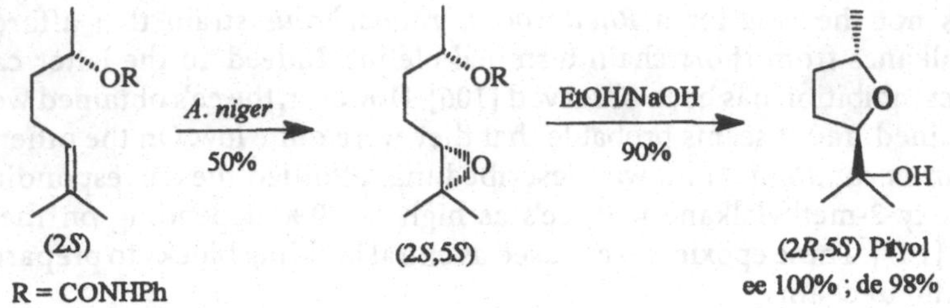

Fig. 6. Epoxidation of a (S)-sulcatol derivative by the fungus *A. niger*. Synthesis of the biologically active pheromone pityol [112]

tiomer of such a diol [113]. As was previously emphasized, vicinal diols themselves can either lead to the corresponding epoxide without loss of enantiomeric integrity or be used as their cyclic sulfates or sulfites [1] (see below).

Other fungal strains, such as *Cunninghamella elegans* and *Syncephalastrum racemosum*, also epoxidize aromatic substrates, but again the intermediate epoxides are directly processed into the less reactive (and less toxic) corresponding *trans*-diols. The enzymes implied in these oxidations were reported to be cytochrome P-450 type enzymes, which possessed stereo and regioselectivities different to the mammalian enzymes [114, 115]. An interesting application of such a fungal epoxidation has been described for 20 fungi that carry out the epoxidation of a precursor to fosfomycin, a broad-spectrum antibiotic [116]. This can thus be obtained (although in low concentration, i.e. 0.5 g L^{-1}) in 90% enantiomeric purity using this very simple procedure [117] (Fig. 7).

As exemplified above, direct epoxidation of olefinic double bonds is possible by using various monooxygenases from different types and origins, either as wild or genetically engineered strains or as purified enzymes. Furthermore, these can in certain cases lead to interesting, although generally not sufficient, enantiomeric purities of the produced epoxides. However, it is clear that severe practical difficulties are linked to the fact that these enzymes are essentially cofactor dependent and that one often encounters substrate and/or product inhibition, so only low concentrations of product can be accumulated in the medium. In spite of some biotechnological improvements based on the use of biphasic systems implying organic solvents [52, 118–120], no practical breakthrough process seems to have been achieved, thus hampering the use of this

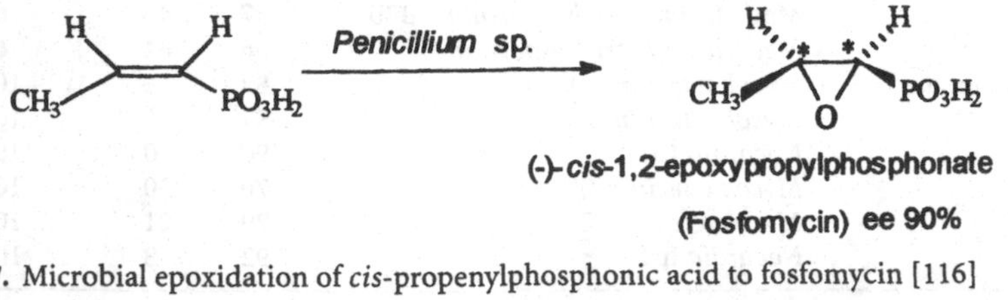

Fig. 7. Microbial epoxidation of *cis*-propenylphosphonic acid to fosfomycin [116]

type of enzymes for large scale laboratory and, more important, industrial production. Therefore, various other strategies for preparation of enantiopure epoxides have been explored. These include indirect ways either starting from different chemical precursors to the epoxide targets or as resolution processes starting from racemic compounds.

2.2
Microbial Reduction of α-Haloketones

Owing to the enormous amount of work devoted to the study of alcohol dehy-drogenase-mediated reduction of various substituted carbonyl compounds, it appears a priori that a strategy that would imply stereoselective reduction of an α-halogenated ketone and further cyclization of the halohydrine obtained from the corresponding epoxide would be an alternative way of synthesising enantiopure epoxides (Fig. 8). Surprisingly enough, only scant information is available about such an approach. One possible explanation may be the intrin-sic reactivity of α-haloketones, which may act more like enzyme inhibitors than as "normal" substrates.

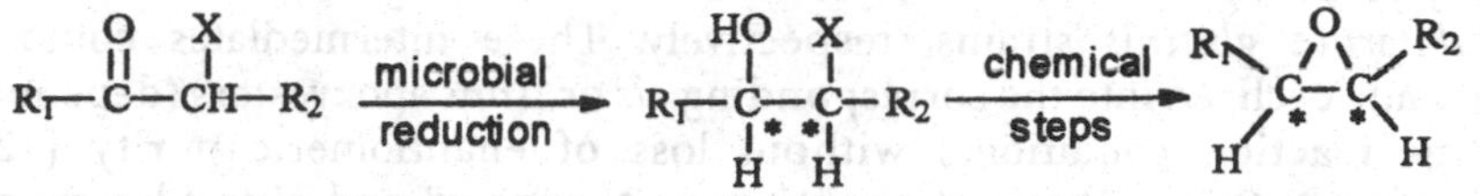

$$R_1-\overset{\overset{O}{\|}}{C}-\overset{\overset{X}{|}}{C}H-R_2 \xrightarrow[\text{reduction}]{\text{microbial}} R_1-\overset{\overset{HO}{|}}{\underset{\overset{|}{H}}{C}}-\overset{\overset{X}{|}}{\underset{\overset{|}{H}}{C}}-R_2 \xrightarrow[\text{steps}]{\text{chemical}} \overset{R_1}{\underset{H}{}}\diagdown\overset{O}{\underset{}{C-C}}\diagup\overset{R_2}{\underset{H}{}}$$

Fig. 8. Chemoenzymatic synthesis of chiral epoxides via the α-haloketone reduction route

Nevertheless, examples have been described that illustrate this approach. Thus, Imuta et al. achieved the reduction of α-bromo- and α-chloroacetophe-none using the bacteria *Cryptococcus macerans*, which leads to the correspond-ing (R)-2-halohydrines as intermediates which cyclize further into enantiomeri-cally pure styrene oxide [121]. More recently, Weijers et al. studied the reduction of α-chloroacetone into 1-chloro-2-propanol with the aim of preparing enan-tiopure 1,2-epoxypropane [122]. This could be obtained using several bacterial or yeast cells. It has been shown, however, that racemic 3-chloro-2-butanone led to a 1:1 mixture of the (2S,3R) and (2S,3S) diastereoisomers. Thus, the reduction of this substrate occurred in a non-enantioselective manner (i.e. each of the two enantiomers was reduced), but since the reduction itself was stereoselective, the two corresponding diastereoisomeric halohydrines were obtained in high enan-tiomeric purity (ee >97%). As a consequence, separation of these two diaster-eoisomers was able to provide a route to the *cis* or the *trans* epoxide in high optical purity. A similar approach has been developed by Besse and Veschambre, who screened numerous microorganisms for the reduction of α-haloketones such as 3-bromo-2-octanone. Here again, a mixture of diastereoisomers was formed systematically [6]. Interestingly, these authors observed that it was pos-sible to obtain a mixture of either the (2S,3SR) or the (2R,3SR) diastereoisomers depending on the microbial strain used. These diastereoisomers could be further cyclized into the corresponding epoxides using conventional synthetic methods (i.e. NaH/benzene) without noticeable loss of stereochemical integrity. In certain cases, one of these diastereoisomers was obtained selectively. Reduction of 4-phenyl-3-bromo-2-butanone by using the fungus A. *niger* led to a 48% yield of the single (2R,3R) stereoisomer in excellent enantiomeric purity (ee >98%). In this case, the remaining (unreacted) α-haloketone substrate was also optically pure. Thus, in this particular case, the reduction was both enantio- and stereoselective. Further studies led these authors to develop this result to a more general method, showing that, from various α-haloketones, it was possible to prepare – although through an intermediate chromatographic separation – all four stereoisomers of a given substrate such as for 4-phenyl-2,3-epoxybutane or *trans*-β-methylstyrene oxide. The yields were moderate, but the enantiomeric purities were excellent (ee >95%) [123]. Such a strategy has also been developed starting from 2-chloro-3-oxoesters, which thus allow the synthesis of valuable glycidic ester derivatives [124]. For example, using a reduction by *Mucor plumbeus*, the *anti* (2R,3R)-ethyl-2-chloro-3-hydroxyhexanoate enantiomer was obtained in 50% yield and 92% ee from the corresponding chlorooxoester. On the other hand, both *syn* (2S,3R) and *anti* (2R,3R) enantiomers could also be obtained from ethyl-2-chloro-3-oxo-3-phenylproprionates in 45–50% yield (ee >95%) using *Mucor racemosus* or *Rhodotorula glutinis* strains, respectively. These intermediates could be efficiently cyclized into the corresponding *cis* or *trans* epoxyesters (depending on the reaction conditions) without loss of enantiomeric purity [125]. This allowed the synthesis of enantiopure N-benzoyl and N-*tert*-butoxycarbonyl(2R,3S)-3-phenylisoserine, i.e. the side chains of taxol and taxotere (Fig. 9).

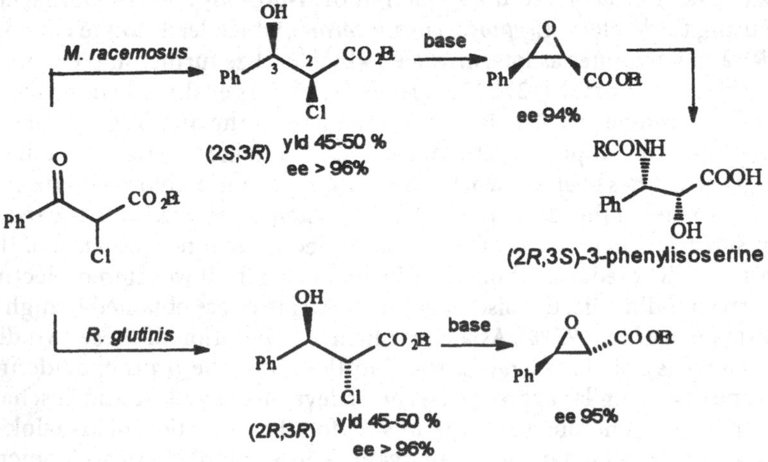

Fig. 9. Stereoselective microbial reduction of 2-chloro-3-oxoester. Synthesis of (2R,3S)-3-phenylisoserine [125]

3
Resolution of Epoxide-Bearing Substrates

In addition to the above described procedures implying either direct oxidation of an olefinic double bond or stereoselective reduction of a ketone precursor, which, as discussed above, do not really provide very efficient ways for the large scale synthesis of enantiopure epoxides, some indirect strategies have also been explored. These are essentially based on the resolution of epoxide-ring bearing substrates as exemplified below. As will be seen, these approaches imply the use of cofactor-independent enzymes, which are in practice much easier to work with, and lead to very interesting results. As a matter fact, some of these processes are already used on an industrial scale, and it can be predicted that future industrial applications will continue to be essentially based on the use of these very promising "easy-to-use" biocatalysts.

3.1
Resolutions Using Lipases

The first, and most efficient, synthetic process is based on the use of lipases. These enzymes have been extensively studied because of their commercial availability and surprising stability, even in the presence of organic solvents. For instance, lipase-catalyzed hydrolysis of various glycidic esters has been studied because the enantiomerically pure products thus obtainable are valuable chiral synthons for the pharmaceutical industry [126–128]. Thus, porcine pancreatic lipase was shown to hydrolyze several glycidyl esters enantioselectively, with the butyrate derivative leading to the best results [129]. Further improvement of this method using a multiphase enzyme membrane reactor has been described recently [130], and has been patented by a Japanese company, using whole cell cultures of *Rhodotorula*, *Pseudomonas* or *Rhizopus* sp. [131]. Similar approaches have also been applied to the synthesis of other glycidyl derivatives, starting from either racemic or prochiral substrates. In the case of racemic 2-(hydroxy-methylene)-1,2-epoxybutan-4-ol, a transesterification with vinyl acetate was performed by the *Pseudomonas cepacia* lipase in the presence of organic solvents. This allowed the recovery of unreacted (*S*)-diol with an ee of 86% at 60% conversion, which implies a low yield of the desired product [132]. Similar results were obtained using the *Pseudomonas fluorescens* lipase-catalyzed reaction (ee 90%) [133] but better yields were obtained starting from a prochiral glycidyl derivative. Thus, the optically active epoxy alcohol (*R*)-2-butyryloxy-methylglycidol was obtained in high enantiomeric purity (ee >98%) using a phosphate buffer and an organic co-solvent system. The best results were obtained with *Pseudomonas* sp. lipase (95% yield; ee >98%) [134].

The use of lipase-catalyzed ester hydrolyses or transesterification reactions has led, interestingly, to an efficient method for synthesizing the *N*-benzoyl-(2*R*,3*S*)-3-phenyl isoserine moiety, the C-13 side chain of taxol. The *Mucor miehei* lipase proved to be uniquely suited for the stereospecific transesterification of racemic methyl *trans*-β-phenyl glycidate. Further improvement of this reaction led to the choice of an isobutanol-hexane mixture (1:1; vol/vol) as the reaction medium. Thus, this enzymatic process, coupled with substrate recycling, afforded both the recovered levorotatory-starting substrate and the isobutyl ester formed in 42 and 43% yield, respectively, with both compounds showing ee's higher than 95%. This experiment was able to be carried out without any problems on a 20-g lab scale. Interestingly, product and substrate could be separated by fractional distillation, and the enzyme could be recycled without much loss of activity (85% recovered activity) [135] (Fig. 10).

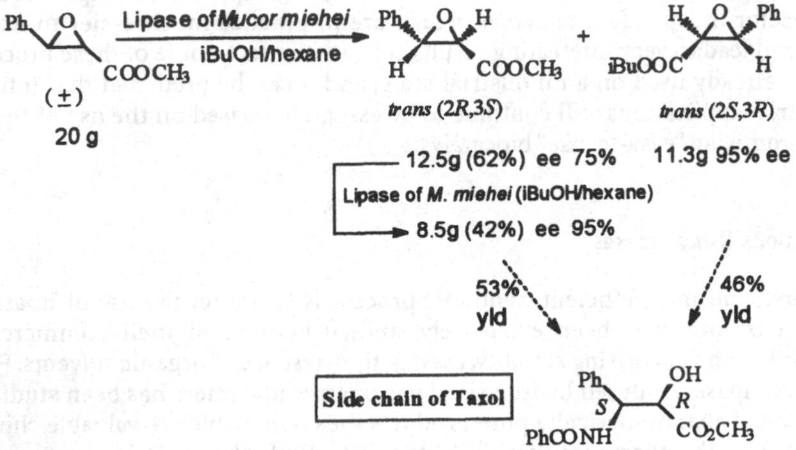

Fig. 10. Lipase. Stereospecific transesterification of methyl *trans-β*-phenyl glycidate, precursor of the side chain of Taxol® [135]

A similar methodology has also been applied to the large scale synthesis of (2R,3S) p-methoxyphenyl glycidic acid esters, the key chiral building block of Diltiazem, a drug widely used as an antihypertensive agent because of its calcium antagonist activity [136]. Although this optically active intermediate is currently available through conventional synthesis – i.e. chemical resolution [137] or asymmetric synthesis [138] – an enzymatic approach has been devised, which in fact is used on the industrial scale. Thus, the racemic p-methoxyphenyl glycidic methyl ester [139, 140] was resolved using a lipase mediated hydrolysis, affording the unreacted (and desired) *trans-*(2R,3S) enantiomer to the exclusion of the other stereoisomer. Although the corresponding acid formed during this hydrolysis decomposes in the reaction conditions, and thus cannot be recycled by racemization, this synthetic scheme appears to be the least expensive and most efficient. The costs resulting from the loss of half the racemic substrate (which is in fact relatively inexpensive to prepare via conventional chemistry) is balanced by the fact that this resolution is achieved in an early stage of the process, avoiding the so-called "enantiomeric ballast" during the further synthetic steps (Fig. 11).

The kinetic resolution of epoxide-bearing substrates has also been achieved via a biotransformation different from hydrolysis. Thus, Weijer et al. [141]

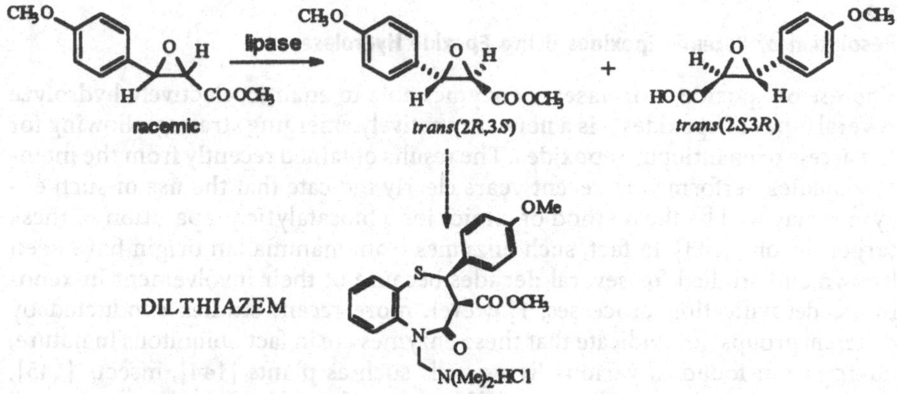

Fig. 11. Lipase. Enzymatic preparation of (2R,3S)-phenyl glycidic esters, the key building-block of Diltiazem® [139, 140]

described the enantioselective degradation of short-chain *cis* and *trans* 2,3-epoxyalkanes by *Xanthobacter* sp. bacteria able to grow on propene. Only the (2S) enantiomers of these substrates were degraded, leading to an accumulation of (2R)-2,3-epoxyalkanes (ee >98%) in the reaction medium. On the other hand, the 1,2-epoxyalkane isomers were degraded without enantioselectivity. Other bacterial strains – i.e. some *Nocardia* species – did, fortunately, make it possible to prepare the remaining epoxide with excellent enantiomeric purity. However, the high ee value was paid for by low yields. Again, *trans*-2,3-epoxy-butane gave much better results.

A similar approach has also been developed recently by a Dutch team for the preparation of glycidyl derivatives (Fig. 12) [142]. Thus, the strain *Acetobacter pasteurianus* was shown to achieve the kinetic resolution of racemic glycidol, affording the unreacted (R)-glycidol enantiomer. From the calculated E value (E = 16), it can be predicted that a 99.5% ee value for this product should be attainable at 64% conversion. The reactive (S)-antipode was in fact oxidized via the corresponding aldehyde into the glycidic acid. The preliminary studies conducted in the course of this work indicate that this method could be attractive when compared with chemical approaches (although the yield of product will be quite low). However, more work must be carried out to judge the economic feasibility of this process.

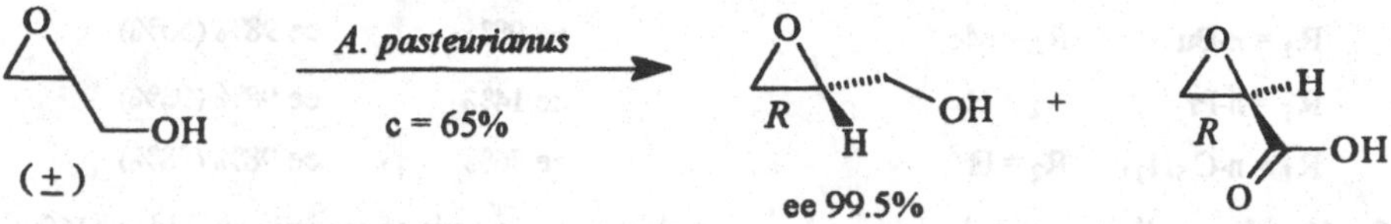

Fig. 12. Production of (R)-glycidol by *Acetobacter pasteurianus* [142]

3.2
Resolution of Racemic Epoxides Using Epoxide Hydrolases

The use of epoxide hydrolases – enzymes able to enantioselectively hydrolyze several types of epoxides – is a new, very actively emerging strategy allowing for the access to enantiopure epoxides. The results obtained recently from the intensive studies performed in recent years clearly indicate that the use of such enzymes may well be the method of choice for a biocatalytic preparation of these target chirons [143]. In fact, such enzymes from mammalian origin have been known and studied for several decades because of their involvement in xenobiotic detoxification processes. However, more recent studies conducted by different groups now indicate that these enzymes are in fact ubiquitous in nature, having been found in various living cells such as plants [144], insects [145], bacteria [146, 147], yeasts [148] and filamentous fungi [149].

As far as mammalian enzymes are concerned, interesting fundamental studies have been conducted by Hammock et al. [150] as well as by Knehr et al. [151], from which important knowledge of their biological properties was gained. Several of these enzymes were recently purified, cloned and overexpressed, and elegant studies have been devoted to the determination of the detailed mechanism involved in these reactions [152–157]. Their possible use for practical application to fine organic synthesis has been intensively explored by Berti and al for years [158]. This pioneering work highlighted the fact that several substrates, including racemic aromatic or aliphatic compounds, can be efficiently processed by epoxide hydrolases which often lead to enantiomerically enriched – or even enantiomerically pure – epoxides (the unreacted enantiomer) and/or to the corresponding vicinal diols [159] (Fig. 13). Moreover, in the cases of α/β–dialkyl substituted epoxides, each of the two enantiomers was attacked with an opposite regioselectivity (i.e. always at the (S)-carbon atom), leading to an almost quantitative yield of the corresponding (R,R) diol [160–161].

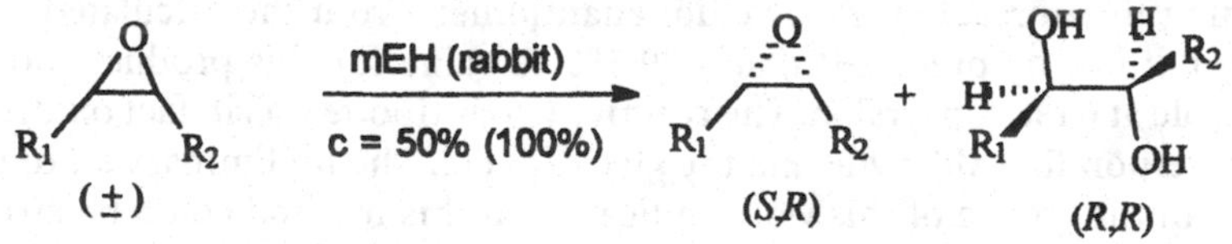

			(S,R)	(R,R)
$R_1 = Ph$	$R_2 = Me$		ee 98%	ee 98% (0%)
$R_1 = Ph$	$R_2 = Et$		ee 98%	ee 98% (90%)
$R_1 = Ph$	$R_2 = n\text{-Pr}$		ee 35%	ee 98% (90%)
$R_1 = Ph$	$R_2 = n\text{-Bu}$		ee 5%	ee 98% (90%)
$R_1 = n\text{-Bu}$	$R_2 = Me$		ee 98%	ee 98% (50%)
$R_1 = n\text{-Pr}$	$R_2 = Et$		ee 14%	ee 98% (98%)
$R_1 = n\text{-C}_5H_{11}$	$R_2 = Et$		ee 56%	ee 98% (98%)

Fig. 13. Mammalian epoxide hydrolase. Enantioselective hydrolysis of racemic epoxides. [160–161]

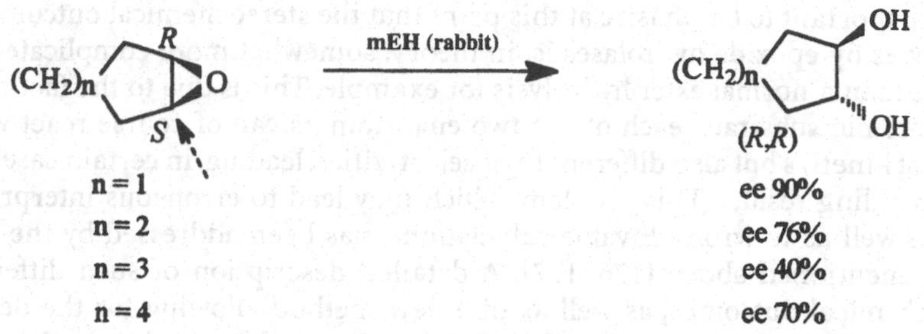

n = 1	ee 90%
n = 2	ee 76%
n = 3	ee 40%
n = 4	ee 70%

Fig. 14. Mammalian epoxide hydrolase. Stereoselective hydrolysis of *meso*-epoxides [162]

Most interestingly, it has been observed by the same authors that some prochiral substrates can also be processed by these enzymes [162]. Thus, in the case of such compounds, the regioselective attack at one single carbon atom of the oxirane ring affords the corresponding diol in optically enriched (or even enantiopure) form. The outstanding interest in these two last cases is of course the fact that, in contrast to a classical resolution process which is limited to a 50% yield, the theoretically possible product yield is 100% (Fig. 14).

In spite of these very interesting results, the use of mammalian epoxide hydrolases is still severely hampered – or even impossible – for large-scale industrial production. Even overexpressed enzymes are not currently available in large enough quantities at a reasonable price.

Therefore, the use of enzymes from other origins, and in particular from microbial sources which can be cultivated in almost unlimited amounts, was another very promising track for such applications. In fact, the use of a bacterial epoxide hydrolases was patented as long ago as 1975 for the industrial synthesis of L- [163] and *meso*-tartaric acid [164] from the precursor epoxide, but surprisingly enough no further application of this type of biocatalyst has been described since. This enzyme, isolated from a *P. putida* strain, was isolated and crystallized as early as 1969 by Allen and Jakoby [164]. Similarly, Niehaus and Schroepfer [165] studied the enantioselective hydrolysis of *cis*- and *trans*-9,10-epoxystearic acids by a *Pseudomonas* sp., whereas Michaels et al. [166] observed the hydrolysis of epoxypalmitate by *Bacillus megaterium*. Since then, several such epoxide hydrolase containing bacteria have been found. For example, De Bont et al. [167], Escoffier and Prome [168], Jacobs et al. [169], Nakamura et al. [170, 171] have described the enzyme-catalyzed hydrolysis of either short chain epoxides or of an epoxysteroid. All these epoxide hydrolases originated from bacterial strains. Surprisingly enough, all these studies were only achieved on the analytical scale, and no real applications of these enzymes to organic synthesis have been developed by these authors.

In fact, the real breakthrough concerning the use of an epoxide hydrolase applied to fine organic chemistry was only recently published, by two groups independently, i.e. by Faber et al. [172, 173] using bacterial enzymes, and by Furstoss et al. [174, 175] using fungal enzymes. In both cases, it was found that these enzymes could be excellent biocatalysts for achieving the resolution of several racemic epoxides.

It is important to emphasize at this point that the stereochemical outcome of hydrolyses by epoxide hydrolases is, in theory, somewhat more complicated to analyze than a normal ester hydrolysis for example. This is due to the fact that, for a racemic substrate, each of the two enantiomers can of course react with different kinetics but also different regioselectivities, leading, in certain cases, to quite puzzling results. This problem, which may lead to erroneous interpretations as well as to wrong E value calculations, has been addressed by the two groups mentioned above [176, 177]. A detailed description of such different stereochemical outcomes, as well as of a new method allowing for the determination of the regioselectivity of the enzymatic attack on both enantiomers, has been described very recently [178].

3.2.1
Use of Bacterial Enzymes

In their preliminary studies, Faber et al. [173] observed that an immobilized enzyme preparation derived from *Rhodococcus* sp. (sold by Novo Industry, Denmark) and designed for the enzymatic hydrolysis of nitriles, also contained an epoxide hydrolase activity. This *Rhodococcus* enzyme was further purified and characterized as being a cofactor-independent, soluble protein [179]. Straight-chain terminal epoxides as well as glycidyl derivatives were well accepted as substrates, but the enantioselectivities were low. The best results were obtained with 2-methyl-2-pentyl oxirane, which was obtained in 72% ee [180]. Further work, using various 2-alkyl substituted straight-chain oxiranes of different lengths, suggested that chiral recognition by the enzyme depends on the difference in size between the two alkyl groups [181]. When sodium azide was added to the medium, enantiomerically enriched (R)-azido alcohol generated from the unreacted (R)-epoxide was also obtained [182]. Although the authors claim that this should be due to an enzyme-catalyzed reaction, it is more reasonable to consider this reaction to be a pure chemical addition of the azide to the substrate in the course of the reaction. Therefore it might well be that these reactions are catalyzed by the chiral protein surface instead of being a real enzymatic reaction. This interpretation is in accordance with the recently published mechanism described for cytosolic or microsomal rat liver epoxide hydrolases [153, 154]. It is worth mentioning that similar results have been observed for the aminolysis of aryl glycidyl ethers by hepatic microsomes from the rat [183] as well as from lipase-catalyzed epoxide aminolysis [184]. Further screening of alkene utilizing bacteria led to the discovery of other bacterial strains similarly equipped with interesting epoxide hydrolase activities [185]. One of these, *Nocardia* EH1, showed almost absolute enantioselectivity on 2-methyl-1,2-epoxyheptane, leading to both the enantiopure (R)-diol and epoxide [186].

Interestingly, this enzyme also accepted *cis*- and *trans*-2,3-disubstituted olefins. Moreover, it appeared that racemic *cis*-2,3-epoxyheptane led to almost total deracemisation, thus affording the corresponding (R,R) diol with 79% chemical yield and 91% ee [187] (Fig. 15). This is due to the fact that the regioselectivity of the enzymatic attack on each of the two enantiomers was different. Recent work described that this enzyme could be immobilized on DEAE-cellulose, thus

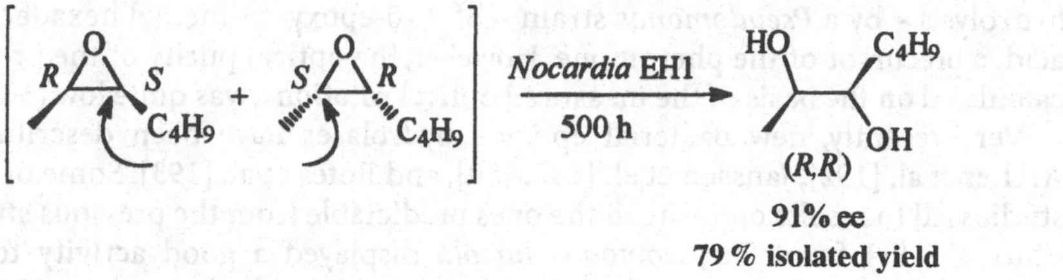

Fig. 15. Bacterial epoxide hydrolase. Deracemization of (±)-*cis*-2,3-epoxyheptane via enantio-convergent biohydrolysis using *Nocardia* EH1 [187]

affording a more stable biocatalyst which could be used in repeated batch reactions [188].

Unexpectedly, as compared with microsomal mammalian epoxide hydrolases, all the *meso*-epoxides tested with the two above mentioned bacterial enzymes were not substrates.

A few synthetic applications for obtaining biologically active compounds have been described, based on the use of these bacterial enzymes. For instance, the pheromone (*S*)-frontalin was synthesized in five steps in 94% ee (but rather low overall yield) via a chemoenzymatic route implying epoxide resolution using lyophilized cells of *Rhodococcus equi* [189] (Fig. 16).

Similarly, the synthesis of both enantiomers of linalool oxide was achieved [190]. It is interesting to note that this bacterial enzyme showed an opposite enantio-preference on structurally related substrate as compared to the EH from the fungus *Aspergillus niger* (see below). Indeed, this enzyme preferred to hydrolyze the (*S*)-configurated substrate, whereas the (*R*)-enantiomer was the fastest for the fungal enzyme. Another elegant application of the use of such enzymes for the preparation of bioactive products was described by Otto et al. [191]. Thus, the synthesis of the biologically active enantiomer of disparlure, the sex pheromone of the moth *Lymantria dispar*, which causes severe damage to trees in some parts of the world, was achieved. This implied enantioselective

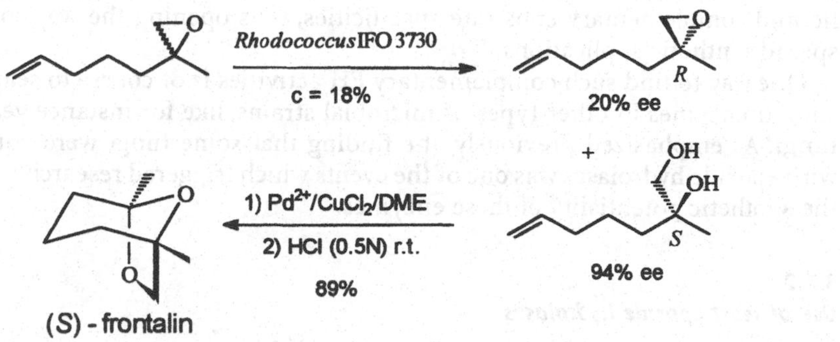

Fig. 16. Bacterial epoxide hydrolase. Key steps of chemoenzymatic synthesis of (*S*)-frontalin [189]

hydrolysis – by a *Pseudomonas* strain – of 9,10-epoxy-15-methyl hexadecanoic acid, a precursor of the pheromone. However, the optical purity of the product, calculated on the basis of the measured optical rotations, was quite low (≈25%).

Very recently, new bacterial epoxide hydrolases have been described by Archer et al. [192], Janssen et al. [147, 169], and Botes et al. [193]. Some of these studies led to results opposite to the ones predictable from the previous studies. Thus, the EH from *Chryseomonas luteola* displayed a good activity toward terminal olefins, whereas 2-substituted derivatives were bad substrates (Fig. 17).

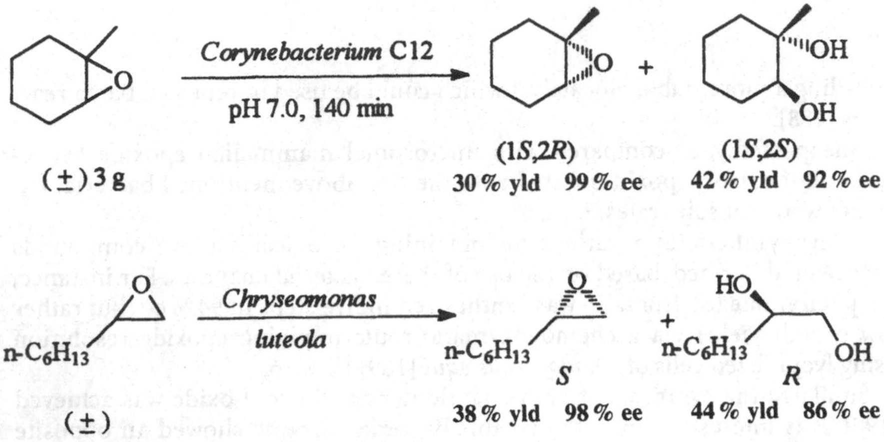

Fig. 17. Bacterial epoxide hydrolase. Resolution of (±)-1-methyl-1,2-epoxycyclohexane [192] and (±)-1,2-epoxyoctane [193] using respectively whole cells of *Corynebacterium* C12 and *Chryseomonas luteola*

On the other hand, the *Agrobacterium radiobacter* enzyme exhibited excellent activity toward aromatic substrates, which was not the case for the one from *Rhodococcus* or *Nocardia* sp. known to have bad (if any) activity toward aromatic compounds. These findings clearly illustrate the fact that, in the future, it will undoubtedly be possible to find various sources of different EHs showing specific and complementary substrate specificities, thus opening the way to widespread synthetic applications.

One way to find such complementary EH activities is of course to search for similar enzymes in other types of microbial strains, like for instance yeasts or fungi. As emphasized previously, the finding that some fungi were equipped with epoxide hydrolases was one of the events which triggered research work on the synthetic potentiality of these enzymes.

3.2.2
Use of Yeast Epoxide Hydrolases

Yeasts showing EH activity have only been described very recently [193, 194]. In particular, a *Rhodotorula glutinis* strain was shown to possess remarkable EH

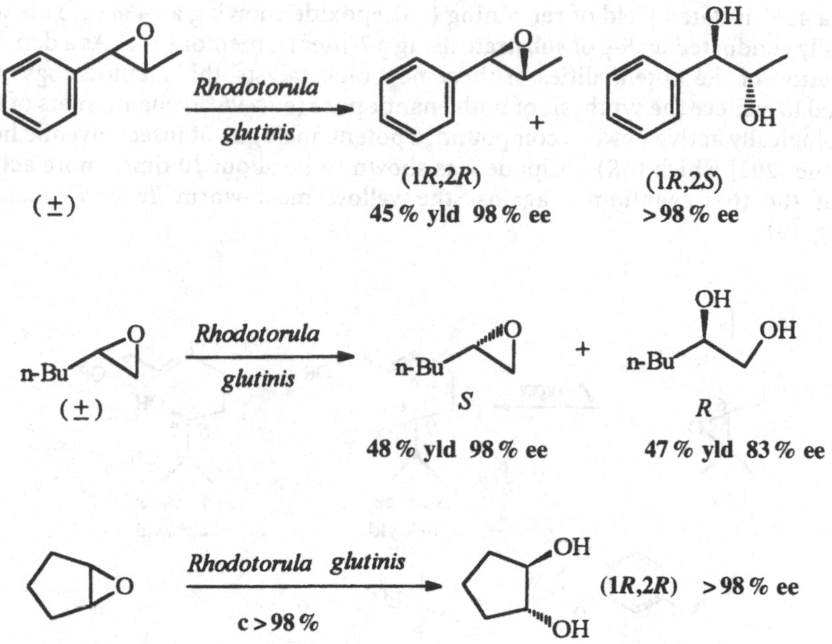

Fig. 18. Yeast epoxide hydrolase. Enantioselective hydrolyses of aromatic [194] and aliphatic [194, 195] epoxide using whole cells of *Rhodotorula glutinis*. Stereoselective hydrolysis of cyclopentene oxide [194]

activity and to have a high preference for aromatic substrates (Fig. 18). Thus, a very satisfactory resolution of *trans*-methyl substituted styrene oxide was achieved using a washed cells suspension of this strain. Further work with this strain was to explore its potentialities for aliphatic terminal olefins, which also proved to be good substrates. Enantioselectivity and reaction rate were strongly influenced by the chain length of the epoxide used [195]. In all the cases studied, the (R) epoxide was preferentially hydrolyzed to the (R)-diol, indicating that the enzymatic attack essentially occurred at the terminal, less substituted, carbon atom. It ought to be stressed that, quite surprisingly as compared to the other epoxide hydrolases from microbial origin, this EH is able to hydrolyze some *meso* compounds very efficiently.

3.2.3
Use of Fungal Epoxide Hydrolases

The presence of epoxide hydrolase activities of fungal cells has been described by several authors [196–201]. However, the real breakthrough in the use of such enzymes for preparative scale synthesis was developed by the group of Furstoss et al. One of their first results was the description of the efficient resolution of racemic geraniol *N*-phenylcarbamate by a culture of the fungus *A. niger*, leading to a 42% isolated yield of remaining (6S)-epoxide showing a 94% ee. This was easily conducted on 5-g of substrate using a 7-liter fermentor [174]. As a demonstration of the potentialities of these new biocatalysts, this methodology was used to achieve the synthesis of both enantiopure (ee > 96%) enantiomers of the biologically active Bower's compound, a potent analogue of insect juvenile hormone [202] which (6R)-antipode was shown to be about 10 times more active that the (6S)-enantiomer against the yellow meal worm *Tenebrio molitor* (Fig. 19).

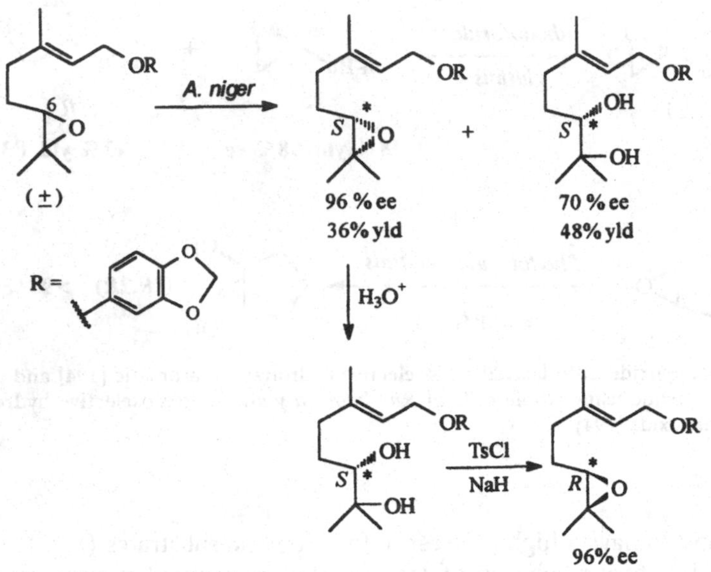

Fig. 19. Fungal epoxide hydrolase. Application to the synthesis of (R)- and (S)-Bower's compound [202]

In the same context, it was shown that this fungus was also capable of achieving the diastereoselective hydrolysis of the exocyclic limonene epoxides, thus opening the way to the synthesis of either enantiopure bisabolol stereoisomer [174]. One of these enantiomers, i.e. (4S,8S)-α-bisabolol, is used on an industrial scale for the preparation of various skin-care creams, lotions and ointments.

Similar results were obtained with styrene oxide, which was efficiently hydrolyzed by washed cells of *A. niger*, affording the (S)-enantiomer in 96% ee within a few hours [175]. Moreover, another fungus, *Beauveria sulfurescens* (presently *B. bassiana*) showed an opposite enantioselectivity, leading to the (R)-enantiomer in 98% ee. Since both hydrolyses afford the corresponding (R)-diol, an enantioconvergent process, by using a mixture of the two fungi, led to an overall 92% yield of the (R)-diol in 89% ee (Fig. 20). Further work elicited interesting information concerning the mechanism implied in these transformations, and led to the conclusion that these enzymes operate via different mechanisms [203]

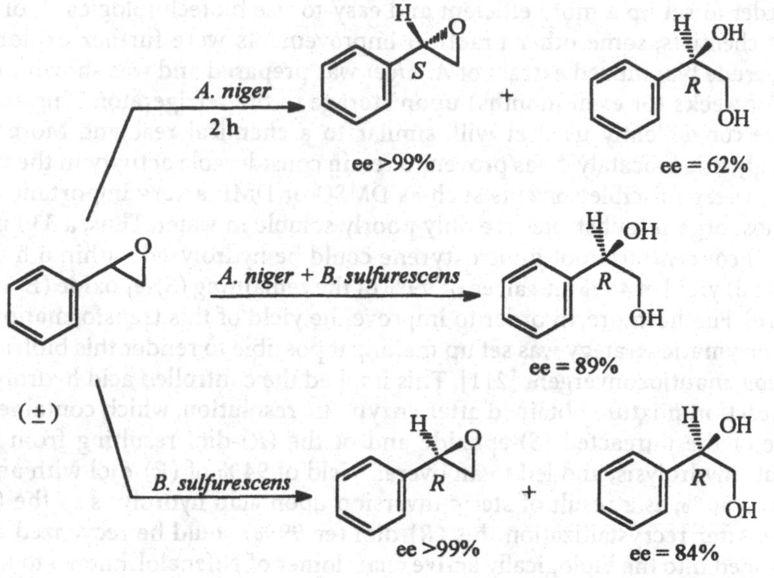

Fig. 20. Resolution of styrene oxide by *A. niger* and *B. sulfurescens* epoxide hydrolases. Deracemization of styrene oxide by using a mixture of these two fungi [175]

Also, the scope of suitable substrates has been explored. Thus, it was shown that substituted styrene derivatives such as various *para*-substituted styrene oxides [204] as well as β-disubstituted derivatives [176] could be accommodated by one or both of these fungi. In the latter case, an interesting enantioconvergent hydrolysis of *cis*-methyl substituted styrene oxide was observed, affording an 85% preparative yield of enantiopure (1R,2R)-diol. Further screening conducted on various other fungal strains indicated that this type of enzyme does indeed seem to be widespread within the fungal world [149, 205].

Another application of these fungal enantioselective hydrolyses was the synthesis of indene oxide and of its corresponding diol, these intermediates being of crucial importance for the synthesis of the orally active HIV protease inhibitor Indinavir. Thus, when submitted to a culture of *B. sulfurescens*, racemic epoxyindene was rapidly hydrolyzed, leading to a 20% yield of recovered enantiomerically pure (ee 98%) (1R,2S) epoxide, and to a 48% yield of the corresponding (1R,2R) *trans*-diol showing a 69% ee [206]. Since this diol has the absolute configuration desired for synthesizing the biologically active HIV protease inhibitor, this approach was quite promising for achieving the preparation of this key-chiral building block by optimizing this biotransformation. Thus, a more extensive study was achieved by the Merck Company which performed a screening on eighty fungal strains to evaluate their ability to enantioselectively hydrolyze racemic epoxyindene [207, 208]. This led to the discovery that some of them were capable of affording both enantiomers of indene oxide in optically pure form, although in rather low yields. Epoxydihydronaphtalene was hydrolyzed similarly to the (1R,2R)-diol in excellent enantiomeric purity [206].

In order to set up a more efficient and easy-to-use biotechnological tool for organic chemists, some other practical improvements were further explored. Thus, a crude lyophilized extract of *A. niger* was prepared and was shown to be stable for weeks (or even months) upon storage in the refrigerator. This, then, could be conveniently used at will, similar to a chemical reactant. More interestingly, this biocatalyst has proven to retain considerable activity in the presence of water-miscible solvents such as DMSO or DMF, a very important fact since most organic substrates are only poorly soluble in water. Thus, a 330 mM (54 g L^{-1}) concentration of *p*-nitrostyrene could be hydrolyzed within 6 h to a (analytical) yield of 49% and an ee of 99% of the remaining (*S*)-epoxide (E ≈ 40) [209, 210]. Furthermore, in order to improve the yield of this transformation, a chemoenzymatic strategy was set up making it possible to render this biotransformation enantioconvergent [211]. This implied the controlled acid hydrolysis of the reaction mixture obtained after enzymatic resolution, which contained a mixture of the unreacted (*S*)-epoxide and of the (*R*)-diol resulting from the enzymatic hydrolysis, and led to an overall yield of 94% of (*R*)-diol with an ee as high as 80%, as a result of steric inversion upon acid hydrolysis of the (*S*)-epoxide. After recrystallization this (*R*)-diol (ee 99%) could be recyclized and transformed into the biologically active enantiomer of Nifenalol, known to have *β*-blocker activity [212] (Fig. 21).

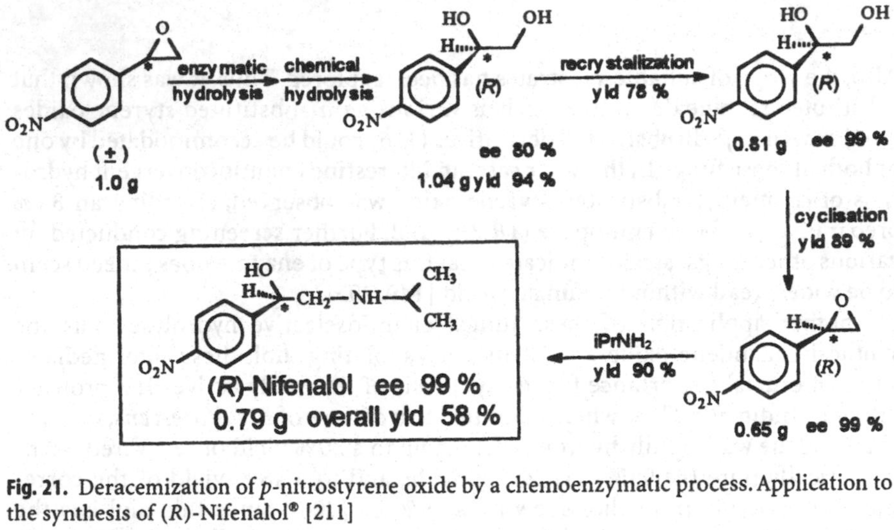

Fig. 21. Deracemization of *p*-nitrostyrene oxide by a chemoenzymatic process. Application to the synthesis of (*R*)-Nifenalol® [211]

Another way to improve the yield of such transformations is to combine a chemical asymmetric oxidation and a biocatalytic approach. This has been illustrated on 2-methyl-epoxyheptane as shown in Fig. 22. Thus, a 33% overall yield of the corresponding (*S*)-epoxide was obtained with an ee of 97% [149]. It is to be emphasized that, in this particular case the best corresponding chemical method for obtaining this epoxide, i.e. the Sharpless asymmetric dihydroxylation, only led to an ee of 71%.

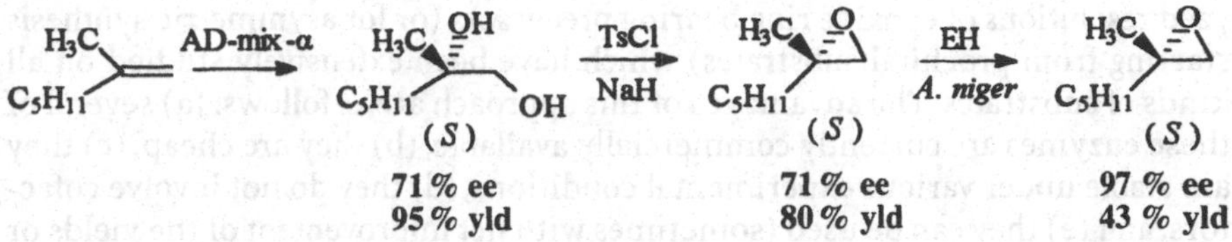

Fig. 22. Fungal epoxide hydrolysis. Chemoenzymatic synthesis of (S)-2-methyl-heptene oxide [149]

The use of a higher substrate concentration for a given amount of enzyme is also a very important goal for preparative scale applications. Such an achievement has been successful very recently [213]. Thus, a two-liquid-phase process allowing for the preparative scale resolution of *para*-bromo styrene oxide (6 g, 0.38 mole L^{-1}) by a crude *A. niger* extract (350 mg) has been achieved. By running this reaction at 4 °C the stability of the enzyme was greatly improved. Surprisingly, the use of this procedure led to a considerable enhancement of the reaction enantioselectivity (E value), thus leading to a 39% yield of optically pure epoxide (ee 99.7%) and a 49% yield of diol (ee 96%) (Fig. 23).

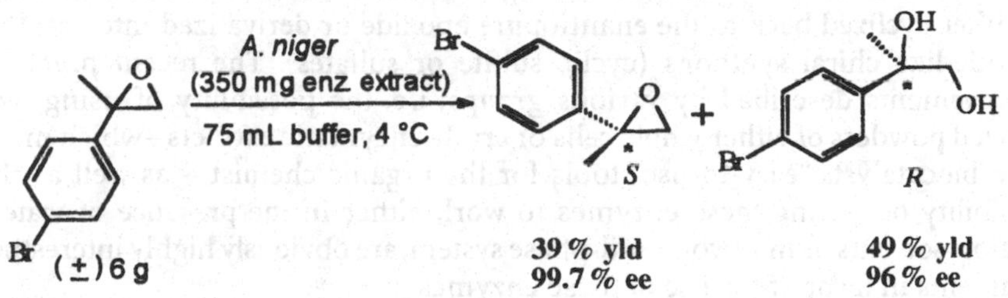

Fig. 23. Fungal epoxide hydrolase. Preparative scale resolution of *para*-bromo styrene oxide using a two-liquid-phase process [213]

4
Summary and Outlook

It can be concluded from the various results described above that numerous continuous efforts have been devoted to the synthesis of enantiopure epoxides over the last twenty to thirty years which has led to important fundamental knowledge on this topic. Unfortunately, many of these approaches still suffer from severe limitations as far as large-scale (industrial) applications are concerned. This is particularly the case for direct epoxidation of alkenes using monooxygenases. Indeed, most of these biocatalytic routes imply multienzymatic processes, and/or are hampered by substrate and/or product inhibition leading to low productivity. Highly sophisticated processes had therefore to be set up in order to partially overcome these drawbacks.

Some more recently investigated indirect ways are, on the other hand, more attractive from this point of view. For instance, this is the case for lipase-catalyzed resolutions of epoxide ring bearing precursors (or for asymmetric synthesis starting from prochiral substrates) which have been extensively studied on all kinds of substrates. The advantages of this approach are as follows: (a) several of these enzymes are currently commercially available, (b) they are cheap, (c) they are stable under various experimental conditions, (d) they do not involve cofactors, and (e) they can be used (sometimes with net improvement of the yields or ee's) in the presence of organic solvents. However, the obvious drawback of this approach is the fact that suitable substrates have to be compounds bearing an epoxide ring and another function – ester or alcohol – near the oxirane ring.

In this respect, the most promising method up till now seems to be the use of microbial epoxide hydrolases, a new aspect of research which is presently blooming. Indeed these enzymes, which act directly on the epoxide ring, independent of any other functionality, seem to offer the same advantages as lipases: (a) they have been shown recently to be ubiquitous in nature, (b) they are cofactor-independent enzymes, (c) they can be produced easily from various microorganisms, (d) they can be partly purified and used as an enzymatic powder without noticeable loss of enzymatic activity upon storage, (e) they can act in the presence of organic solvents, allowing the use of water-insoluble substrates, and (f) they often lead to excellent ee's for the remaining epoxide, but also in certain cases for the diol formed, which can itself be used as such or can be either cyclized back to the enantiopure epoxide or derivatized into reactive epoxide-like chiral synthons (cyclic sulfite or sulfates). The recent practical improvements described by various groups, i.e. the possibility of using lyophilized powders of either whole cells or crude enzymatic extracts –which make these biocatalysts "easy-to-use" tools for the organic chemist – as well as the possibility of getting these enzymes to work either in the presence of water-miscible solvents or in a two-liquid-phase system, are obviously highly interesting arguments in favor of the use of these enzymes.

As a conclusion, examination of the present literature clearly indicates that, depending on the circumstances, any of the methods described in this review may be "the best" for the preparation of a given enantiopure epoxide. In particular, the recent progress achieved by using metal-catalyzed chemical processes obviously has to be taken into account. As far as biocatalytic methods are concerned, one can anticipate that, in the near future, lipases or, better, epoxide hydrolases, will prove to be "the best" choice, particularly as far as industrial applications are concerned. Research is ongoing in diverse laboratories to explore the scope and limitations of these very promising enzymes.

5
References

1. Lohray BB (1992) Synthesis 1035
2. Fedresel HF (1993) Drug chirality-scale-up, manufacturing, and control. A comprehensive review of all the things you have to consider when making the drug you want. In: Ojima I (ed), Verlag Chemie, New York, 23:24
3. Stinson SC (1993) Chem Eng News 27:38

4. Katsuki T (1995) J Synth Org Chem Jpn 53:940
5. Schurig V, Betschinger F (1992) Chem Rev 92:873
6. Besse P, Veschambre H (1994) Tetrahedron 50:8885
7. Katsuki T (1995) Coord Chem Rev 140:189
8. Kolb HC, van Nieuwenhze MS, Sharpless KB (1994) Chem Rev 94:2483
9. Pedragosa-Moreau S, Archelas A, Furstoss R (1995) Bull Soc Chim Fr 132:769
10. Swaving J, de Bont JAM (1998) Enz Microb Technol 22:19
11. Archelas A, Furstoss R (1997) Ann Rev Microbiol 51:491
12. Katsuki T, Sharpless KB (1980) J Am Chem Soc 102:5976
13. Klunder JM, Ko SY, Sharpless KB (1986) J Org Chem 51:3710
14. Pfenninger A (1986) Synthesis 2:89
15. Kolb HC, Sharpless KB (1992) Tetrahedron 48:10515
16. Tokunaga M, Larrow JF, Kakiuki F, Jacobsen EN (1997) Science 277:936
17. For a review see: Hodgson DM, Gibbs AR, Lee GP (1996) Tetrahedron 46:14361
18. Brandes BD, Jacobsen EN (1997) Tetrahedron: Asymmetry 23:3927
19. Archelas A, Furstoss R (1998) TIBTECH 16:108
20. Besse P, Veschambre H (1993) Tetrahedron: Asymmetry 4:1271
21. De Bont JAM (1993) Tetrahedron: Asymmetry 4:1331
22. Hawkins DR (ed) (1994) A survey of the biotransformations of drugs and chemicals in animals. The Royal Society of Chemistry, Cambridge, Vols 1–5
23. Guengerich FP (1991) J Biol Chem 266:10019
24. Poulos TL, Raag R (1992) Faseb J 6:674
25. Sono M, Roach MP, Coulter ED, Dawson JH (1996) Chem Rev 96:2841
26. Holland HL (1992) In: Holland HL (ed) Organic synthesis with oxidative enzymes VCH, New York, 55–199
27. Ortiz de Montellano PR (1987) Cytochromes P-450. Plenum, New York
28. Raag R, Poulos TL (1991) Biochemistry 30:2674
29. Lu AYH, West SB (1980) Pharmacol Rev 31:277
30. Ryan DE, Levin W, Reik LM, Thomas PE (1982) Xenobiotica 12:727
31. Ruettinger RT, Fulco AJ (1981) J Biol Chem 256:5728
32. Ortiz de Montellano PR, Fruetel JA, Collins JR, Camper DL, Loew GH (1991) J Am Chem Soc 113:3195
33. Zeldin DC, Dubois RN, Falck JR, Capdevila JH (1995) Arch Biochem Biophys 322:76
34. For a review see: Wong LL, Westlake CG, Nikerson DP (1997) Struct Bond 88:175
35. Levinger DC, Stevenson JA, Wong LL (1995) J Chem Soc, Chem Commun 22:2305
36. May SW, Abbott BJ. (1973) J Biol Chem 248:1725
37. Abbott BJ, Hou CT (1973) Appl Microbiol 26:86
38. May SW, Steltenkamp MS, Schwartz RD, McCoy CJ (1976) J Am Chem Soc 98:7856
39. Fu H, Shen GJ, Wong CH (1991) Rec Trav Chim Pays Bas 110:167
40. Ruettinger RT, Griffith GR, Coon MJ (1977) Arch Biochem Biophys 183:528
41. Ueda T, Coon JM (1972) J Biol Chem 247:5010
42. May SW, Schwartz RD, Abbott BJ, Zaborsky OR (1975) Biochim Biophys Acta 403:245
43. Shanklin J, Achim C, Schmidt H, Fox BG, Munck E (1997) Proceedings Nat Acad Sciences of the USA 94:2981
44. Johnstone SL, Phillips GT, Robertson BW, Watts PD, Bertola MA, Koger HS, Marx AF (1987) Stereoselective synthesis of S-(-)-β-blockers via microbially produced epoxide intermediates. In: Laane C, Tramper J, Lilly MD (eds) Biocatalysis in organic media. Elsevier, Amsterdam, 387–392
45. Wubbolts MG, Panke S, Beilen JB, Witholt B (1996) Chimia 50:436
46. Nöthe C, Hartmans S (1994) Biocatalysis 10:219
47. Onumonu AN, Colocoussi A, Matthews C, Woodland MP, Leak DJ (1994) Biocatalysis 10:211
48. Rigby SR, Matthews CS, Leak DJ (1994) Bioorg Med Chem 2:553–56
49. Wubbolts MG, Noordman R, van Beilen JB, Witholt B (1995) Recl Trav Chim Pays-Bas 114:139

50. Brown HC, Pai GG (1983) J Org Chem 48:1784
51. Coote SJ, Davies SG, Middlemiss D, Naylor A (1989) J Chem Soc Perkin Trans 1 2223
52. Wubbolts MG, Hoven J, Melgert B, Witholt B (1994) Enzyme Microb Technol 16:887
53. Wallar BJ, Lipscomb JD (1996) Chem Rev 96:2625
54. Fox BG, Borneman JG, Wackett LP, Lipscomb JD (1990) Biochemistry 29:6419
55. Leak DJ, Dalton H (1987) Biocatalysis 1:23
56. Hou CT, Patel RN, Laskin AI, Barnabe N (1979) Appl Environn Microbiol 38:127
57. Elliot SJ, Zhu M, Tso L, Nguyen HHT, Yip JHK, Chan SI (1997) J Am Chem Soc 119: 9949
58. Gallagher SC, Cammack R, Dalton H (1997) Eur J Biochem 247:635
59. Small FJ, Ensign SA (1997) J Biol Chem 272:24913
60. Ohno M, Okura I (1990) J Mol Catal 61:113
61. Seki Y, Shimoda M, Sugimori D, Okura I (1994) J Mol Catal 87:17
62. Geigert J, Neidleman SL, Dalietos DJ, deWitt SK (1983) Appl Environ Microbiol 45:366
63. Fetzner S, Lingens F (1994) Microbiol Rev 58:641
64. Krenn BE, Plat H, Wever R (1988) Biochim Biophys Acta 952:255
65. Vilter H (1983) Bot Mar 26:451
66. Wiesner W, van Pee KH, Lingens F (1988) J Biol Chem 263:13725
67. Van Pee KH (1995) B.C.G. halogenation. In: Drauz K, Waldmann H (eds) Enzyme catalysis in organic synthesis. A comprehensive handbook. VCH, Weinheim, 2:783–807
68. Hager LP, Morris DR, Brown FS, Eberwein H (1966) J Biol Chem 241:1769–77
69. Morris DR, Hager LP (1966) J Biol Chem 241:1763
70. Bangcharoenpaurpong O, Champion PM, Hall KS, Hager LP (1986) Biochemistry 25:2374
71. Libby RD, Thomas JA, Kaiser LW, Hager LP (1982) J Biol Chem 257:5030
72. Morrison M, Schonbaum GR (1976) Annu Rev Biochem 45:861
73. Ramakrishnan K, Oppenhuizen ME, Saunders S, Fisher J (1983) Biochemistry 22:3271
74. Neidleman SL, Levine SD (1968) Tetrahedron Lett 37:4057
75. Griffin BW (1983) Biochem Biophys Res Commun 116:873
76. Den Wijngaard AJ, Reuvekamp PTW, Janssen DB (1991) J Bacteriol 173:124
77. Castro CE, Bartnicki EW (1968) Biochemistry 7:3213
78. Neidleman SL (1980) Hydrocarb Process pp 135–38
79. Itoh M, Kusaka M, Hirota T, Nomura A (1995) Appl Microbiol Biotechnol 43:394
80. Kasai N, Tsujimura K, Unoura K, Suzuki T (1992) J Indust Microbiol 9:97
81. Nagasawa T, Nakamura T, Yu F, Watanabe I, Yamada H (1992) Appl Microbiol Biotechnol 36:478
82. Nakamura T, Yu F, Mizunashi W, Watanabe I (1991) Agric Biol Chem 55:1931
83. Allain EJ, Hager LP (1993) J Am Chem Soc 115:4415
84. Zaks A, Dodds DR (1995) J Am Chem Soc 117:10419
86. Colonna S, Gaggero N, Casella L, Carrea G, Pasta P (1993) Tetrahedron: Asymmetry 4:1325
86. Dexter AF, Lakner FJ, Campbell RA, Hager LP (1995) J Am Chem Soc 117:6412
87. Lakner FJ, Cain KP, Hager LP (1997) J Am Chem Soc 119:443
88. Geigert J, Lee TD, Dalietos DJ, Hirano DS, Neidleman SL (1986) Biochem Biophys Res Comm 136:778
89. McCarthy MB, White RE (1983) J Biol Chem 258:9153
90. Ortiz de Montellano PR, Choe YS, Depillis G, Catalano CE (1987) J Biol Chem 262:11641
91. Lakner FJ, Hager LP (1996) J Org Chem 61:3923
92. Lakner FJ, Hager LP (1997) Tetrahedron: Asymmetry 8:3547
93. Dexter AF, Hager LP (1995) J Am Chem Soc 117:817
94. Blanke SR, Yi S, Hager LP (1989) Biotechnol Lett 11:769
95. Van Deurzen MPJ, Seelbach K, van Rantwijk F, Kragl U, Sheldon RA (1997) Biocat Biotrans 15:1
96. Seelbach K, Vandeurzen MPJ, Vanrantwijk F, Sheldon RA, Kragl U (1997) Biotech Bioeng 55:283
97. Hager LP, Allain EJ (1994) US Patent No 5358860

98. Habets-Crutzen AQH, Carlier SJN, de Bont JAM, Wistuba D, Schurig V, Hartman S, Tramper J (1985) Enzyme Microb Technol 7:17
99. Weijers CAGM, de Haan A, de Bont JAM (1988) Microbiol Sci 5:156
100. De Bont JAM, Attwood MM, Primrose SB, Harder W (1979) Fems Microbiol Lett 6:183
101. Archelas A, Hartmans S, Tramper J (1988) Biocatalysis 1:283
102. Van Ginkel CG, Welten HGJ, de Bont JAM (1986) Appl Microbiol Biotechnol 24:334
103. Habets-Crutzen AQH, Brink LES, van Ginkel CG, de Bont JAM, Tramper J (1984) Appl Microbiol Biotechnol 20:245
104. Mahmoudian M, Michael A (1992) Appl Microbiol Biotechnol 37:23
105. Mahmoudian M, Michael A (1992) Appl Microbiol Biotechnol 37:28
106. Woods NR, Murrell JC (1991) Biotechnol Lett 12:409
107. Takahashi O, Umezawa J, Furuhashi K, Takagi M (1989) Tetrahedron Lett 30:1583
108. Ohta H, Tetsukawa H (1979) Agric Biol Chem 43:2099
109. Weijers CAGM, van Ginkel CG, de Bont JAM (1988) Enzyme Microb Technol 10:214
110. Furuhashi K (1989) Jpn Patent No 01075479 A
111. Hou CT, Patel RN, Laskin AI (1980) US Patent No 4368267
112. Archelas A, Furstoss R (1992) Tetrahedron Lett 33:5241
113. Zhang XM, Archelas A, Furstoss R (1991) J Org Chem 56:3814
114. Cerniglia CE, Yang SK (1984) Appl Environ Microbiol 47:119
115. McMillan DC, Cerniglia CE, Fu PP (1987) Appl Environ Microbiol 53:2560
116. White RF, Birnbaum J, Meyer RT, Broeke JT, Chemerda JM, Demain AL (1971) Appl Microbiol 22:55
117. White RF (1980) US Patent No 2054310
118. Brink LES, Tramper J (1985) Biotechnol Bioeng 27:1258
119. Furuhashi K, Shintani M, Takagi M (1986) Appl Microbiol Biotechnol 23:218
120. Prichanont S, Leak DJ, Stuckey DC (1998) Enzyme Microb Technol 22:471
121. Imuta M, Kawai KI, Ziffer H (1980) J Org Chem 45:3352
122. Weijers CAGM, Litjens MJJ, de Bont JAM (1992) Appl Microbiol Biotechnol 38:297
123. Besse P, Renard MF, Veschambre H (1994) Tetrahedron: Asymmetry 5:1249
124. Cabon O, Buisson D, Larcheveque M, Azerad R (1995) Tetrahedron: Asymmetry 6:2199
125. Cabon O, Buisson D, Larcheveque M, Azerad R (1995) Tetrahedron: Asymmetry 6:2211
126. Avignon-Tropis M, Treilhou M, Pougny JR, Frechard-Ortuno I, Linstrumelle G (1991) Tetrahedron 47:7279
127. Geerlof A, Jongejan JA, van Dooren TJGM, Raemakers-Franken PC, den Tweel WJJ, Duine JA (1994) Enzyme Microb Technol 16:1059
128. Kloosterman M, Elferink VHM, van Iersel J, Roskam JH, Meijer EM, Hulshof LA, Sheldon RA (1988) Trends Biotechnol 6:251
129. Ladner WE, Whitesides GM (1984) J Am Chem Soc 106:7250
130. Wu DR, Cramer SM, Belfort G (1993) Biotechnol Bioeng 41:979
131. Agase Agrochemicals (1995) Jpn Patent No Jp 07099-993
132. Ferraboschi P, Grisenti P, Manzocchi A, Santaniello E (1994) Tetrahedron: Asymmetry 5:691
133. Ferraboschi P, Casati S, Grisenti P, Santaniello E (1994) Tetrahedron 50:3251
134. Seu YB, Lim TK, Kim CJ, Kang SC (1995) Tetrahedron: Asymmetry 6:3009
135. Gou DM, Liu YC, Chen CS (1993) J Org Chem 58:1287
136. Elks J, Ganellin CR (1990) Diltiazem in dictionary of drugs. Chapmann and Hall, London, 426
137. Wynberg H, ten Hoeve W (1989) Eur Patent Appl 0342903A1
138. Palmer JT (1985) US Patent No 4552695
139. Gentile A, Giordano C, Fuganti C, Ghirotto L, Servi S (1992) J Org Chem 57:6635
140. Huylshof LA, Hendrik J (1989) Eur Patent No 0343714A1
141. Weijers CAGM, de Haan A, de Bont JAM (1988) Appl Microbiol Biotechnol 27:337
142. Machado SS, Wandel U, Straathof AJJ, Jongejan JA, Duine JA (1996) Production of (R)-glycidol by Acetobacter pasteurianus. International Conference on Biotechnology for Industrial Production of Fine Chemicals. Zermatt, Switzerland

143. For a recent very exhaustive review see: Archer IVJ (1997) Tetrahedron 53:15617
144. Blee E, Schuber F (1995) Eur J Biochem 230:229
145. Linderman RJ, Walker EA, Haney C, Roe RM (1995) Tetrahedron 51:10845
146. Orru R, Kroutil W, Faber K (1997) Tetrahedron Lett 38:1753
147. Spelberg JHL, Rink R, Kellog RM, Janssen DB (1998) Tetrahedron: Asymmetry 9:459
148. Botes AL, Weijers CAGM, van Dyk MS (1998) Biotechnol Lett 20:421
149. Moussou P, Archelas A, Furstoss R (1998) Tetrahedron 54:1563
150. Hammock BD, Grant DF, Storms DH (1996) Epoxide hydrolases. In: Sipes I, McQueen C, Gandolfi A (eds) Comprehensive toxicology. Pergamon, Oxford, chap18.3
151. Knehr M, Thomas H, Arand M, Gebel T, Zeller HD, Oesch F (1993) J Biol Chem 268:17623
152. Arand M, Wagner H, Oesch F (1996) J Biol Chem 271:4223
153. Borhan B, Jones DA, Pinot F, Grant DF, Kurth MJ, Hammock BD (1995) J Biol Chem 270:26923
154. Lacourciere GM, Armstrong RN (1993) J Am Chem Soc 115:10466
155. Tzeng HF, Laughlin LT, Lin S, Armstrong RN (1996) J Am Chem Soc 118:9436
156. Laughlin LT, Tzeng HF, Lin S, Armstrong RN (1998) Biochemistry 37:2897
157. Tzeng HF, Laughlin LT, Armstrog RN (1998) Biochemistry 37:2905
158. Berti G (1986) Enantio- and diastereoselectivity of microsomal epoxide hydrolase: potential applications to the preparation of non-racemic epoxides and diols. In: Schneider MP (ed) Enzymes as catalysts in organic synthesis. D Reidel, Dordrecht, Holland, 178:349
159. Bellucci G, Chiappe C, Ingrosso G, Rosini C (1995) Tetrahedron: Asymmetry 6:1911
160. Chiappe C, Cordoni A, Moro GL, Palese CD (1998) Tetrahedron: Asymmetry 9:341
161. Bellucci G, Chiappe C, Cordoni A (1996) Tetrahedron: Asymmetry 7:197
162. Belluci G, Capitani I, Chiappe C, Marioni F (1989) J Chem Soc Chem Commun 16:1170
163. Sato H (1975) Jpn Patent No 75140684
164. Allen RH, Jakoby WB (1969) J Biol Chem 244:2078
165. Niehaus WG, Schroepfer GJ (1967) J Biol Chem 89:4227
166. Michaels BC, Fulco AJ, Ruettinger RT (1980) Biochem Biophys Res Commun 92:1189
167. De Bont JAM, van Dijken JP, van Ginkel KG (1982) Biochim Biophys Acta 714:465
168. Escoffier B, Prome JC (1989) Bioorg Chem 17:53
169. Jacobs MH, van den Wijngaard AJ, Pentenga M, Janssen DB (1991) Eur J Biochem 202:1217
170. Nakamura T, Nagasawa T, Yu F, Watanabe I, Yamada H (1992) J Bacteriol 174:7613
171. Nakamura T, Nagasawa T, Yu F, Watanabe I, Yamada H (1994) Appl Environ Microbiol 60:4630
172. Faber K, Mischitz M, Kroutil W (1996) Acta Chem Scand 50:249
173. Hechtberger P, Wirnsberger G, Mischitz M, Klempier N, Faber K (1993) Tetrahedron: Asymmetry 4:1161
174. Chen X-J, Archelas A, Furstoss R (1993) J Org Chem 58:5528
175. Pedragosa-Moreau S, Archelas A, Furstoss R (1993) J Org Chem 58:5533
176. Pedragosa-Moreau S, Archelas A, Furstoss R (1996) Tetrahedron 52:4593
177. Mischitz M, Mirtl C, Saf R, Faber K (1996) Tetrahedron: Asymmetry 7:2041
178. Moussou P, Archelas A, Baratti J, Furstoss R (1998) Tetrahedron: Asymmetry 9:1539
179. Mischitz M, Faber K, Willetts A (1995) Tetrahedron: Asymmetry 17:893
180. Mischitz M, Kroutil W, Wandel U, Faber K (1995) Tetrahedron: Asymmetry 6:1261
181. Wandel U, Mischitz M, Kroutil W, Faber K (1995) J Chem Soc Perkin Trans 1 735
182. Mischitz M, Faber K (1994) Tetrahedron Lett 35:81
183. Kamal A, Rao AB, Rao MV (1992) Tetrahedron Lett 33:4077
184. Kamal A, Damayanthi Y, Rao MV (1992) Tetrahedron: Asymmetry 3:1361
185. Osprian I, Kroutil W, Mischitz M, Faber K (1997) Tetrahedron: Asymmetry 8:65
186. Kroutil W, Mischitz M, Faber K (1997) J Chem Soc Perkin Trans 1 3629
187. Kroutil W, Mischitz M, Plachota P, Faber K (1996) Tetrahedron Lett 37:8379
188. Kroutil W, Orru RVA, Faber K (1998) Biotechnol Lett 20:373
189. Kroutil W, Osprian I, Mischitz M, Faber K (1997) Synthesis 156
190. Mischitz M, Faber K (1996) Synlett 978

191. Otto PPJHL, Stein F, van der Willigen CA (1988) Agric Ecosys Environ 21:121
192. Archer IVJ, Leak DJ, Widddowson DA (1996) Tetrahedron Lett 37:8819
193. Botes AL, Steenkamp JA, Letloenyane MZ, van Dyck MS (1998) Biotechnol Lett 20:427
194. Weijers CAGM (1997) Tetrahedron: Asymmetry 8:639
195. Weijers CAGM, Botes AL, van Dyk MS, de Bont JAM (1998) Tetrahedron: Asymmetry 9:467
196. Hartmann GR, Frear DS (1963) Biochem Biophys Res Commun 10:366
197. Imai K, Marumo S, Mori K (1974) J Am Chem Soc 96:5925
198. Kolattukudy PE, Brown L (1975) Arch Biochem Biophys 166:599
199. Suzuki Y, Marumo S (1972) Tetrahedron Lett 19:1887
200. Wackett LP, Gibson DT (1982) Biochem J 205:117
201. Choi WJ, Huh EC, Park HJ, Lee EY, Choi CY (1998) Biotechnology Techniques 12:225
202. Archelas A, Delbecque JP, Furstoss R (1993) Tetrahedron: Asymmetry 4:2445
203. Pedragosa-Moreau S, Archelas A, Furstoss R (1994) Bioorg Med Chem 2:609
204. Pedragosa-Moreau S, Morisseau C, Zylber J, Archelas A, Baratti JC, Furstoss R (1996) J Org Chem 61:7402
205. Moussou P, Archelas A, Furstoss R (1998) J Molecular Catalysis B, Enzymatic 5:447
206. Pedragosa-Moreau S, Archelas A, Furstoss R (1996) Tetrahedron Lett 37:3319
207. Chartrain M, Senanayake CH, Rosazza JPN, Zhang J (1996) Int Patent No WO96/12818
208. Zhang J, Reddy J, Roberge C, Senanayake C, Greasham R, Chartrain M (1995) J Ferment Bioeng 80:244
209. Nellaiah H, Morisseau C, Archelas A, Furstoss R, Baratti JC (1996) Biotechnol Bioeng 49:70
210. Morisseau C, Nellaiah H, Archelas A, Furstoss R, Baratti JC (1997) Enzyme Microb Technol 20:446
211. Pedragosa-Moreau S, Morisseau C, Zylber J, Baratti JC, Archelas A, Furstoss R (1997) Tetrahedron 53:9707
212. Murmann W, Rumore G, Gamba A (1967) Bull Chim Farm 106:251
213. Cleij M, Archelas A, Furstoss R (1998) Tetrahedron: Asymmetry 9:1839

Oxynitrilases:
From Cyanogenesis to Asymmetric Synthesis

Michael Schmidt · Herfried Griengl

Spezialforschungsbereich Biokatalyse and Institut für Organische Chemie der Technischen Universität Graz, Stremayrgasse 16, A-8010 Graz, Austria
E-mail: Sekretariat@orgc.tu-graz.ac.at

Oxynitrilases are enzymes which catalyse the formation and cleavage of cyanohydrins. The cyanohydrin formation reaction proceeds by stereoselective addition of hydrogen cyanide to aldehydes or ketones to give enantiopure α-hydroxynitriles. This simple method of C-C bond formation has become a promising method to obtain a number of biologically active compounds. Cyanohydrin fission plays an important role in nature and is involved in plant defence where hydrogen cyanide is liberated upon plant damage. Among the known oxynitrilases only the (R)-oxynitrilase from *Prunus amygdalus* and the (S)-oxynitrilases from *Hevea brasiliensis* and *Manihot esculenta* are available in sufficient quantities which allow cyanohydrin formation on a larger scale. *Prunus amygdalus* oxynitrilase can easily be isolated from natural sources (bitter almond bran) and for two (S)-oxynitrilases functional overexpression allows their production in sufficient amounts for broad preparative applications. The three dimensional structure of the (S)-oxynitrilase from *Hevea brasiliensis* has been determined, and suggestions concerning the reaction mechanism have been discussed. Several procedures employing oxynitrilases have been developed to date which enable cyanohydrin formation on a preparative scale, particularly the use of buffer solutions as the reaction medium, organic solvents with immobilised enzymes, as well as biphasic reaction systems. Possible follow up reactions of the generated hydroxy and nitrile functionality, as well as the conversion of unsaturated cyanohydrins into valuable asymmetric compounds are outlined.

Keywords: Cyanogenesis, Enantioselective syntheses, Cyanohydrins, Enzyme catalysed reactions.

Topics in Current Chemistry, Vol. 200
© Springer Verlag Berlin Heidelberg 1999

1
Introduction

When developing new pharmaceutical compounds with chiral centres it is necessary to prepare all possible stereoisomers in enantiopure form in order to explore their effects in therapy and to study their metabolic pathways [1]. In general, only one enantiomer has the desired therapeutical effect, while the other one is ineffective or even harmful [2]. Therefore, the preparation of enantiopure compounds is advantageous and, in some cases, a necessity. As an alternative to common resolution techniques which can be sometimes laborious, several methods for direct asymmetric syntheses have been developed [3, 4]. Theoretically, these techniques can give a quantitative yield of the desired product with an economical use of raw materials, energy and less waste material production. Chiral cyanohydrins can play an important role in this area since their synthesis leads to a C-C bond and introduces both a hydroxy and a nitrile functionality which can be transformed further. Chiral cyanohydrin formation is a rather uncomplicated reaction in which hydrogen cyanide is reacted with carbonyl compounds (i.e. aldehydes or ketones) by the use of a chiral catalyst. A comparison of the advantages and disadvantages of chemical versus enzymatical methods indicates that the latter approach is the method of choice, because biocatalysts can produce products in excellent enantiopurities as well as high isolated yields [2]. Furthermore, waste treatment can be reduced to a minimum since enzymes are biodegradable. Oxynitrilases can also be used as chiral biocatalysts. In the last decade a number of oxynitrilases have been isolated and characterised [5]. These enzymes exhibit different stereoselectivities and also vary in the nature of carbonyl compounds accepted as substrates. In almost all cases oxynitrilases have been isolated from plant sources [6] where they play an important role as a defence mechanism of the plant against herbivores or microbial attack [7-9] and in amino acid anabolism [8, 10]. This review concentrates on the progress made in this area concerning biological characterisation, genetic elaboration as well as procedures used for the biotransformation and the synthetic applications of cyanohydrins as useful chiral intermediates.

2
Oxynitrilases

2.1
Natural Occurrence

To date about 3000 plant species and also a small number of other organisms are known to be able to release hydrogen cyanide spontaneously from their tissues [6, 11-14]. This ability is commonly known as cyanogenesis. In approximately 300 cyanogenic plants, the source of HCN was identified to be a cyanogenic glycoside or cyano lipid [5, 11] (Fig. 1). In these HCN storage compounds the α-hydroxy group of the cyanohydrin is attached to a sugar moiety via a glycosidic bond or it is protected as a fatty acid ester.

These cyanide donors can be cleaved either spontaneously or by the action of enzymes which enhances the velocity of HCN liberation significantly. One class of enzymes involved in the cyanogenic pathway was thought to be oxynitrilases [15]. However, oxynitrilases could only be found in a few cyanogenic plants, among them important food plants like manjok (*Manihot esculenta*) [16] and millet (*Sorghum bicolor*) [17]. Several oxynitrilases from the *Rosaceae* family [18-26] and from the plant families of *Linaceae* [27, 28], *Filitaceae* [29], *Olacaceae* [30, 31] *and Euphorbiaceae* [16, 32, 33] have also been isolated and investigated in more detail.

Among the 11 oxynitrilases isolated from six plant families and purified to homogeneity to date (Table 1), only four attained preparative interest due to their sufficient availability from natural plant sources or successful functional overexpression.

2.2
Function in Cyanogenic Plants

The presence of high concentrations of cyanohydrin cleavage products during cyanogenesis, i.e. aldehydes, ketones and hydrogen cyanide, is usually highly toxic to animals and microorganisms [9]. This mechanism plays a key role in the self defence of cyanogenic plants [34]. If the subcellular structures of the plant are destroyed, several enzymes gain access to the cyanogenic glycosides or lipids which were previously localised in separate cell compartments [5]. The liberation of HCN is a two-step reaction. First, the action of a β-glycosidase cleaves the cyanogenic glycoside into a sugar residue and a cyanohydrin moiety [7]. Second,

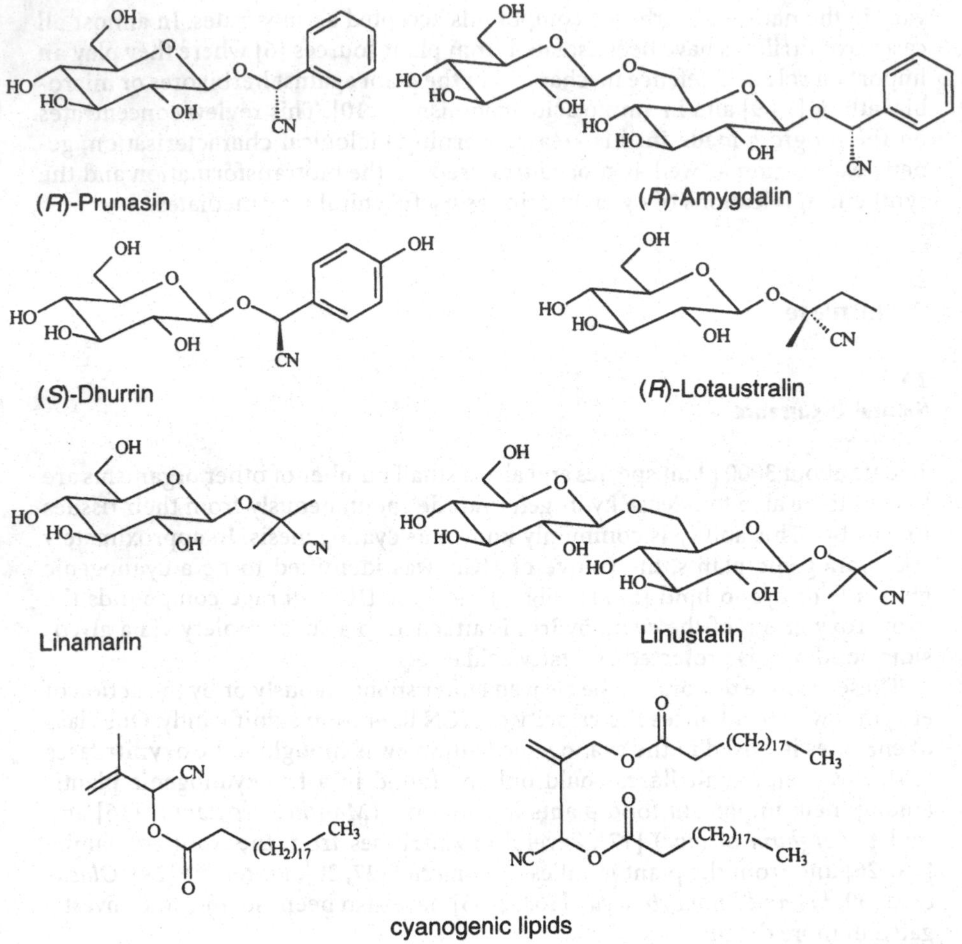

(R)-Prunasin

(R)-Amygdalin

(S)-Dhurrin

(R)-Lotaustralin

Linamarin

Linustatin

cyanogenic lipids

Fig. 1. Naturally occurring cyanogenic glycosides and cyanogenic lipids from various plant sources

the released cyanohydrin is cleaved either spontaneously by base catalysis or enzymatically by the action of oxynitrilases to release the corresponding carbonyl compound and hydrogen cyanide [15]. Since cyanohydrins are unstable at pH values exceeding five, the necessity of an oxynitrilase for HCN liberation was not clearly understood for a long time. The role of oxynitrilases has become more obvious only recently [34, 35]. It was shown that the protective effect and the efficiency thereof did not depend on the total amount of cyanogenic glycosides in the plant but on the rate at which HCN is liberated from the plant tissue. In this context, oxynitrilases play a key role in cyanogenesis because it was shown that they significantly enhance the rate of cyanohydrin fission [6, 10, 15, 34] (Scheme 1).

Table 1. Biochemical properties of purified oxynitrilases from various plant sources

Plant	Family	Native molecular mass [kDa]	Iso-enzymes	Stereo selectivity	pI optimum	pH	FAD	Natural substrate	Glyco-syla-tion	Ref.
Prunus amygdalus	Rosaceae	72±2	3–4	(R)	4.3–4.5	5–6	+	(R)-mandelonitrile	+	[23, 25, 26, 30, 49, 56, 65]
Prunus serotia	Rosaceae	55.6	5	(R)	4.58–4.63	6–7	+	(R)-mandelonitrile	+	[20, 24, 42]
Prunus laurocerasus	Rosaceae	60	3	(R)	4.2–4.4	5.5–6.0	+	(R)-mandelonitrile	+	[22]
Prunus lyonii	Rosaceae	50	no	(R)	4.75	5.5	+	(R)-mandelonitrile	–	[21]
Sorghum bicolor	Poaceae	108±3	3	(S)	4.0		–	(S)-4-hydroxy-mandelonitrile	+	[25, 26, 28, 30, 57, 58, 59]
Sorghum vulgare	Poaceae	180		(S)		5–6	–	(S)-4-hydroxy-mandelonitrile		[55, 56]
Mainhot esculenta	Euphor-biaceae	92–124	3	(S)	4.1–4.6	5.4	–	acetone cyanohydrin, (R)-butanone cyano-hydrin	–	[16, 56, 62]
Hevea brasiliensis	Euphor-biaceae	58		(S)	4.1	5.5–6.0	–	acetone cyanohydrin	–	[33, 61]
Linum usitatissimum	Linaceae	82–87	yes	(R)	4.5–4.85	5.5	–	acetone cyanohydrin, (R)-butanone cyano-hydrin	–	[27, 57, 58, 60]
Phlebodium aureum	Filitaceae	168±30–40	3	(R)		6.5	–	(R)-mandelonitrile	–	[29]
Ximenia americana	Olacaceae	110	3	(S)	3.9–4.0	5.5	–	(S)-mandelonitrile	+	[30, 31]

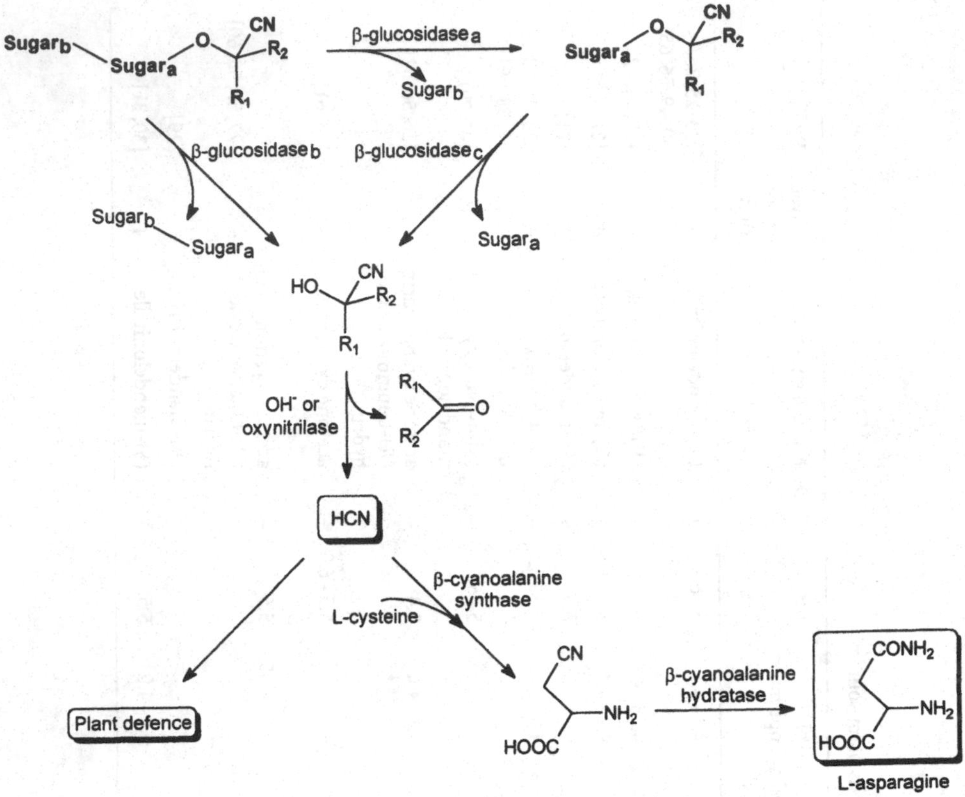

Scheme 1. Role of cyanogenic glycosides and oxynitrilases in the metabolism of cyanogenic plants

Another important role of cyanogenic glycosides in nature is that of nitrogen storage and their subsequent function as a nitrogen source [8, 10, 36]. Huge amounts of the cyanogenic glycoside linamarin could be found in the seeds of the rubber tree *Hevea brasiliensis*. After germination, this compound is metabolised to non-cyanogenic compounds without HCN liberation [10, 37]. More detailed investigations showed that hydrogen cyanide is reacted with cysteine to give β-cyanoalanine which is a precursor of asparagine. The enzyme involved in this conversion, the β-cyanoalanine synthetase, can be found in all cyanogenic plants. The asparagine formed is then channelled into the anabolism of amino acids [36, 38 – 41].

2.3
Biochemical Properties

To classify the known oxynitrilases, two broad groups have been defined. Namely, flavine adenine dinucleotide (FAD) containing oxynitrilases and FAD free oxynitrilases [7].

The FAD containing oxynitrilases can be found exclusively in the *Rosaceae* plant family, namely in the *Prunoideae* and in the *Maloideae* [18]. Therefore it is not surprising that these enzymes show similar biochemical properties [19, 22]. They are N-glycosylated single chain proteins with a carbohydrate content up to 30% [22]. Their molecular weight ranges from 50 to 80 kDa [5] and several different isoforms have been described, which may differ by the degree of glycosylation and in their primary structure [23, 42]. The natural substrates of all these oxynitrilases are either (R)-prunasin or (R)-amygdalin which contains (R)-(+)-mandelonitrile as the chiral aglycon [11, 14]. The analogy between of these groups of oxynitrilases can also be found in their similar immunological behaviour [19]. Their isoelectric points have been reported to be in the range 4.2–4.8 [5] and, in every case, a flavine adenine dinucleotide (FAD) is attached non-covalently as prosthetic group in a hydrophobic region next to the active centre of the protein. The presence of this FAD in an oxidised form seems to be essential for the oxynitrilase activity of these enzymes [43–51]. The FAD group is not directly involved in the cyanogenic pathway [48] since cyanohydrin fission does not occur via a net oxidation or reduction reaction. Therefore, a structure stabilising effect of this group has been suggested, since oxynitrilase activity is lost if FAD is removed [43, 48, 49, 52, 53]. A catalytically active serine has been identified to be involved in the action of these enzymes by inhibition experiments with diisopropylfluorophosphate [28]. In addition, a free cysteine residue also seems to be important for oxynitrilase activity, since common SH reagents as well as the chemical modification with α, β-unsaturated propiophenones [50, 54] lead to inactivation. The oxynitrilase of *Prunus serotina* has been reported to show significant similarity to FAD-dependent oxidoreductases [24]. Corresponding to this model, FAD containing oxynitrilases might have developed from a common precursor flavine containing oxidoreductase which has subsequently lost its ability to catalyse redox reactions. The FAD remaining as a structurally important domain has consequently been preserved during the evolution [48].

Oxynitrilases which do not contain FAD can be found in various plant families [16, 27, 29, 31, 55–57]. Therefore, completely different biochemical properties have to be expected. They are not related to the FAD containing oxynitrilases at all and differ greatly in molecular mass, glycosylation content, isoelectic point and also in structure. The oxynitrilases from *Sorghum bicolor* [25, 26, 28, 57–59], *Sorghum vulgare* [55, 56] and *Ximenia americana* [30, 31] employ the chiral cyanogenic glycosides (S)-dhurrin and (S)-sambunigrin as substrates which contain (S)-4-hydroxymandelonitrile and (S)-mandelonitrile as chiral aglycons. In the case of the oxynitrilases from *Linum usitatissimum* [27, 58, 60], *Hevea brasiliensis* [32–34, 61] and *Manihot esculenta* [16, 62], linamarine is the natural cyanogenic glycoside substrate which contains the achiral acetone cyanohydrin. These enzymes are not glycosylated. Additionally, oxynitrilases from *Linum* and *Manihot* contain lotaustraline with (R)-2-butanone cyanohydrin as aglycon. The molecular masses of the native protein vary in the range 58–180 kDa and the isoelectric points are between 3.9 and 4.6 [5]. Recently, another FAD-independent (R)-oxynitrilase which employes (R)-mandelonitrile as natural substrate has been isolated from *Phlebodium aureum* leaves [29].

2.4
Molecular Cloning and Overexpression

Among the known oxynitrilases from cyanogenic plants, only four are available in sufficient quantities to allow synthetic application. *Prunus amygdalus* oxynitrilase can very easily be isolated from its natural source, the almond tree [23, 25, 26, 49, 56, 63–66] and is even commercially available. Defatted almond meal (bran), prepared by grinding almonds and treatment with an organic solvent to remove organically soluble products, gives an enzyme preparation in which the oxynitrilase is associated with its natural matrix and therefore does not require immobilisation to enhance its stability against denaturation [66].

Sorghum bicolor oxynitrilase was purified and characterised, but a functional overexpression has failed to date. The core protein was expressed correctly, but the posttranslational processing (glycosylation) could not be performed in the host organisms used. Since plants differ in the manner of glycosylation compared to simple eucaryotic organisms, functional overexpression of the oxynitrilase from *Sorghum bicolor* seems not to be possible in the foreseeable future [67].

In contrast, good progress has been made in the overexpression of the oxynitrilase from *Hevea brasiliensis* [32]. This enzyme can easily be isolated from the plant material of the rubber tree by homogenisation of the frozen leaves and has been purified to homogeneity in a five-step purification procedure [32, 33, 61]. Subsequently, this enzyme has been cloned and overexpressed in *Escherichia coli* K 12 and *Sacharomyces cerevisiae* [32]. Overexpression of this oxynitrilase in *E. coli* resulted in the formation of inclusion bodies. Their renaturation is laborious and only low specific activities are obtained. Overexpression in *S. cerevisiae* yielded a highly active soluble oxynitrilase with a specific activity of 22 IU ml^{-1}. Recently the highly productive overexpression of this oxynitrilase in the methylotrophic yeast *Pichia pastoris*, in which up to 60% of the cellular protein consist of recombinant oxynitrilase [68, 69], has been achieved. The specific activity has been determined to be 40 IU ml^{-1} which is about twice the value of that found in the isolated and purified natural enzyme. Remarkably, this enzyme was the first oxynitrilase produced by genetic means and is now available in any quantity, since high cell density fermentation allows its extremely economical production [68].

Progress has also been made in the overexpression of the oxynitrilase from *Manihot esculenta* (cassava) [67, 70] in *Escherichia coli*. As mentioned previously, this oxynitrilase is very similar to the *Hevea* enzyme [16, 32], because both plants belong to the same plant family, the *Euphorbiaceae*. A fermentation on the 40 l scale gave, after simple purification, a total amount of about 40,000 IU of oxynitrilase activity and allowed the exploration of its ability to catalyse the formation of (*S*)-cyanohydrins [67].

2.5
Three-Dimensional Structure

X-ray structure determination of crystallised enzymes is of paramount importance in obtaining a better understanding of the reaction mechanism of the enzyme in question and to give the basis for enzyme engineering. Recently, crystallisation of the (R)-oxynitrilase from *Prunus amygdalus* has been reported [71]. Four isomeric forms of this enzyme could be determined. However, a high resolution crystal structure of this oxynitrilase has not been published to date.

Better progress has been made in the structural characterisation of the *Hevea brasiliensis* oxynitrilase, the first three dimensional structure of an oxynitrilase being obtained with excellent resolution [72, 73] (Fig. 2).

A comparison of the secondary structure of this enzyme with others led to the conclusion that this oxynitrilase belongs to the family of the α, β-hydrolase fold enzymes [74–76] (Fig. 3).

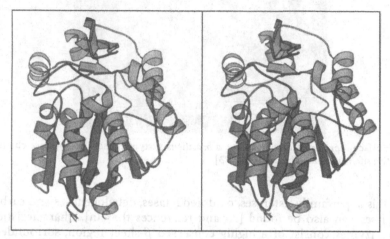

Fig. 2. Stereoview ribbon presentation of the chain fold of the oxynitrilase from *Hevea brasiliensis* [73]

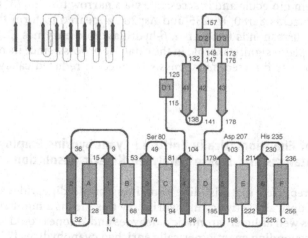

Fig. 3. Topology of the oxynitrilase molecule from *Hevea brasiliensis*. Helices are represented by *rectangles* and β strands by *arrows*, the position of the key amino acid residues are marked [73]. The *upper left insert* shows the topology of the 'prototypic hydrolase fold' [74]

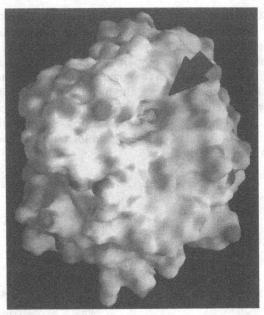

Fig. 4. Surface representation of the *Hevea brasiliensis* oxynitrilase. The entrance channel to the active site is indicated by an *arrow* [73]

In this superfamily, esterases, oxidoreductases, dehalogenases and carboxy-peptidases can also be found [73 and references therein]. Characteristically, these enzymes consist of a highly conserved β-sheet region, surrounded by α-helical domains. A variable so called "cap region" completes their structural appearance. The active site of *Hevea* oxynitrilase has been found to lie deep inside the protein molecule and is accessible via a narrow tunnel [73] (Fig. 4).

The amino acids Ser80, His235 and Asp207 appeared to form the catalytic triade which corresponds to other α, β-hydrolase fold enzymes [32, 76]. Cys81 also seems to play a significant role in the catalytic process, since its replacement with serine by site directed mutagenesis resulted in reduced catalytic activity [32, 76].

3
Formation of Enantiomerically Enriched Cyanohydrins Employing Chiral Chemical Catalysts and Lipase Catalysed Kinetic Resolution

It has been reported that histidine containing cyclic dipeptides catalyse the formation of optically active cyanohydrins. By this means a number of aromatic, aliphatic as well as heteroaromatic aldehydes and ketones could be converted into the corresponding enantiomerically enriched cyanohydrins [77–87]. Interestingly, cyanohydrin formation using this class of catalysts is suppressed by small amounts of benzoic acid. This finding suggests that the mechanism of the dipeptide catalysed cyanohydrin formation could have common features to the reaction course catalysed by oxynitrilases [77, 79], because benzoic acid is also known to be a strong inhibitor of these enzymes [49, 88]. Danda et al. [89] observed that the addition of a small amount of an optically pure cyanohydrin to the cyclic dipeptide catalyst gave significantly higher enantiomeric purities. This finding has been explained by the formation of another, more selective catalytic species. Although in some cases excellent enantiomeric purities could be obtained, a broader application is hindered by a narrow substrate range and insufficient optical purity.

Furthermore, several chiral Lewis acid complexes of titanium [90–95], tin [96], rhenium [97, 98], boron [99], and magnesium [100] have been employed for the preparation of chiral cyanohydrins. However, only in some cases are the enantiomeric excesses satisfactory.

Using optically active aldehydes, ketones or chiral complexes of prochiral carbonyl compounds can induce the diastereoselective addition of hydrogen cyanide to the carbonyl carbon. In this case the substrate range is also very narrow and only in some cases are the obtainable optical yields satisfactory [101–106].

Other possibilities to prepare chiral cyanohydrins are the enzyme catalysed kinetic resolution of racemic cyanohydrins or cyanohydrin esters [107 and references therein], the stereospecific enzymatic esterification with vinyl acetate [108–111] (Scheme 2) and transesterification reactions with long chain alcohols [107, 112]. Many reports describe the use of lipases in this area. Although the action of whole microorganisms in cyanohydrin resolution has been described [110–116], better results can be obtained by the use of isolated enzymes. Lipases from *Pseudomonas* sp. [107, 117–119], *Bacillus coagulans* [110, 111], *Candida cylindracea* [112, 119, 120] as well as lipase AY [120], Lipase PS [120] and the mammalian porcine pancreatic lipase [112, 120] are known to catalyse such resolution reactions.

However, a general disadvantage of kinetic resolution is that the theoretical maximum yield is limited to 50%. To overcome this restriction, racemisation techniques during the resolution have been developed. However, only in a few cases are the optical purities comparable to those obtained with oxynitrilases [121–123].

4
Oxynitrilase Catalysed Formation of Chiral Cyanohydrins

4.1
Substrate Range: Scope and Limitations

4.1.1
(R)-Oxynitrilases

In 1908 Rosenthaler [124] reported the enantioselective synthesis of (R)-mandelonitrile catalysed by an enzyme preparation from almond meal called emulsin. This remarkable result did not attract any scientific attention for a considerable time [125–127]. In the early 1960s the investigation of this interesting

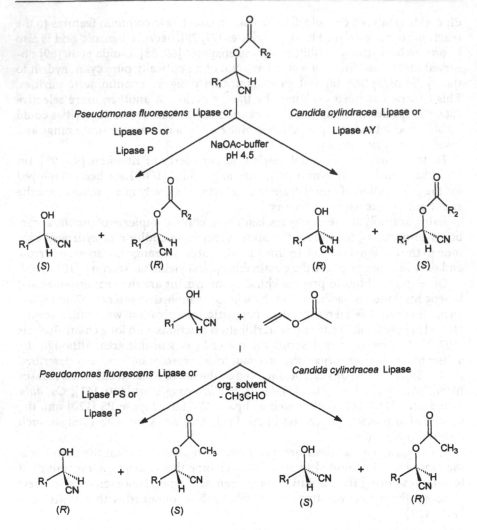

Scheme 2. Kinetic resolution of racemic cyanohydrins by lipase catalysed hydrolysis and by lipase catalysed irreversible esterification with vinyl acetate

route to chiral cyanohydrins was continued. It was found that the aforementioned *Prunus amygdalus* oxynitrilase is responsible for this catalytic effect and this enzyme was employed in the synthesis of several aliphatic, unsaturated, aromatic and heteroaromatic cyanohydrins [49, 65, 88, 128] (Scheme 3). However only with a few substrates were good optical purities obtained and, in particular, slow reacting aldehydes gave unsatisfactory optical yields due to the aqueous ethanol containing reaction system used [49, 65, 88, 128]. In this solvent, at a pH value of five, the spontaneous addition of hydrogen cyanide is obviously not sufficiently suppressed to yield higher enantiopurities [128–131]. In contrast to

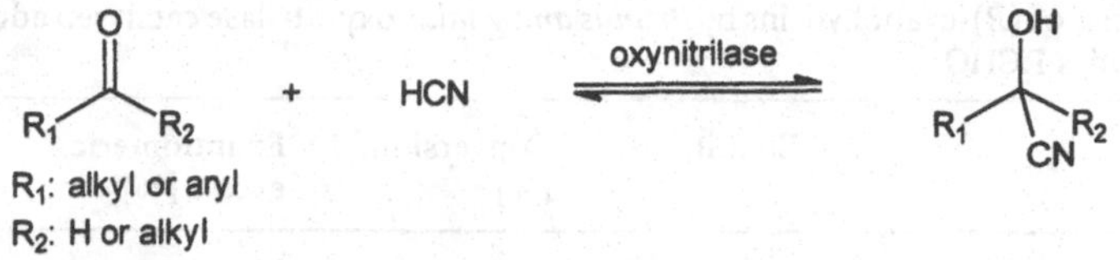

R₁: alkyl or aryl
R₂: H or alkyl

Scheme 3. Principal reaction sequence of enzyme catalysed cyanohydrin formation

that, Kula et al. were able to suppress the chemical cyanohydrin reaction in aqueous medium by reducing the pH value to below 4.0 [132, 133]. Unfortunately, the stability of the oxynitrilase at such low pH values is not very high and rapid deactivation took place. This phenomenon can be overcome by the use of higher enzyme quntities but in the case of slow reacting aldehydes the enantiomeric excess still remained unsatisfactory. A major breakthrough was the observation that the spontaneous addition of hydrogen cyanide can be greatly suppressed by the employment of an organic solvent which is not miscible with water [129–131]. Ethyl acetate or diisopropyl ether have been reported to be such suitable organic solvents although a number of other solvent systems have been investigated for this purpose [134]. In these solvents even slow reacting carbonyl compounds can be converted with a negligible amount of chemical HCN addition, which would lower the optical purity of the product. Fortunately, the range of carbonyl compounds accepted as substrates by *Prunus amygdalus* oxynitrilase is very broad, which allows the formation many of aldehyde [65, 88, 128–131, 135–137] and ketone cyanohydrins [138] with excellent enantiopurity (Table 2).

Among the suitable organic solvents for cyanohydrin synthesis, diisopropyl ether is especially advantageous, since when using it the enzyme activity remains almost constant over a long period of time [139]. In the case of ethyl acetate as reaction medium, a more or less rapid inactivation of the oxynitrilase was observed. However, it seemed to be necessary to immobilise the enzyme to enhance its stability [140], which can be advantageous with respect to reisolation and reuse of the enzyme preparation [2].

Ketones have also been converted into the corresponding cyanohydrins by the use of *Prunus amygdalus* oxynitrilase (Table 3). Although chiral ketone cyanohydrins are valuable intermediates in the preparation of several bioactive compounds [113, 141–144], their preparation by other methods is only poorly described. Only when this oxynitrilase was used for cyanohydrin formation was a broader substrate range accessible. Most of the publications in this area describe the conversion of substituted methyl ketones (2-alkanones) [138]. This class of ketones gives the corresponding cyanohydrins in yields and selectivities comparable to those obtained with aldehydes. 3-Alkanones have also been investigated but they do not seem to be ideal substrates for the *Prunus amygdalus* oxynitrilase, since the chemical and optical yields of the corresponding cyanohydrins in general are lower [145].

Other (*R*)-oxynitrilases have also been used for the purpose of cyanohydrin formation. One of them is the oxynitrilase from *Linum usitatissimum* (flax).

Table 2. Synthesis of (R)-cyanohydrins by *Prunus amygdalus* oxynitrilase catalysed addition of HCN to aldehydes RCHO

R	Conditions	Conversion [%]	Enantiomeric excess [%]	Ref.
C_6H_5	A	95	99	[129]
	B	100	99	[66]
$4\text{-}CH_3C_6H_4$	A	75	98.5	[131]
$4\text{-}CH_3OC_6H_4$	C	55	93	[135]
	B	47	99	[66]
$3\text{-}PhOC_6H_4$	A	99	98	[129]
$3\text{-}NO_2C_6H_4$	A	89	89	[131]
$4\text{-}ClC_6H_4$	A	94	97	[131]
$3,5\text{-}Cl_2\text{-}3NH_2C_6H_2$	C	0	0	[135]
$3\text{-}CF_3C_6H_4$	A	76	75	[161]
$4\text{-}OHC_6H_4$	A	0	0	[161]
$4\text{-}(CH_3)_2NC_6H_4$	A	0	0	[161]
piperonyl	C	50	93	[135]
2-furyl	D	96	99	[137]
3-furyl	D	96	99	[137]
2-thienyl	D	71	99	[137]
3-thienyl	D	95	99	[137]
$2\text{-}(5\text{-}CH_3\text{-}furyl)$	B	70	99	[66]
3-pyridyl	A	97	82	[2]
3-indolyl	A	0	0	[161]
$CH_3(CH_2)_2$	E	100	95	[170]
$(CH_3)_2CH$	E	99	83	[170]
$CH_3(CH_2)_3$	E	100	97	[170]
$(CH_3)_2CHCH_2$	E	100	94	[170]
$CH_3(CH_2)_6$	E	98	87	[170]
$CH_3(CH_2)_8$	E	94	63	[170]
$(CH_3)_3C$	A	84	83	[2]
cyclohexyl	A	90	98.5	[131]
3-cyclohexenyl	C	86	55	[135]
$CH_3S(CH_2)_2$	A	98	96	[2]
$C_6H_5CH_2$	A	88	74	[131]
$C_6H_5(CH_2)_3$	A	94	90	[2]
$C_6H_5OCH_2$	C	83	0	[135]
$(E)\text{-}CH_3CH=CH$	C	94	95	[135]
$(E,E)\text{-}CH_3CH=CHCH=CH$	F	36	95	[154]
$(E)\text{-}C_6H_5CH=CH$	G	54	99	[216]

A: Avicel, ethyl acetate/HOAc-buffer (pH 5.4), HCN, rt. B: almond meal, ethyl acetate/citrate buffer (pH 5.5), HCN, 4°C. C: EtOH/HOAc-buffer (pH 5.4), HCN, 0°C. D: Avicel, diisopropyl ether/HOAc-buffer (pH 3.3), HCN, RT. E: almond meal, diisopropyl ether/citrate buffer (pH 5.5), acetone cyanohydrin, RT. F: Et$_2$O/HOAc-buffer (pH 5.0), acetone cyanohydrin, RT. G: almond meal, methyl *t*-butyl ether/citrate buffer (pH 5.5), HCN, 5°C.

Table 3. Synthesis of (R)-ketone cyanohydrins by *Prunus amygdalus* oxynitrilase catalysed addition of HCN to A methyl ketones (RCOCH$_3$) and B ethyl ketones (RCOCH$_2$CH$_3$)

RCOCH$_3$	Conditions	Conversion [%]	Enantiomeric excess [%]	Ref.
A				
R = CH$_3$CH$_2$	A	80	76	[138]
CH$_3$(CH$_2$)$_2$	A	70	97	[138]
CH$_3$(CH$_2$)$_3$	A	90	98	[138]
CH$_3$(CH$_2$)$_4$	A	88	98	[138]
(CH$_3$)$_2$CH	A	54	90	[138]
(CH$_3$)$_2$CHCH$_2$	A	57	98	[138]
(CH$_3$)$_2$CH(CH$_2$)$_2$	A	64	98	[138]
CH$_2$=CHCH$_2$	A	68	94	[138]
CH$_2$=CH(CH$_2$)$_2$	A	80	97	[138]
Cl(CH$_2$)$_3$	A	87	84	[138]
B				
R = CH$_3$(CH$_2$)$_2$	B	33	85	[145]
CH$_3$(CH$_2$)$_3$	B	21	90	[145]
CH$_3$(CH$_2$)$_4$	B	7	66	[145]
(CH$_3$)$_2$CHCH$_2$	B	0	0	[145]

A: Avicel, diisopropyl ether/HOAc-buffer (pH 4.5), HCN, 0°C or 20°C. B: Avicel, diisopropyl ether/HOAc-buffer (pH 4.5), HCN, 0°C or 20°C.

This enzyme is obviously able to convert aliphatic aldehydes and ketones into the corresponding cyanohydrins although detailed information on the enantiomeric excess was not given [60]. The (R)-oxynitrilase from *Phlebodium aureum* also was also used for the formation of chiral cyanohydrins [29]. While aliphatic carbonyl compounds are not suitable substrates for this enzyme, a few aromatic and heteroaromatic aldehydes could be converted with good optical purities. In search of additional (R)-oxynitrilases, Kiljunen investigated apple, cherry, plum and apricot seeds as oxynitrilase sources and compared the performance of *Prunus amygdalus* oxynitrilase with the enzyme isolated from homogenised apple seeds (apple meal) [146, 147]. The result of this investigation was that this enzyme preparation behaves in a manner similar to the almond oxynitrilase. Long chain aliphatic as well as bulky substituted aromatic aldehydes, in addition to several ketones, were reacted to give the corresponding enantiomerically enriched cyanohydrins.

4.1.2
(S)-Oxynitrilases

In 1961 Bové and Conn isolated and characterised the first (S)-oxynitrilase from *Sorghum vulgare* seedings (millet) [55]. Subsequently, another (S)-oxynitrilase from *Sorghum bicolor* [26] was isolated. These enzymes differ from the (R)-oxynitrilases of the *Rosaceae* family with regard to their biochemical and catalytic properties [25, 26, 56, 57, 59, 65, 148]. The major structural difference is the absence of a prosthetic FAD-group. The isolation and purification from the natural plant source is much more laborious than in the case of the *Prunus* oxynitrilase and yields only small amounts of purified enzyme in three isoforms. Presumably this was the reason why this oxynitrilase was only recently employed in cyanohydrin formation [133, 149]. Another striking difference to other oxynitrilases is the relatively high substrate specificity of this enzyme. *Sorghum* oxynitrilase only catalyses the formation of aromatic cyanohydrins. Aliphatic aldehydes as well as ketones cannot be used as substrates for this oxynitrilase [133, 149, 150]. Therefore, in the last years it has been attempted to identify (*S*)-oxynitrilases which can catalyse the conversion of these substrates (Tables 4–6).

The first described (*S*)-oxynitrilase which fulfils these requirements was isolated from the rubber tree *Hevea brasiliensis* [16, 32, 33, 61]. This oxynitrilase accepts aliphatic as well as aromatic aldehydes of varying structure and chain length [151–153] in addition to ketones [68].

Good progress was also made for the (*S*)-oxynitrilase from *Manihot esculenta* [16, 62, 70]. Overexpression in *E. coli* gave sufficient amounts of enzyme to characterise its properties [67] and it was found that this oxynitrilase catalyses the formation of aliphatic, aromatic and heteroaromatic aldehydes and ketones [67]. The substrate range is very similar compared to that of the aforementioned oxynitrilase from *Hevea brasiliensis* [151–153] which is not surprising since these two enzymes belong to the same plant family and resemble each other in secondary and three dimensional structure (see Sect. 2.4).

Another (*S*)-oxynitrilase was isolated from *Ximenia americana* leaves and seems to accept aromatic aldehydes since it catalyses mandelonitrile fission [31]. However, the investigation of this oxynitrilase probably suffers from its inadequate natural occurrence.

4.2
Reaction Mechanism

In general, little is known about the mechanisms of the oxynitrilase catalysed cyanohydrin reaction. In case of the (*R*)-oxynitrilases mechanistic approaches were reported [50, 154]. However, these preliminary studies suffer partly from insufficient experimental data [51]. Among the (*S*)-oxynitrilases, the enzymes from *Hevea brasiliensis* and *Manihot esculenta* have been explored in more detail very recently. Site directed mutagenesis led to the identification of the amino acids involved in the catalytic process (catalytic triade) and revealed that they belong to the α,β-hydrolase fold class of enzymes [32, 69, 76, 155] which corresponds to the structural data [72, 73]. Inhibition studies showed that both oxynitrilases are strongly deactivated by the action of serine modifying reagents such as diisopropylfluorophosphate [61, 62]. Histidine modifying reagents also heavily inhibit the *Hevea brasiliensis* oxynitrilase [61]. Inhibition could also be detected by the action of thiol specific reagents indicating that a cysteine is involved in the catalytic mechanism. All these findings led to the conclusion that

Table 4. Formation of aromatic (*S*)-cyanohydrins by addition of HCN to aldehydes RCHO catalysed by different (*S*)-oxynitrilases

R	Sorghum bicolor			Manihot esculenta			Hevea brasiliensis		
	Conversion [%]	Enantiomeric excess [%]	Ref.	Conversion [%]	Enantiomeric excess [%]	Ref.	Conversion [%]	Enantiomeric excess [%]	Ref.
C$_6$H$_5$	91	97	[2]	100	98	[67]	96	99	[68]
3-CH$_3$C$_6$H$_4$	80	96	[133]						
4-CH$_3$C$_6$H$_4$	78	87	[2]						
2-CH$_3$OC$_6$H$_4$							61	77	[152]
3-CH$_3$OC$_6$H$_4$	93	89	[2]	82	98	[67]	80	99	[152]
4-CH$_3$OC$_6$H$_4$							49	95	[152]
3-PhOC$_6$H$_4$	93	96	[2]				98	99	[68]
3-OHC$_6$H$_4$	90	98	[133]						
4-OHC$_6$H$_4$	87	99	[133]						
2-Cl C$_6$H$_4$				100	92	[67]			
3-Cl C$_6$H$_4$	95	91	[2]						
4-Cl C$_6$H$_4$	87	54	[2]						
3-BrC$_6$H$_4$	94	92	[2]						
3-CF$_3$C$_6$H$_4$	87	52	[2]						
2-furyl	96	99	[137]	98	92	[67]	95	94	[68]
3-furyl	96	99	[137]	85	96	[67]	96	97	[68]
2-thienyl	71	99	[137]	98	98	[67]	52	99	[152]
3-thienyl	95	99	[137]	84	86	[67]	49	99	[152]
piperonyl									
2-pyrrolyl							0	0	[152]
2-pyridyl							n.d.	0	[152]
3-pyridyl							n.d.	0	[152]
4-pyridyl							n.d.	0	[152]
3-indolyl							0		[152]

Table 5. Formation of aliphatic (S)-cyanohydrins by addition of HCN to aldehydes RCHO catalysed by different (S)-oxynitrilases

R	Manihot esculenta			Hevea brasiliensis		
	Conversion [%]	Enantiomeric excess [%]	Ref.	Conversion [%]	Enantiomeric excess [%]	Ref.
CH_3CH_2	86	91	[67]			
$CH_3(CH_2)_2$	70	88	[62]	n.d.	80	[153]
$CH_3(CH_2)_3$	100	91	[67]			
$CH_3(CH_2)_4$				81	96	[68]
$CH_3(CH_2)_7$				n.d.	85	[153]
$(CH_3)_2CH$	91	95	[67]	n.d.	81	[153]
$(CH_3)_3C$	80	94	[67]	n.d.	67	[153]
Cyclohexyl	100	92	[67]	95	99	[68]
$CH_2=CH$	100	47	[67]	98	94	[151]
$(E)-CH_3CH=CH$	100	92	[67]	80	86	[151]
$(E)-CH_3(CH_2)_2CH=CH$	82	97	[67]	46	95	[151]
$(Z)-CH_3(CH_2)_2CH=CH$				35	80	[151]
$CH_3(CH_2)_2C\equiv C$				88	80	[151]
3-cyclohexenyl				87	99	[152]
$(E)-PhCH=CH$				77	92	[68]
$PhCH_2$				44	99	[152]
$Ph(CH_2)_2$				88	93	[152]
$PhOCH_2$				n.d.	0	[152]
$BnOCH_2$				92	12	[68]

Table 6. Formation of (S)-cyanohydrins by addition of HCN to ketones $RCOCH_3$ catalysed by different (S)-oxynitrilases

R	Manihot esculenta			Hevea brasiliensis		
	Conversion [%]	Enantiomeric excess [%]	Ref.	Conversion [%]	Enantiomeric excess [%]	Ref.
CH_3CH_2	91	18	[67]			
$CH_3(CH_2)_2$	36	69	[67]	51	75	[68]
$CH_3(CH_2)_3$	58	80	[67]			
$CH_3(CH_2)_4$	39	92	[67]			
$(CH_3)_2CHCH_2$	69	91	[67]			
$(CH_3)_3C$	81	28	[67]	49	78	[68]
C_6H_5	13	78	[67]			
$C_6H_5CH_2$				74	95	[68]

Ser80 plays a key role in both enzymes. In addition His236 and Asp208 as well as His235 and Asp207 were found to be essential for the enzyme activity of *Manihot esculenta* and *Hevea brasiliensis*, respectively. Based upon these findings, two different models of the reaction mechanism have been developed. In the case of *Hevea brasiliensis* oxynitrilase Wagner et al. and Hasslacher et al. [73, 156] suggest the covalent binding of the prochiral carbonyl compound to the active site of the enzyme to give a hemiacetal or hemiketal structure forming the tetrahedral intermediate. More recently, a modified reaction mechanism for this oxynitrilase has been proposed [157].

Recently, Wajant and Pfitzenmaier [155] have presented a different mechanistic model for cyanohydrin fisson catalysed by *Manihot esculenta* oxynitrilase. In this approach the cyanohydrin is orientated in the active site of the enzyme by hydrogen bonding.

However, since the structures of enzyme-substrate complexes have not been widely investigated yet, these mechanistic proposals have to be considered with care.

4.3
Methods of Biotransformation

4.3.1
Reaction in Aqueous Medium

Oxynitrilase catalysed chiral cyanohydrin formation generally requires the employment of hydrogen cyanide. This reagent can be generated in situ by reacting an aqueous alkali cyanide solution with an appropriate acidic component [64, 66, 88, 135, 151, 152, 158]. Using such a reaction system, one has to ensure that the spontaneous uncatalysed addition of HCN is sufficiently suppressed to obtain chiral cyanohydrins in good enantiomeric purities [129]. This goal can be achieved by adjusting the pH value to below 4.0 [132, 133]. This procedure avoids the employment of highly toxic free hydrogen cyanide. One disadvantage may, however, be the reduced enzyme stability at such low pH values [159, 160]. Employing this method, a number of good water soluble substrates could be converted. However, in the case of poorly water soluble carbonyl compounds such as *m*-phenoxybenzaldehyde [152], the chemical yields were particularly unsatisfactory [132, 133].

The following procedure is a typical example [152]. To a stirred solution of 1 mmol aldehyde in 1.7 ml of 0.1 mol l^{-1} sodium citrate buffer (pH 4.5), 2000 IU of (S)-oxynitrilase (1000 IU/ml) were added and the mixture was cooled down to ice bath temperature. Subsequently, 2.5 mmole equivalents of potassium cyanide adjusted to pH 4.5 with cold 0.1 mol l^{-1} citric acid (17 ml), were added in one portion. After stirring for 1 h at 0–5°C, the reaction mixture was extracted with methylene chloride (3 × 50 ml). The combined organic layers were dried over anhydrous sodium sulfate and the solvent was removed by evaporation to give the crude cyanohydrin, which was purified by column chromatography using petroleum ether/ethyl acetate (5/1 or 9/1) acidified with trace amounts of anhydrous HCl as the eluent.

4.3.2
Reaction in Organic Solvents with Immobilised Oxynitrilases

A major breakthrough in oxynitrilase catalysed cyanohydrin formation was the finding that the spontaneous addition of hydrogen cyanide to the carbonyl compound is almost completely suppressed by the employment of a non-water miscible organic solvent [64,66,129,131]. This approach offers several advantages. Due to the negligible chemical addition of HCN in such reaction systems, high optical purities could be obtained even in the case of slow reacting bulky carbonyl compounds. Furthermore, both the starting material and the chiral cyanohydrin are soluble in the reaction medium. Among the solvents used for the oxynitrilase catalysed preparation of chiral cyanohydrins [134], diisopropyl ether seems to be especially advantageous [139]. In contrast to ethyl acetate, the oxynitrilase activity remains almost constant for weeks in diisopropyl ether [134]. Obviously, the low solubility of water in diisopropyl ether is responsible for the enhanced activity and stability of oxynitrilases [139]. It was reported to be practical to immobilise the enzyme on a solid support, which allows a simple work up by filtration and enables the reuse of the enzyme preparation [2]. The simplest way of 'immobilisation' is the association with the natural matrix. This was shown to be the case for *Prunus amygdalus* oxynitrilase. Defatted almond meal obtained by the homogenisation of almonds in which the enzyme is bound was used without further purification for cyanohydrin formation [66]. Several other supports such as cellulose based ion exchangers like ECTEOLA-cellulose [88,128,129,161], DEAE cellulose [129,161], Sepharose 4B [161], controlled pore glass beads [129] and silica gel [162] were investigated for cyanohydrin syntheses. In addition, immobilisation was also carried out with microcrystalline cellulose (Avicel) [129, 138, 161], Eupergit C [161], nitrocellulose [67] and with Celite [139, 163]. The latter has been reported to be especially advantageous because of its hydrophobic nature which is obviously beneficial for the enzyme activity [164]. Another crucial point is the water content of the organic solvent employed. If the latter is not saturated with water (buffer) the enzyme will be removed from the carrier and loses its activity [163]. Alternatively, the covalent immobilisation and the fixation of oxynitrilases in liquid crystals have been described [165]. On the basis of these results a method for the continuous production of optically active cyanohydrins was developed which demonstrates remarkable productivities [166].

However, in such a reaction system the employment of hydrogen cyanide in a neat form is unavoidable. Alternatively, HCN can be created from an aqueous solution of alkali cyanide by acidification followed by extraction with an organic solvent which can be directly used for the enzyme catalysed reaction [66].

A representative example for cyanohydrin formation in organic solvents with immobilised oxynitrilases is the following [149]. A suspension of Avicel-cellulose (0.5 g) in 0.05 mol l^{-1} phosphate buffer (pH 5.4, 10 ml), containing ammonium sulfate (4.72 g), was stirred for 1 h and a solution of (*S*)-oxynitrilase from *Sorghum bicolor* (50 μl, 1000 units ml^{-1}, specific activity 70 units mg^{-1}) was added. The mixture was stirred at room temperature for 10 min, filtered, and the immobilised enzyme suspended in diisopropyl ether (10 ml). After addition of aldehyde (2 mmol) and HCN (300 µl, 7.5 mmol), the mixture was stirred until all aldehyde had reacted. After removal of the immobilised enzyme, the filtrate was concentrated to yield the crude cyanohydrin.

4.3.3
Reaction in Biphasic Solvent Mixtures

Biphasic solvent mixtures can also be used for chiral cyanohydrin formation. Loos et al. reported optimisation studies during the development of an industrial process for the production of (R)-mandelonitrile [167, 168]. A number of crucial reaction parameters such as pH-value, temperature, type of solvent and the phase ratio (organic/aqueous) were varied and optimised to give a highly productive method.

Employing the (S)-oxynitrilase from *Hevea brasiliensis* in biphasic solvent mixtures led to enantiomerically enriched cyanohydrins in good yields and selectivities [68]. Using high concentrations of several starting carbonyl compounds (1 – 1.5 mol l⁻¹) in the desired organic solvent gave the corresponding aldehyde cyanohydrins in > 77 % chemical yield and > 92 % e.e. with the exception of benzyloxyacetaldehyde, which is obviously not an ideal substrate for this enzyme. In all cases, a reaction time of approximately 1 h at 0 – 5 °C was sufficient for a complete conversion. These two highly optimised reaction systems seem to allow the application of cyanohydrin formation on an industrial scale.

A typical procedure follows [167]. Freshly distilled, benzoic acid free, benzaldehyde (37.1 g = 0.35 mol), HCN (12.2 g = 0.45 mol and (R)-oxynitrilase (78 mg) were dissolved in 225 ml of methyl *t*-butyl ether (MTBE) and 250 ml of 50 mmol l⁻¹ citric acid buffer (pH 5.5) at 22 °C. After stirring for 20 min the MTBE layer was separated and the aqueous layer was extracted with 25 ml of MTBE. The combined organic layers were dried over $MgSO_4$, filtered and concentrated under reduced pressure (200 mbar) using a water bath (30 °C). Yield: 45.2 g (97 %), purity 98 %, e.e. 98 %. The aqueous layer was reused in a series of four consecutive experiments using the same amounts of reagents in the organic phase. A total of 185.5 g of benzaldehyde was converted into 226 g of (R)-mandelonitrile using 78 mg of (R)-oxynitrilase (0.035 % by weight).

4.3.4
Transhydrocyanation

As pure hydrogen cyanide is highly toxic, alternative methods for enzyme catalysed cyanohydrin formation have been investigated. Ognyanov et al. [154] reported the use of acetone cyanohydrin in a *Prunus amygdalus* catalysed cyanohydrin synthesis. This achiral cyanohydrin has several advantages. It is commercially available, well soluble in water and readily decomposes to HCN which is subsequently reacted to give the corresponding cyanohydrin. The byproduct acetone can easily be removed by distillation.

Transhydrocyanation is a two-step equilibrium reaction [169] (Scheme 4). First, acetone cyanohydrin is cleaved by the action of the oxynitrilase which is

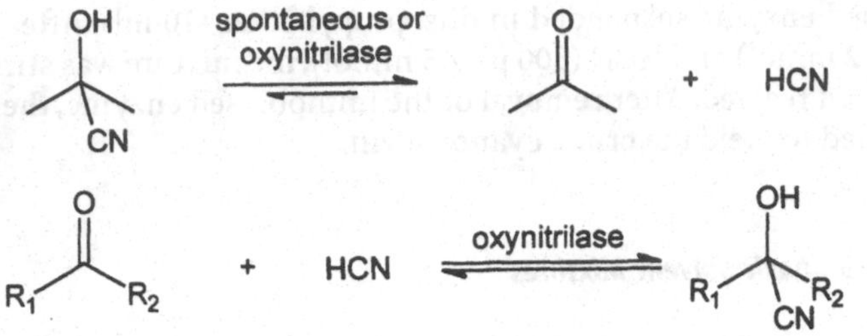

Scheme 4. The principal reaction sequence of transhydrocyanation

facilitated because of the favourable equilibrium position of ketone cyanohydrin fission. Second, the HCN formed is consumed in an enzyme catalysed reaction to give the enantiomerically enriched cyanohydrin. The slow rate of hydrogen cyanide evolution in such a reaction system is advantageous as it suppresses the uncatalysed chemical HCN addition which lowers the optical purity. Careful adjustment of the reaction conditions led to cyanohydrins with good to excellent enantiopurities [154]. The influence of partition coefficients in byphasic systems was also investigated. It was shown that the substrates which are only poorly soluble in water give high optical purities whilst those which are easily soluble in water led to cyanohydrins with decreased e.e. values [154]. Increasing the chain length of the employed aldehydes resulted in a decrease of optical purity [170]. Since under the conditions employed transhydrocyanation reactions seem to be slower than the direct asymmetric synthesis (hydrocyanation), especially with slow reacting carbonyl compounds, only moderate optical purities were obtained due to the predominance of the chemical HCN addition. This reaction system was also applied to the oxynitrilases from *Hevea brasiliensis* [153], *Sorghum bicolor* [150, 171] and for the cyanohydrin synthesis starting from racemic aldehydes using the *Prunus amygdalus* oxynitrilase [172].

Some information is available concerning the kinetics of transhydrocyanation reactions employing benzaldehyde as the substrate [173]. In this study the time dependence of the concentrations of all educts and products were determined by NMR spectroscopy. However, this complex kinetic system is still not fully understood.

An example of cyanohydrin formation using acetone cyanohydrin as the cyanide source is given in the following procedure [154]. To a solution of 120 mg (1 mmole) of phenylacetaldehyde and 110 mg (1.3 mmole) of acetone cyanohydrin in 11 ml of diethyl ether at 23 °C 0.50 ml of the oxynitrilase buffer solution (10 mg/ml, 0.4 mol l^{-1} acetate buffer, pH 5.0) was added. The mixture was stirred for 18 h at 23 °C and diluted with 50 ml of ether. The aqueous phase was extracted with 2x10 ml of ether and the combined organic phases were dried over anhydrous magnesium sulfate. Evaporation of solvent gave a pale amber liquid that was chromatographed on a flash silica gel column in 1:30:50 ethyl acetate-benzene-dichloromethane to afford 122 mg (83%) of cyanohydrin, e.e. 88%.

4.3.5
Kinetic Resolution of Racemic Cyanohydrins

Since in Nature oxynitrilases catalyse cyanohydrin fission it should be possible to employ this class of enzymes in kinetic resolution such as has already been described for lipases and esterases. (R)-cyanohydrins should be obtainable via the action of an (S)-oxynitrilase since the (R)-enantiomer does not seem to be a substrate for this oxynitrilase. Conversely, (S)-cyanohydrins should be available by the employment of an (R)-oxynitrilase. In fact, several investigations were made in this area. It was reported that the oxynitrilase from Sorghum bicolor was able to resolve a racemic mixture of p-hydroxybenzaldehyde cyanohydrin, the natural substrate of this enzyme [174]. The oxynitrilase from Prunus amygdalus was used for the resolution of racemic cyanohydrins [175–177]. However, using this reaction system high yields of the desired compound can only be obtained if the reaction products, namely the liberated carbonyl compound and the evolving HCN, are removed from the reaction mixture in a suitable manner. This technique is often accompanied by reduced yields due to the difficult work up procedures. Several techniques have been developed to circumvent these problems. The removal of the above-mentioned reaction byproduct HCN can be achieved by either gas-membrane extraction [175] or by capture with ω-bromoaldehydes [178] or acetaldehyde [176]. Recently Effenberger and Schwämmle reported the removal of the formed carbonyl compound byproduct by the action of sodium hydrogen sulfite, hydroxylamine and semicarbazide [177]. In some cases this method gave good optical purities but it seems to be limited to the resolution of aromatic and heteroaromatic aldehyde cyanohydrins. Furthermore, these carbonyl modifying reagents harm the enzyme upon prolonged exposure.

Although these resolution techniques can be successfully employed in some cases, they cannot compete with the direct asymmetric synthesis of optically active cyanohydrins.

A typical procedure may be as follows [177]. Prunus amygdalus oxynitrilase (1500 IU, activity 2200 IU/ml) was added to a solution of (R,S)-mandelonitrile (10 mmol) in 200 ml citrate buffer (50 mmol l^{-1}, pH 3.5) and the reaction mixture stirred for 0.5 h. Subsequently, 2.5 ml of a NaHSO$_3$ solution (2 mol l^{-1}) were added, followed by another 2.5 ml NaHSO$_3$ solution after 3 h. After 10 h the precipitate was filtered off and washed with dichloromethane. The aqueous filtrate was extracted three times with dichloromethane (100 ml). The combined organic phases were washed with NaHSO$_3$ solution (2 mol l^{-1}), dried (MgSO$_4$) and concentrated. Yield of (S)-mandelonitrile was 44%, e.e. 96%.

5
Enantiopure Cyanohydrins as Building Blocks in Chiral Syntheses

5.1
Synthetic Relevance of Cyanohydrins

Chiral cyanohydrins can be valuable chiral building blocks in organic syntheses [2, 179] (Fig. 5). The interconversion of the hydroxyl functionality for example

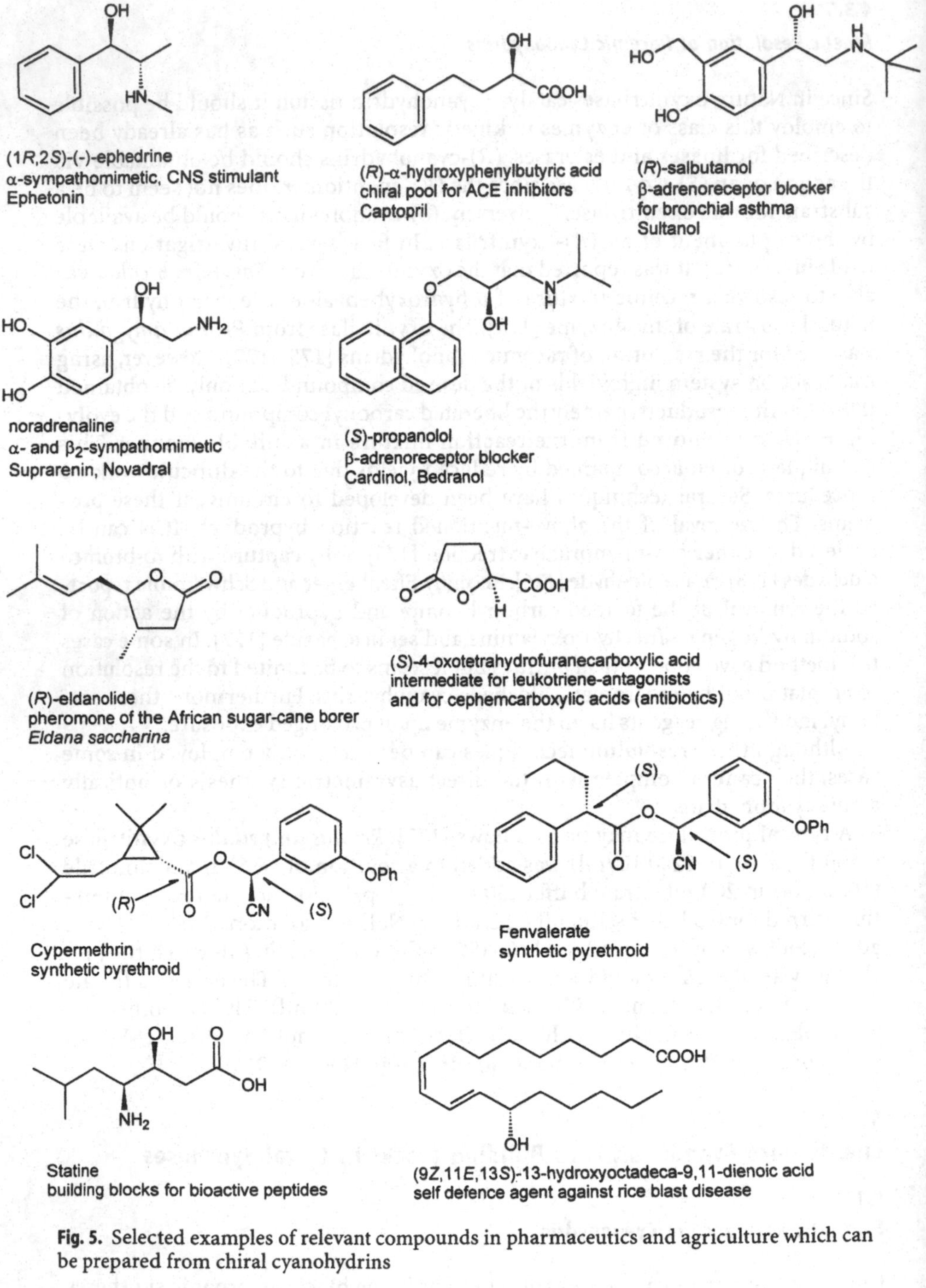

(1R,2S)-(-)-ephedrine
α-sympathomimetic, CNS stimulant
Ephetonin

(R)-α-hydroxyphenylbutyric acid
chiral pool for ACE inhibitors
Captopril

(R)-salbutamol
β-adrenoreceptor blocker
for bronchial asthma
Sultanol

noradrenaline
α- and β$_2$-sympathomimetic
Suprarenin, Novadral

(S)-propanolol
β-adrenoreceptor blocker
Cardinol, Bedranol

(R)-eldanolide
pheromone of the African sugar-cane borer
Eldana saccharina

(S)-4-oxotetrahydrofuranecarboxylic acid
intermediate for leukotriene-antagonists
and for cephemcarboxylic acids (antibiotics)

Cypermethrin
synthetic pyrethroid

Fenvalerate
synthetic pyrethroid

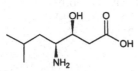

Statine
building blocks for bioactive peptides

(9Z,11E,13S)-13-hydroxyoctadeca-9,11-dienoic acid
self defence agent against rice blast disease

Fig. 5. Selected examples of relevant compounds in pharmaceutics and agriculture which can be prepared from chiral cyanohydrins

can give α-azidonitriles, α-aminonitriles, α-acetoxynitriles and α-fluoronitriles with inverted configuration. The nitrile group can be converted into valuable chiral species such as α-hydroxy acids, α-hydroxy aldehydes, α-hydroxy ketones and β-aminoalcohols. Furthermore, this moiety can be attacked by metalorganic reagents. Chiral cyanohydrins bearing other functionalities, such as a double bond, can be used for other valuable synthetic applications.

5.2
Chemical Transformations

5.2.1
Reaction of the Hydroxy Group

The conversion of the hydroxy group of an enantiomerically enriched cyanohydrin into a good leaving group allows an S_N-type displacement reaction with inversion of configuration [180, 181] (Scheme 5). α-Sulfonyloxynitriles have been reported to be cyanohydrins with suitable leaving groups. Aliphatic sulfonyl activated cyanohydrins can be prepared without loss of optical purity, whilst aromatic sulfonyloxynitriles are configurationally less stable [180, 181]. The mesylate of mandelonitrile can be prepared but the corresponding tosylate decomposes even at room temperature [180]. Furthermore, trifluoromethansul-

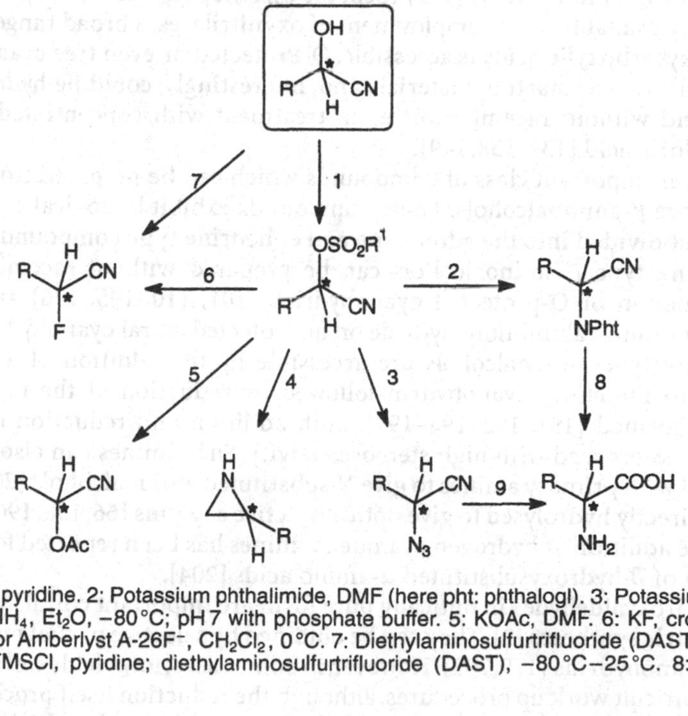

1: R¹SO₂Cl, pyridine. 2: Potassium phthalimide, DMF (here pht: phthalogl). 3: Potassium azide, DMF. 4: LiAlH₄, Et₂O, –80°C; pH 7 with phosphate buffer. 5: KOAc, DMF. 6: KF, crownether, 45–100°C or Amberlyst A-26F⁻, CH₂Cl₂, 0°C. 7: Diethylaminosulfurtrifluoride (DAST), –80°C –25°C or TMSCl, pyridine; diethylaminosulfurtrifluoride (DAST), –80°C–25°C. 8: N₂H₄,Δ. 9: Pd-C, H₂.

Scheme 5. Possible follow-up reactions of the hydroxy group of chiral cyanohydrins

phonates of various cyanohydrins have been prepared [181]. In the displacement reaction aromatic and α, β-unsaturated sulfonyloxynitriles can racemise to a large extent even under mild reaction conditions [181]. Aliphatic sulphonated cyanohydrins gave the desired substitution product cleanly under mild conditions [181]. Inversion of aromatic sulfonyloxynitriles under Mitsunobu conditions yielded, at least with O-nucleophiles, a complete inversion of configuration [182]. α-Sulphonyloxynitriles can be converted to α-azidonitriles, α-aminoacids, α-aminonitriles and aziridines in high chemical and optical yields. Furthermore, displacement with alkali acetates gave the inverted α-acetoxynitriles which can be cleaved to give the corresponding chiral cyanohydrin with inverted configuration [180, 181]. α-Fluoronitriles can also be obtained with inversion of configuration by the action of diethylaminosulfurtrifluoride (DAST) [183]. Fluorinated nitriles can be transformed into α-fluoro carboxylic acids [184, 185] and to β-fluoroamines [186, 187] without racemisation.

5.2.2
Reaction of the Nitrile Group

Optically active α-hydroxycarboxylic acids are valuable synthons for synthetic purposes [188–192] and for optical resolutions [193]. Since they are not widely distributed in Nature, several methods of preparation have been developed [2, 161]. However, all these methods cannot compete with the acid catalysed hydrolysis of chiral cyanohydrins [149] (Scheme 6). Since (R)- and (S)-cyanohydrins are readily available by the employment of oxynitrilases, a broad range of chiral α-hydroxycarboxylic acids is accessible. O-Protected or even free cyanohydrins have been used as starting materials and, interestingly, could be hydrolysed in good yield without racemisation upon treatment with concentrated aqueous hydrochloric acid [131, 138, 149].

Another important class of compounds which can be prepared from cyanohydrins are β-aminoalcohols. These compounds exhibit biological activity and can be subdivided into the adrenaline and ephedrine type compounds [2, 194]. Adrenaline type β-aminoalcohols can be prepared without racemisation by hydrogenation of O-protected cyanohydrins [101, 110, 195, 196] and by the action of lithium aluminium hydride on unprotected chiral cyanohydrins [131]. Ephedrine type aminoalcohols are accessible by the addition of a Grignard reagent to the chiral cyanohydrin followed by reduction of the imine intermediate formed [191, 192, 196–199]. Both addition and reduction have been reported to proceed with high stereoselectivity. Such imines can also be transiminated with primary amines to give N-substituted aminoalcohols [200, 201] or can be directly hydrolysed to give optically active acyloins [66, 158, 196, 198, 202, 203]. The addition of hydrogen cyanide to imines has been reported for the preparation of β-hydroxysubstituted α-amino acids [204].

α-Hydroxyaldehydes in optically pure form are important chiral synthons in asymmetric syntheses [2]. They can be obtained by selective partial reduction of chiral cyanohydrins [179, 205]. However, in some cases, low yields were obtained due to difficult work up procedures, although the reduction itself proceeds without loss of optical purity [206]. α, β-Dihydroxy acids can be obtained by the

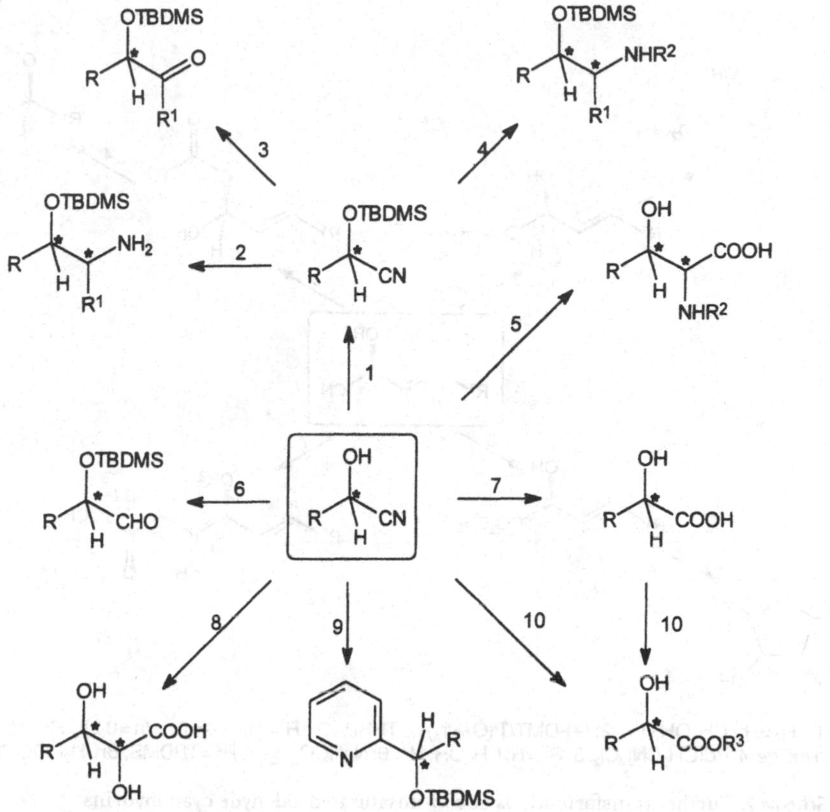

1: TBDMS-Cl, imidazole. 2: R^1MgX; NaBH$_4$; H$_3$O$^+$. 3: R^1MgX; H$_3$O$^+$. 4: R^1MgX; R$_2$NH$_2$; NaBH$_4$; H$_3$O$^+$. 5: 2-methoxypropene, POCl$_3$; Et$_3$N, H$_2$O; DIBAL-H; NH$_4$Br, CH$_3$OH; R^{2}NH$_2$; HCN; dil. HCl; 1,1'-carbonyldiimidazole, Et$_3$N; K$_2$CO$_3$, EtOH; dil. HCl; 2M KOH. 6: TBDMS-Cl, imidazole; DIBAL-M; H$_3$O$^+$. 7: HCl conc., Δ. 8: H$_2$/Ni; NaHSO$_3$, NaCN; 2,2-dimethoxypropane, H$^+$; KOH, EtOH; H$_3$O$^+$. 9: TBFMS-Cl, imidazole; acetylene, p, Co-catalyst. 10: R^{3}OH, HCl$_{(g)}$.

Scheme 6. Possible follow up reactions of the nitrile functionality of chiral cyanohydrins

addition of HCN to the aforementioned α-hydroxyaldehydes [188, 192, 206]. Cyclisation reactions of cyanohydrins with acetylene to give chiral 2-pyridyl alcohols have also been described [207].

5.2.3
Other Transformations

Unsaturated chiral cyanohydrins bear another synthetically interesting functionality, the C-C double bond. Among the opportunities of transforming unsaturated cyanohydrins, oxidative cleavage [208, 215], epoxidation [209, 210], iodolactonisation [209], addition reactions [211, 212, 216] and metal assisted

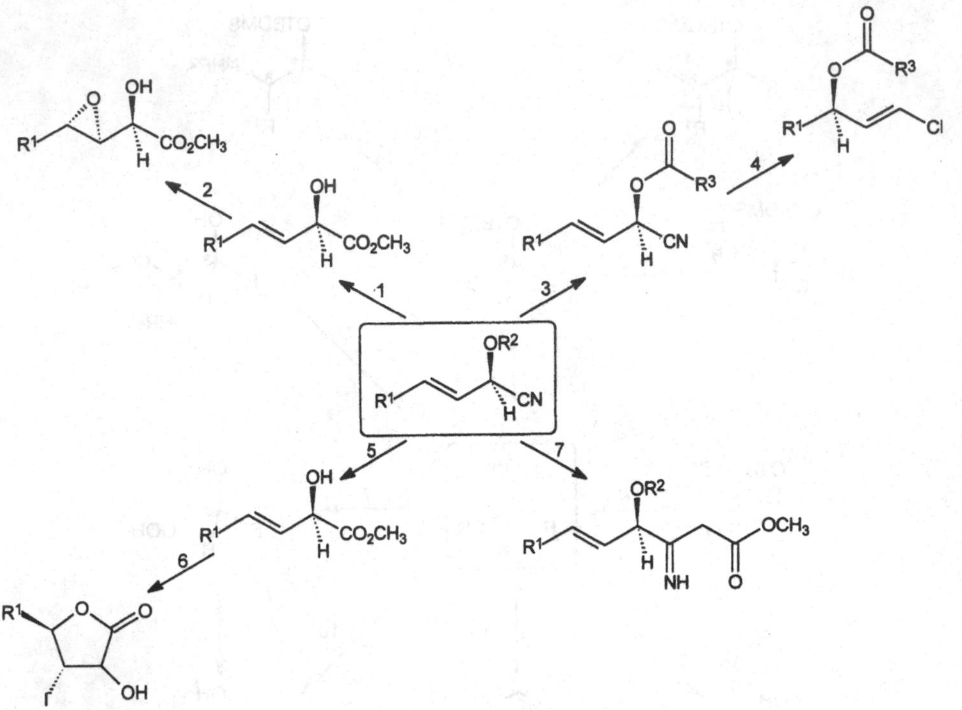

1: R_2=H; CH$_3$OH,H$^+$. 2: (+)-DMT/Ti(Oi-Pr$_4$)$_4$, TBHP. 3: R^1=CH$_3$(CH$_2$)$_n$, (n=0,2); R^2=H; Bu$_2$O, pyridine. 4: Pd(CH$_3$CN)$_2$Cl$_2$. 5: R^2=H; CH$_3$OH, H$^+$. 6: NaHCO$_3$, I$_2$. 7: R^2=TBDMS; BrCH$_2$COOCH$_3$, Zn.

Scheme 7. Further transformations of α,β-unsaturated aldehyde cyanohydrins

rearrangement reactions [213, 214] have been reported. Consequently, these highly functionalised chiral synthons can obviously be of great synthetic value to the chemist [215–219] (Scheme 7).

6
Future Perspectives

Nowadays, the oxynitrilase catalysed formation of chiral cyanohydrins is well established. This technology is obviously superior to other methods with respect to the obtained chemical and optical yields. A better understanding of this class of C-C bond forming enzymes on a molecular level and also the resolution of their 3D structures should help to explore further the reaction mechanism of the oxynitrilase catalysed cyanohydrin formation in more detail. Genetic engineering can yield modified enzymes which might enlarge the range of carbonyl compounds that are accepted as substrates. The ongoing development of new synthetic transformation methods would facilitate the application of this useful class of compounds in science and industry.

Acknowledgement. Research of the Graz group in this area has been supported by the Eurpean Community, Fonds zur Förderung der Wissenschaftlichen Forschung and Forschungs-förderungsfonds der Gewerblichen Wirtschaft. Help of Mag. Astrid Preisz and Dr. Anna de Raadt-Stütz in proof-reading of the manuscript is gratefully acknowledged.

7
References

1. FDA, Department of Health and Human Services of the USA (1992) FDA's Policy Statement for the Development of New Stereoisomeric Drugs, Federal Register, 2 May 1992 57:22, 102
2. (a) Effenberger F (1994) Angew Chem 106:1609; (b) Angew Chem Int Ed Engl 33:1555
3. Stinson SC (1995) Chem Eng News 73:44
4. Collins AN, Sheldrake GN, Crosby J (1997) Chirality in industry II. Wiley, New York
5. Hickel A, Hasslacher M, Griengl H (1996) Physiol Plant 98:891
6. Conn EE (1981) The biochemistry of plants: a comprehensive treatise. Academic Press, New York
7. Poulton JE (1988) In: Evered D, Hernett S (eds) Cyanide compounds in biology. Ciba foundation symposium no 140. Wiley, Chichester, p 67
8. Selmar D (1993) ACS Symp Ser 13:191
9. Nahrstedt A (1985) Plant Syst Evol 150:35
10. Lieberei R, Selmar D, Biehl B (1985) Plant Syst Evol 150:49
11. Seigler DS (1991) In: Rosenthal GA, Berenbaum MR (eds) Herbivores: their interaction with secondary plant metabolites, 2nd edn. Academic Press, New York, p 35
12. Duffey SS (1981) In: Vennesland B, Conn EE, Knowles CJ, Westley J, Wissing F (eds) Cyanide in biology. Academic Press, New York, p 385
13. Müller E, Nahrstedt A (1990) Planta Med 56:612
14. Conn EE (1981) In: Stumpf PK, Conn EE (eds) Secondary plant products, vol 7. Academic Press, New York, p 479
15. Poulton JE (1990) Plant Physiol 94:401
16. Hughes J, De Carvalho FJP, Hughes MA (1994) Arch Biochem Biophys 311:496
17. Seigler DS (1975) Phytochem 14:9
18. Gerstner E, Mätzke V, Pfeil E (1968) Naturwissenschaften 55:561
19. Gerstner E, Pfeil E (1972) Hoppe-Seyler's Z Physiol Chem 353:271
20. Wu H-C, Poulton JE (1991) Plant Physiol 96:1329
21. Xu L-L, Singh BK, Conn EE (1986) Arch Biochem Biophys 250:322
22. Gerstner E, Kiel U (1975) Hoppe-Seyler's Z Physiol Chem 356:1853
23. Aschoff H-J, Pfeil E (1970) Hoppe-Seyler's Z Physiol Chem 351:818
24. Cheng I-P, Poulton JE (1993) Plant Cell Physiol 34:1139
25. Jansen I, Woker R, Kula MR (1992) Biotechnol Appl Biochem 15:90
26. Smitskamp-Wilms E, Brussee J, van der Gen A, van Scharrenburg GJM, Sloothaak JB (1991) Recl Trav Chim Pays-Bas 110:209
27. Xu L-L, Singh BK, Conn EE (1988) Arch Biochem Biophys 263:256
28. Wajant H, Mundry K-W, Pfizenmaier K (1994) Plant Mol Biol 26:735
29. Wajant H, Förster S, Selmar D, Effenberger F, Pfizenmaier K (1995) Plant Physiol 109:1231
30. van Scharrenburg GJM, Sloothaak JB, Kruse CG, Smitskamp-Wilms E, Brussee J (1993) Ind J Chem 32 B:16
31. Kuroki GW, Conn EE (1989) Proc Natl Acad Sci USA 86:6978
32. Hasslacher M, Schall M, Hayn M, Griengl H, Kohlwein SD, Schwab H (1996) J Biol Chem 271:5884
33. Schall M (1996) PhD Thesis, University of Graz
34. Selmar D, Lieberei R, Biehl B, Conn EE (1989) Physiol Plant 75:97
35. Bernays EA (1983) In: Lee JA, McNeill S, Rorison IH (eds) Nitrogen as an ecological factor. Blackwell Scientific, Oxford, p 321

36. Rosenthal GA, Bell EA (1979) In: Rosenthal GA, Janzen DH (eds) Herbivores: their inter-
    actions with secondary plant metabolites, 1st edn. Academic Press, New York, p 353
37. Lieberei R, Nahrstedt A, Selmar D, Gasparotto L (1986) Phytochemistry 25:1573
38. Castric PA, Farnden JF, Conn EE (1972) Arch Biochem Biophys 152:62
39. Blumenthal-Goldschmidt S, Butler GW, Conn EE (1963) Nature 197:718
40. Fowden L, Bell EA (1965) Nature 206:110
41. Oaks A, Johnson FJ (1972) Phytochemistry 11:3465
42. Yemm RS, Poulton JE (1986) Arch Biochem Biophys 247:440
43. Bärwald K-R, Jaenicke L (1978) FEBS Lett 90:255
44. Vargo D, Jorns MS (1979) J Am Chem Soc 101:7625
45. Vargo D, Pokora A, Wang S-W (1981) J Biol Chem 256:6027
46. Pokora A, Jorns MS, Vargo D (1982) J Am Chem Soc 104:5466
47. Jorns MS, Ballenger C, Kinney G, Pokora A, Vargo D (1983) J Biol Chem 258:8561
48. Jorns MS (1979) J Biol Chem 254:12,145
49. Becker W, Benthin U, Eschenhof E, Pfeil E (1963) Biochem Z 337:156
50. Jaenicke L, Preun J (1984) Eur J Biochem 138:319
51. Jorns MS (1985) Eur J Biochem 146:481
52. Hochuli E (1983) Helv Chim Acta 66:489
53. Kaul R, Mattiasson B (1987) Biotechnol Appl Biochem 9:294
54. Petrounia IP, Goldberg J, Brush EJ (1994) Biochemistry 33:2891
55. Bové C, Conn EE (1961) J Biol Chem 236:207
56. Seely MK, Criddle RS, Conn EE (1966) J Biol Chem 241:4457
57. Wajant H, Mundry KW (1993) Plant Sci 89:127
58. Wajant H, Riedel D, Benz S, Mundry K-W (1994) Plant Sci 103:145
59. Wajant H, Böttinger H, Mundry K-W (1993) Biotechnol Appl Biochem 18:75
60. Albrecht J, Jansen I, Kula MR (1993) Biotechnol Appl Biochem 17:191
61. Wajant H, Förster S (1996) Plant Sci 115:25
62. Wajant H, Förster S, Böttinger H, Effenberger F, Pfizenmaier K (1995) Plant Sci 108:1
63. Vernau J, Kula M-R (1990) Biotechnol Appl Biochem 12:397
64. van den Nieuwendijk AMCH, Warmerdam EGJC, Brussee J, van der Gen A (1995) Tetra-
    hedron: Asymmetry 6:801
65. Becker W, Pfeil E (1966) Biochem Z 346:301
66. Zandbergen P, van der Linden J, Brussee J, van der Gen A (1991) Synth Comm 21:1387
67. (a) Förster S, Roos J, Effenberger F, Wajant H, Sprauer A (1996) Angew Chem 108:493;
    (b) Förster S, Roos J, Effenberger F, Wajant H, Sprauer A (1996) Angew Chem Int Ed Engl
    35:437
68. Griengl H, Hickel A, Johnson DV, Kratky C, Schmidt M, Schwab H (1997) J Chem Soc
    Chem Commun 1933
69. Hasslacher M, Schall M, Hayn M, Bona R, Rumbold K, Lückl J, Griengl H, Kohlwein SD,
    Schwab H (1997) Protein Express Purif 11:61
70. Wajant H, Förster S, Sprauer A, Effenberger F, Pfizenmaier K (1996) Ann N Y Acad Sci
    (Enzyme Engineering XIII) 799:771
71. Lauble H, Müller K, Schindelin H, Förster S, Effenberger F (1994) Proteins: Struct Funct
    Genet 19:343
72. Wagner UG, Kratky C (1996) Acta Crystallogr D52:591
73. Wagner UG, Hasslacher M, Griengl H, Schwab H, Kratky C (1996) Structure 4:811
74. Ollis DL, Cheah E, Cygler M, Dijkstra B, Frolow F, Franken SM, Harel M, Remington SJ,
    Silman I, Schrag J, Sussman JL, Verschueren KHG, Goldman A (1992) Protein Eng 5:197
75. Cygler M, Schrag JD, Sussman JL, Harel M, Silman I, Gentry MK, Doctor BP (1993) Pro-
    tein Sci 2:366
76. Hasslacher M, Kratky C, Griengl H, Schwab H, Kohlwein SD (1997) Proteins 27:1
77. North M (1993) Synlett 807
78. Oku J, Ito N, Inoue S (1979) Makromol Chem 180:1089
79. Oku J, Inoue S (1981) J Chem Soc Chem Commun 229
80. Oku J, Ito N, Inoue S (1982) Makromol Chem 183:579

81. Mori A, Ikeda Y, Kinoshita K, Inoue S (1989) Chem Lett 2119
82. Tanaka K, Mori A, Inoue S (1990) J Org Chem 55:181
83. Asada S, Kobayashi Y, Inoue S (1985) Makromol Chem 186:1755
84. Kobayashi J, Asada S, Watanabe I, Hayashi H, Motoo Y, Inoue S (1986) Bull Chem Soc Jpn 59:893
85. Jackson WR, Jayatilake GS, Matthews BR, Wilshire C (1988) Austr J Chem 41:203
86. Matthews BR, Jackson WR, Jayatilake GS, Wilshire C, Jocobs HA (1988) Aust J Chem 41:1697
87. Kobayashi Y, Hayashi H, Miyaji K, Inoue S (1986) Chem Lett 931
88. (a) Becker W, Freund H, Pfeil E (1965) Angew Chem 77:1139; (b) Becker W, Freund H, Pfeil E (1965) Angew Chem Int Ed Engl 4:1079
89. Danda H, Nishikawa H, Otaka K (1991) J Org Chem 56:6740
90. Reetz MT, Kyung SH, Bolm C, Zierke T (1986) Chem Ind 824
91. Hayashi M, Matsuda T, Oguni N (1990) J Chem Soc Chem Commun 1364
92. Hayashi M, Matsuda T, Oguni N (1992) J Chem Soc Perkin Trans 1 3135
93. Hayashi M, Miyamoto Y, Inoue T, Oguni N (1993) J Org Chem 58:1515
94. Narasaka K, Yamada T, Minamikawa H (1987) Chem Lett 2073
95. Minamikawa H, Hayakawa S, Yamada T, Iwasawa N, Narasaka K (1988) Bull Chem Soc Jpn 61:4379
96. Kobayashi S, Tsuchiya Y, Mukaiyama T (1991) Chem Lett 541
97. Dalton DM, Garner CM, Fernandez JM, Gladysz JA (1991) J Org Chem 56:6823
98. Garner CM, Fernandez JM, Gladyz JA (1989) Tetrahedron Lett 30:3991
99. Reetz MT, Kunisch F, Heilmann P (1986) Tetrahedron Lett 27:4721
100. Corey EJ, Wang Z (1993) Tetrahedron Lett 34:4001
101. Elliott JD, Choi VMF, Johnson WS (1983) J Org Chem 48:2294
102. Choi VMP, Eliott JD, Johnson WS (1984) Tetrahedron Lett 25:591
103. Reetz MT, Drewes MW, Harms K, Reif W (1988) Tetrahedron Lett 29:3295
104. Reetz MT, Drewes MW, Schmitz A, Holdgruen X, Wuensch T, Binder J (1988) Philos Trans R Soc London A 326:573
105. Herranz R, Castro-Pichel J, Garcia-Lopez T (1989) Synthesis 9:703
106. Garcia Ruano JL, Martin Castro AM, Rodriguez JH (1991) Tetrahedron Lett 32:3195
107. Effenberger F, Gutterer B, Ziegler T, Eckhardt E, Aichholz R (1991) Liebigs Ann Chem 47
108. Wang YF, Chen ST, Liu KKC, Wong CH (1989) Tetrahedron Lett 30:1917
109. Hsu SH, Wu SS, Wang YF, Wong CH (1990) Tetrahedron Lett 31:6403
110. Ohta H, Miyamae Y, Tsuchhashi G (1986) Agric Biol Chem 50:3181
111. Ohta H, Miyamae Y, Tsuchhashi G (1989) Agric Biol Chem 53:215
112. Bevinakatti HS, Banerji AA, Newadkar RV (1989) J Org Chem 54:2453
113. Ohta H, Kimura Y, Sugano Y, Sugai T (1989) Tetrahedron 45:5469
114. Ohta H, Kimura Y, Sugano Y (1988) Tetrahedron Lett 29:6957
115. Ohta H, Miyamae Y, Tsuchhashi G (1989) Agric Biol Chem 53:281
116. Ohta H, Hiraga S, Miyamoto K, Tsuchihashi G (1988) Agric Biol Chem 52:3023
117. Matsuo N, Ohno N (1985) Tetrahedron Lett 26:5533
118. van Almsick A, Buddrus J, Hönicke-Schmidt P, Laumen K, Schneider MP (1989) J Chem Soc Chem Commun 1391
119. Hirohara H, Mitsuda S, Ando E, Komaki R (1985) Stud Org Chem (Amsterdam) 22:119
120. Kanerva LT, Kiljunen E, Huuhtanen TT (1993) Tetrahedron: Asymmetry 4:2355
121. Inagaki M, Hiratake J, Nishioka T, Oda J (1991) J Am Chem Soc 113:9360
122. Inagaki M, Hatanaka A, Mimura M, Hiratake J, Nishioka T, Oda J (1992) Bull Chem Soc Jpn 65:111
123. Inagaki M, Hiratake J, Nishioka T, Oda J (1992) J Org Chem 57:5643
124. Rosenthaler L (1908) Biochem Z 14:238
125. Krieble VK, Wieland WA (1921) J Am Chem Soc 43:164
126. Albers H, Hamann K (1934) Biochem Z 255:44
127. Albers H, Hamann K (1934) Biochem Z 269:14
128. Becker W, Pfeil E (1966) J Am Chem Soc 88:4299

129. (a) Effenberger F, Ziegler T, Förster S (1987) Angew Chem 99:491; (b) Effenberger F, Ziegler T, Förster S (1987) Angew Chem Int Ed Engl 26:458
130. (a) Effenberger F, Ziegler T, Förster S, Degussa AG (1988) DE-B3 701 383; (b) Effenberger F, Ziegler T, Förster S, Degussa AG (1989) Chem Abstr 110:74,845r
131. Ziegler T, Hörsch B, Effenberger F (1990) Synthesis 575
132. Kragl U, Niedermeyer U, Kula MR, Wandrey C (1990) Ann N Y Acad Sci 613:167
133. (a) Niedermeyer U, Kula MR (1990) Angew Chem 102:423; (b) Niedermeyer U, Kula MR (1990) Angew Chem Int Ed Engl 29:386
134. (a) Bauer B, Strathmann H, Effenberger F (1991) DE-B 4041 896; (b) Bauer B, Strathmann H, Effenberger F (1992) Chem Abstr 116:150,138d
135. Brussee J, Loos WT, Kruse CG, van der Gen A (1990) Tetrahedron 46:979
136. Effenberger F, Eichhorn J, Roos J (1995) Tetrahedron: Asymmetry 6:271
137. Effenberger F, Eichhorn J (1997) Tetrahedron: Asymmetry 8:469
138. Effenberger F, Hörsch B, Weingart F, Ziegler T, Kühner S (1991) Tetrahedron Lett 32:2605
139. Wehtje E, Adlercreutz P, Mattiasson B (1990) Biotechnol Bioeng 36:39
140. Klibanov AM, Cambou B (1987) Methods Enzymol 136:117
141. Mori K, Ebata T, Takechi S (1984) Tetrahedron 40:1761
142. Redlich H, Xiang-jun J (1982) Liebias Ann Chem 717
143. Redlich H, Xiang-jun J, Paulsen H, Francke W (1981) Tetrahedron Lett 22:5053
144. Guzzi U, Ciabatti R, Padora G, Battaglia F, Cellentani M, Galliani G, Schiatti P, Spina G (1986) J Med Chem 29:1826
145. Effenberger F, Heid S (1995) Tetrahedron: Asymmetry 6:2945
146. Kiljunen E, Kanerva LT (1996) Tetrahedron: Asymmetry 7:1225
147. Kiljunen E, Kanerva LT (1997) Tetrahedron: Asymmetry 8:1551
148. Woker R, Champluvier B, Kula M-R (1992) J Chromatogr 584:85
149. Effenberger F, Hörsch B, Förster S, Ziegler T (1990) Tetrahedron Lett 31:1249
150. Kiljunen E, Kanerva LT (1996) Tetrahedron: Asymmetry 7:1105
151. Klempier N, Pichler U, Griengl H (1995) Tetrahedron: Asymmetry 6:845
152. Schmidt M, Herve S, Klempier N, Griengl H (1996) Tetrahedron 52:7833
153. Klempier N, Griengl H, Hayn M (1993) Tetrahedron Lett 34:4769
154. Ognyanov VI, Datcheva VK, Kyler KS (1991) J Am Chem Soc 113:6992
155. Wajant H, Pfitzenmaier K (1996) J Biol Chem 271:25,830
156. Hasslacher M, Kratky C, Griengl H, Schwab H, Kohlwein SD (1997) Proteins Struct Funct Genet 27:438
157. Zuegg J, Gugganig M, Wagner UG, Kratky C (1997) Abstract C-24, 3rd International Symposium on Biocatalysis and Biotransformations, 22–26 September Le Grand Motte, France
158. Brussee J, Roos EC, van der Gen A (1988) Tetrahedron Lett 29:4485
159. Hickel A, Graupner M, Lehner D, Hermetter A, Glatter O, Griengl H (1997) Enz Microb Technol 21:361
160. Selmar D, De Carvalho FJP, Conn EE (1987) Anal Biochem 166:208
161. Hörsch B (1990) PhD thesis, University of Stuttgart
162. Wehtje E, Adlercreutz P, Mattiasson B (1988) Appl Microbiol Biotechnol 29:419
163. Wehtje E, Adlercreutz P, Mattiasson B (1994) Biotechnol Bioeng 41:171
164. Reslow M, Adlercreutz P, Mattiasson B (1988) J Biochem 172:573
165. Voß H, Miehte P (1992) In: Tramper J, Vermae MH, Beet HH, Stockar UV (eds) Biocatalysis in non-conventional media: progress in biotechnology. Elsevier, London 8:739
166. Miethe P, Jansen I, Niedermeyer U, Kragl U, Haftendorn R, Kula M-R, Wandrey C, Mohr K-H, Meyer H-W (1992) Biocatal 7:61
167. Loos WT, Geluk HW, Ruijken MMA, Kruse CG, Brussee J, van der Gen A (1995) Biocatal Biotransform 12:255
168. (a) Geluk HW, Loos WT (1993) EP 0 547,655; (b) Geluk HW, Loos WT (1993) Chem Abstr 119:158,357g
169. Hickel A (1996) PhD Thesis, Technical University of Graz
170. Huuhtanen TT, Kanerva LT (1992) Tetrahedron: Asymmetry 3:1223

171. Kiljunen E, Kanerva LT (1994) Tetrahedron: Asymmetry 5:311
172. Danieli B, Barra C, Carrea G, Riva S (1996) Tetrahedron: Asymmetry 7:1675
173. Hickel A, Gradnig G, Griengl H, Schall M, Sterk H (1996) Spectrochim Acta Part A 52:93
174. Mao C-H, Anderson L (1967) Phytochemistry 6:473
175. (a) van Eikeren P (1993) US Patent 5,241,087; (b) van Eikeren P (1994) Chem Abstr 120:8336f
176. (a) Niedermeyer U (1993) DE 4 139 987; (b) Niedermeyer U (1993) Chem Abstr 119:70, 564m
177. Effenberger F, Schwämmle A (1997) Biocatal Biotransform 14:167
178. Menéndez E, Brieva R, Rebolledo F, Gotor V (1995) J Chem Soc Chem Commun 989
179. Kruse CG (1992) In: Collins AN, Sheldrake GN, Crosby J (eds) Chirality in industry. Wiley, New York, p 279
180. (a) Effenberger F, Stelzer U (1991) Angew Chem 103:866; (b) Effenberger F, Stelzer U (1991) Angew Chem Int Ed Engl 30:873
181. Effenberger F, Stelzer U (1993) Chem Ber 126:779
182. Warmerdam EGJC, Brussee J, Kruse CG, van der Gen A (1993) Tetrahedron 49:1063
183. Stelzer U, Effenberger F (1993) Tetrahedron: Asymmetry 4:161
184. Bömelburg J, Heppke G, Ranft A (1989) Z Naturforsch B 44:1127
185. Focella A, Bizzarro F, Exon C (1991) Synth Commun 21:2163
186. Kollonitsch J, Marburg S, Perkins LM (1979) J Org Chem 44:771
187. Alvernha GM, Ennakoua CM, Lacombe SM, Laurent AJ (1981) J Org Chem 46:4938
188. Hanessian S (1983) Total synthesis of natural products. Pergamon Press, New York
189. Effenberger F, Bukhardt U, Willfahrt J (1986) Liebigs Ann Chem 314
190. Lerchen HG, Kunz H (1985) Tetrahedron Lett 26:5257
191. Larcheveque M, Petit Y (1986) Synthesis 60
192. Kobayashi Y, Takemoto Y, Ito Y, Terashima S (1990) Tetrahedron Lett 31:3031
193. Whitesell JK, Reynolds D (1983) J Org Chem 48:3548
194. Kleemann A, Engel J (1982) Pharmazeutische Wirkstoffe, Synthesen, Patente, Anwendungen, 2nd edn and Supplement 1987
195. Matsuo N, Ohno N (1985) Tetrahedron Lett 26:5533
196. Brussee J, Dofferhoff F, van der Gen A (1990) Tetrahedron 46:1653
197. Krepski LR, Jensen KM, Heilmann SM, Rasmussen JK (1986) Synthesis 301
198. Jackson WR, Jacobs HA, Matthews BR, Jayatilake GS, Watson KG (1990) Tetrahedron Lett 31:1447
199. Effenberger F, Gutterer B, Ziegler T (1991) Liebigs Ann Chem 269
200. Zandbergen P, van den Nieuwendijk AMCH, Brussee J, van der Gen A (1992) Tetrahedron 48:3977
201. Brussee J, van der Gen A (1991) Recl Trav Chim Pays-Bas 110:25
202. Jackson WR, Jacobs HA, Jayatilake GS, Matthews BR, Watson KG (1990) Austr J Chem 43:2045
203. Brussee J, van Benthem RATM, Kruse CG, van der Gen A (1990) Tetrahedron: Asymmetry 1:163
204. Zandbergen P, Brussee J, van der Gen A, Kruse CG (1992) Tetrahedron: Asymmetry 3:769
205. Hayashi M, Yoshiga T, Nakatani K, Ono K, Oguni N (1994) Tetrahedron 50:2821
206. Matthews BR, Gountzos H, Jackson WR, Watson KG (1989) Tetrahedron Lett 30:5157
207. Chelucci G, Cabras MA, Saba A (1994) Tetrahedron: Asymmetry 5:1973
208. Warmerdam EGJC, van Rijn RD, Brussee J, Kruse CG, van der Gen A (1996) Tetrahedron: Asymmetry 7:1723
209. Warmerdam EGJC (1995) PhD Thesis, University of Leiden
210. Warmerdam EGJC, van den Nieuwendijk AMCH, Brussee J, Kruse CG, van der Gen A (1996) Tetrahedron: Asymmetry 7:2539
211. Hannich SM, Kishi Y (1983) J Org Chem 48:3833
212. Schmitz J (1992) PhD Thesis, University of Paderborn
213. Oehlschlarger AC, Mishra P, Dhami S (1984) Can J Chem 62:791
214. Abe A, Nitta H, Mori A, Inoue S (1992) Chem Lett 2443

215. Warmerdam EGJC, Brussee J, van der Gen A (1994) Helv Chim Acta 77:252
216. Warmerdam EGJC, van den Niewendijk AMCH, Kruse CG, Brussee J, van der Gen A (1996) Recl Trav Chim Pays-Bas 115:20
217. Johnson DV, Griengl H (1997) Tetrahedron 53:617
218. Johnson DV, Griengl H (1997) Abstract No P I-52, 3rd International Symposium on Biocatalysis and Biotransformations, 22–26 September, Le Grand Motte, France
219. Effenberger F, Jäger J (1997) J Org Chem 62:386

Glycosyltransferase-Catalyzed Synthesis of Non-Natural Oligosaccharides

Reinhold Öhrlein

S-507.702, NOVARTIS Pharma AG, CH-4002 Basle, Switzerland
E-mail: *reinhold.oehrlein@pharma.novartis.com*

Enzymes are indispensable tools in modern organic synthesis. Recent progress in cloning techniques, microbiology and protein purification supply an ever growing number and quantity of preparatively useful biocatalysts. Initially used to synthesize compounds found in nature, enzymes are now probed on non-natural substrates to an increasing degree. They work on highly functional and unprotected substrates under mild and environmentally friendly conditions. In addition, they exhibit an excellent chemo-, regio- and stereoselectivity. The following article will focus on the application of glycosyltransferases in the area of molecular glycobiology. Selected reactions with non-natural substrates will be discussed.

Keywords: Glycosyltransferases, Sugar nucleotides, Non-natural acceptors, Non-natural donors, Non-natural oligosaccharides.

Topics in Current Chemistry, Vol. 200
© Springer Verlag Berlin Heidelberg 1999

1
Introduction

Carbohydrates are the most widely and abundantly distributed class of bio-molecules in Nature [1]. They either form homo- or heteropolymers with them-selves [2] or they may be linked to various lipids and proteins to build O- and N-glycans [3, 4]. This diversity is reflected by their broad biological functions. Besides their importance as storage and supporting compounds, carbohydrates play key-roles in biological recognition and adhesion phenomena [5]. For example, they are involved in egg-sperm fertilization [6], cell differentiation during embryonic development [7, 8], or antigenic events [9, 10]. In addition, parasitic viruses, bacteria and toxins take advantage of various cell surface carbohydrates. They firmly attach to peripheral sugars before penetrating and damaging the cell [11–14]. Also malignant, metastasizing cells present and recognize specific cell surface carbohydrates [15]. Carbohydrates and structural analogues are therefore considered as potential drug candidates [16, 17]. Recent research efforts have been devoted to the leukocyte and influenza virus adhesions [18–21] and the xenotransplantation area [22, 23].

In order to carry out these investigations seriously, large numbers of natural and modified oligosaccharides have to be made available for high-throughput screenings. These numbers of compounds can hardly be supplied by classical glycoside synthesis, although a mature level with numerous protocols has been reached [24, 25a]. An alternative, or better a complementary strategy has been pioneered by Whitesides et al. [26] and Augé et al. [27] who explored the preparation of oligosaccharides with the help of biocatalysts. This review will focus on the use of Leloir-glycosyltransferases for the synthesis of non-natural oligosaccharide derivatives. The reader is referred to other excellent compilations for a more comprehensive survey [28–32].

2
Prerequisites of Enzymatic Glycosylations

The biocatalytic glycosylation of unprotected carbohydrates excels by its exclusive regio- and stereoselectivity. In addition, the absence of any protecting group manipulation simplifies the preparative routes towards oligosaccharides and reduces the workload and hence the synthetic costs for carbohydrate chemists as compared to chemical synthesis [25a]. However, there are two prerequisites which have to be fulfilled in order to be able to make the switch towards a biocatalytic method. The first prerequisite concerns the glycosyltransferases as the biocatalysts. For the synthetic chemist, one has to make oneself familiar with this particular class of biocatalysts. Glycosyltransferases are type II membrane bound glycoproteins consisting of a short N-terminal cytoplasmic domain, a transmembrane domain with a "stem" or "neck" region which is easily cleavable by proteases, and the C-terminal catalytic domain [25b]. Normally, the transferases are anchored to an internal cell membrane by the transmembrane region. However, in body fluids like blood or milk, functional transferases can be found which have been released by certain proteases by cleavage at the "neck" region. Such truncated enzymes are preferentially cloned and overexpressed in a number of cell types, in contrast to the full length transferases. Because a single transferase may show microheterogeneity with regard to chain length and glycosylation pattern due to the individual expression system. Therefore it is best purified by affinity chromatography.

There is no general rule concerning the stability of a transferase; e.g., $\beta(1-4)$GalT can be stored as a lyophilized powder and reconstituted in an appropriate buffer system before use. The transferases show optimum conversion rates at 37 °C and at a pH around 7. Some enzymes like the GalTs and FucTs require Mn^{2+} as a cofactor whereas SiaTs do not. The user should also pay attention to the different stabilities of the various transferases depending on the expression and purification protocols.

The second prerequisite of the Leloir-transferases are the necessary donor substrates which are nucleotide mono- or diphosphate sugars. Particularly in microorganisms, there seems to be an almost unlimited number of donor sugars occurring in Nature. However, mammals use only a highly restricted set of donors. The description of the preparative use of mammalian transferases in combination with non-natural donor sugars and non-natural acceptor substrates is the main topic of this article.

2.1
Leloir-Transferases

In mammalian systems, the Leloir-transferases play the central role in the biosynthesis of glycosidic bonds [33]. These enzymes transfer a monosaccharide unit from a nucleotide activated donor regio- and stereoselectively onto a specific OH-group of an acceptor sugar (Fig. 1).

Apart from the desired saccharide, a nucleotide side-product is produced which acts as a natural inhibitor on the respective transferase. To shift the equilibrium to the oligosaccharide product the nucleotide has to be removed in vitro. This can be accomplished by the addition of an alkaline phosphatase to the incubation mixture [34a], or by recycling the glycosyl donor if it is a natural compound [34b].

Well over a hundred different transferases [E.C.2.4. ...] have been described so far [31]. Only a limited number is commercially available for preparative use, and others have to be isolated from natural sources [28].

2.2
Donor Substrates

Surprisingly, mammalian cells are able to produce the vast diversity of oligosaccharide structures with only eight sugar nucleotide building blocks (Fig. 2). Following transfer, the sugars may be even further modified by other carbohydrate processing enzymes [35].

A few of these natural donors are commercially available in larger quantities [36] but most of them can be synthesized following original protocols. A selection of more recent preparations is given in Table 1.

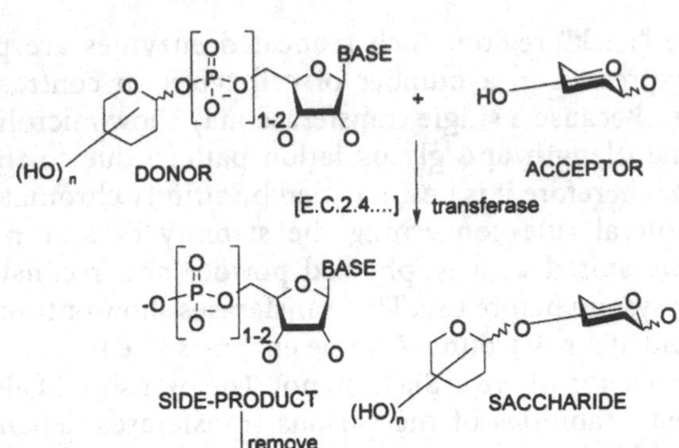

Fig. 1. General scheme for enzymatic glycosylations by Leloir-transferases

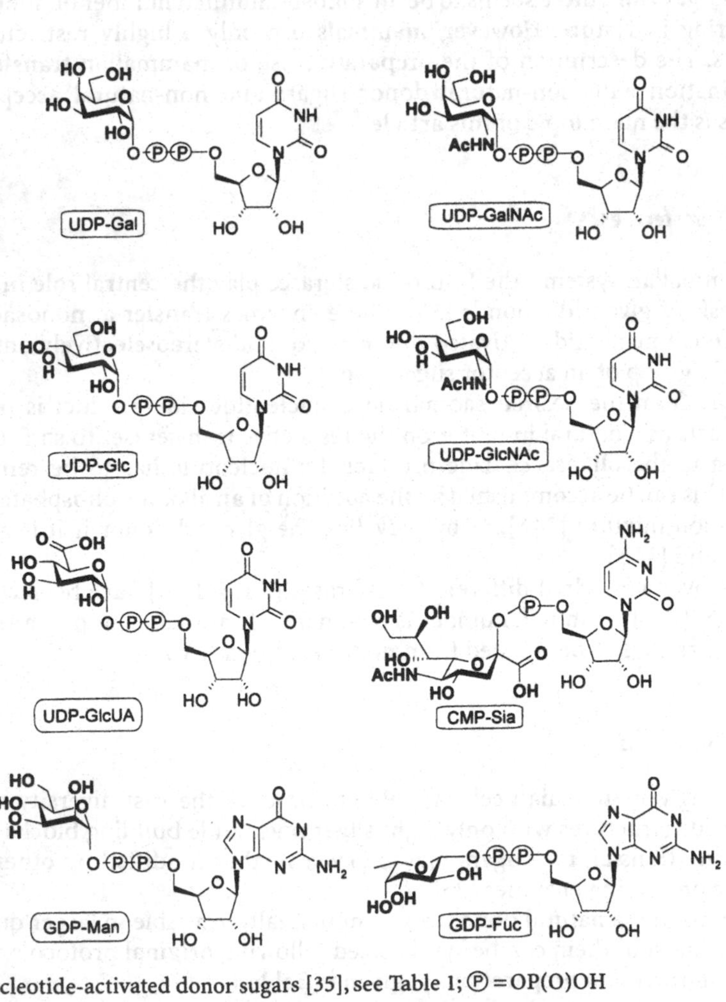

Fig. 2. Nucleotide-activated donor sugars [35], see Table 1; ℗ = OP(O)OH

Table 1. Useful protocols for the synthesis of nucleotide-activated donor sugars

Donor	Reference	Donor	Reference
CMP-Sia	[37–39]	UDP-GlcUA	[35, 44]
GDP-Fuc	[40–42]	UDP-Glc	[45]
GDP-Man	[41, 43]	UDP-GalNAc	[46]
UDP-Gal	[35, 36, 41]	UDP-GlcNAc	[35]

In some cases the natural nucleotide-activated donors can be generated in situ prior to the enzymatic transfer [28, 34b, 47]. A prerequisite for this technology, however, is the availability of an arsenal of enzymes, e.g. aldolases, synthetases, phosphokinases, nucleoside monophosphate kinases, pyruvate kinases, epimerases etc. Although some of these enzymes accept non-natural substrates, their one-pot use for recycling non-natural substrates in a multi-enzyme sequence is still an area of rewarding research. The references to non-natural donors will be given below, together with their use in the respective glycosylation reactions.

3
Enzymatic Glycosylations

3.1
Galactosyltransferases

3.1.1
β(1–4)Galactosyltransferase

To date commercial β(1-4)galactosyltransferase from bovine milk – β(1-4)GalT – is the most thoroughly studied transferase with respect to its scope of the acceptor- and donor-specificities [28, 48]. The enzyme transfers a D-galactose unit from UDP-Gal onto the 4-OH-group of a terminal N-acetylglucosamine acceptor in a β-mode to form N-acetyllactosamine (Fig. 3). In the presence of α-lactalbumine, the preferred acceptor is glucose to give lactose, respectively.

The aglycon moiety is the most extensively varied residue of the acceptor structure. The transferase tolerates almost any derivative as long as it is β-linked to the N-acetylglucosamine and fairly soluble in the incubation mixture. Selected examples are listed in Table 2. Besides compounds with highly lipophilic O-aglycons (entries 2–8), S-aglycons (entry 10) have also been found to be good acceptor substrates for β(1-4)GalT using standard glycosylation conditions [49]. This method is also applicable for the rapid and stereochemically unambiguous synthesis of di- and tri-antennary carbohydrate structures (entries 11–13). The enzymatic galactosylation of various glycopeptide entities (entries 15–17) was also easily achieved on a preparative scale, including that of multiantennary compounds which mimic natural glycopeptide structures [50]. In some cases (see Table 2) two substituents of the natural N-acetylglucosamine

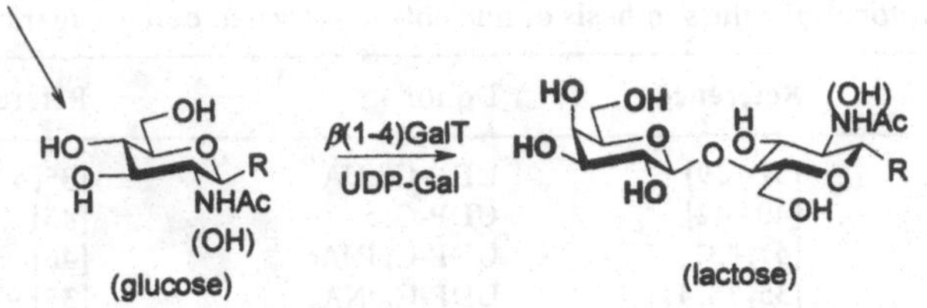

Fig. 3. Galactosylation with $\beta(1-4)$galactosyltransferase

Table 2. $\beta(1-4)$Galactosylations of glucosamides with various β-linked aglycons

Entry	Aglycon	[%]	Ref.	Entry	Aglycon	[%]	Ref.
1	-OH	96 [b]	[51]	2	-O(CH$_2$)$_8$COOCH$_3$	95 [b]	[52]
3	-OCH$_2$CH=CH$_2$	93 [b]	[51]	4	-O(CH$_2$)$_3$CH=CH$_2$	86 [b]	[51]
5	-O(CH$_2$)$_{11}$CH$_3$	42 [b]	[51]	6	-OCH$_2$CH$_2$SiMe$_3$	98 [b]	[53]
7	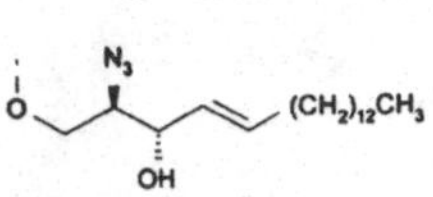 (structure)	39 [b]	[51]	8	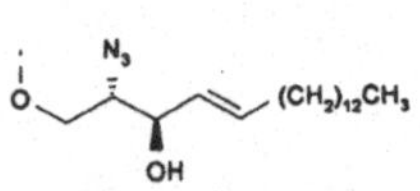 (structure)	87 [a]	[54]
9	-Osugar	68 [b]	[55]	10	-SPh	60 [a]	[49]
11	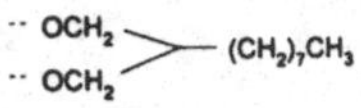 (structure)	53 [b]	[56]	12	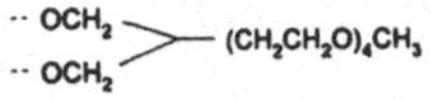 (structure)	77 [a]	[56]
13	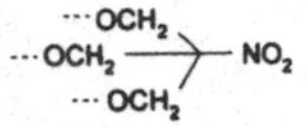 (structure)	38 [b]	[56]	14	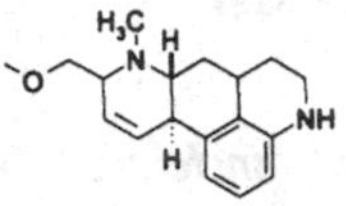 (structure)	65 [a]	[57]
15	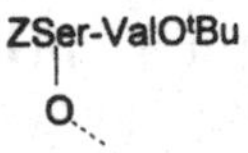 ZSer-ValO^{t}Bu	59 [a]	[58]	16	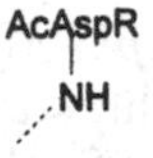 AcAspR	96 [a]	[50]
17	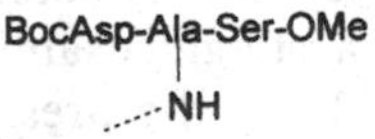 BocAsp-Ala-Ser-OMe	35 [a]	[59]	18	glucosteroids	55 [a]	[60]

[a] acceptor GlcNAc.
[b] acceptor GlcNHCOOCH$_2$CH$_2$=CH$_2$; according to Fig. 3.

acceptor had been exchanged; in addition to the aglycon part the N-acetyl group can be replaced by the synthetically more versatile allyloxycarbonyl residue [51].

Permissible substitutions of the natural N-acetyl group have been further investigated (Table 3) [52]. For example, this group can be substituted by thio-amides or sulfonamides (entries 6, 7, 14 – 16). Unexpectedly, positive or negative charges on the N-acyl chain are also tolerated, such as a guanidinium or sulfate group (entries 5, 10, 20, 22). The natural N-acetyl group can be supplanted by heterocycles that closely resemble the uridine part of the donor component without adverse effects (entry 9). The stereoselective galactosylation of the

Table 3. Enzymatic $\beta(1-4)$galactosylations of various *N*-acyl glycosamides with aglycon-O-$(CH_2)_8COOMe$, according to Fig. 3 [52, 56, 61]

Entry	Acyl residue	Yield % (mg)	Entry	Acyl residue	Yield % (mg)
1	- C(O)CH₃	95 (35)	14	- C(S)NHCH₂CH₃	98 (40)
2	- C(O)H	88 (28)	15	- S(O)₂CH₃	87 (30)
3	- C(O)CH(CH₃)₂	90 (31)	16	- S(O)₂toluyl	82 (16)
4	- C(O)CH₂OH	84 (35)	17	- C(O)Ph	81 (16)
5	- C(O)CH₂NH₂	75 (66)	18	- C(O)CH₂SH	29 (7)
6	- C(S)CH₃	100 (29)	19	- C(O)OCH₂CH=CH₂	100 (18)
7	- C(S)OCH₂CH₃	85 (19)	20	- C(O)CH₂SO₃Na	77 (39)
8		74 (19)	21		78 (16)
9		71 (35)	22		73 (29)
10		65 (30)	23		37 (14)
11		53 (11)	24		58 (17)
12		61 (16)	25		40 (12)
13		96 (37)	26		83 (27)

highly polar and bulky glycuronamide derivatives (entries 11–13, 24, 25) is surprising as well. None of the investigated sugar amides [61] inhibits the $\beta(1-4)$GalT, and galactosyltransfer occurs exclusively as expected onto the 4-OH-group of the glucosamide moiety.

The $\beta(1-4)$GalT has been reported to tolerate modifications at any OH-group of the galactose moiety within the "UDP-Gal donor", including the ring oxygen [28, 48]. Accordingly, the galactose ring oxygen (Fig. 2) has been replaced by sulfur [48], and various hydroxyl groups have been replaced by hydrogen [48] or fluoride [62]. It has also been demonstrated that donors like UDP-GalN, UDP-GalNAc and UDP-Glc are transferred by the enzyme in the expected manner [48, 63]. Very interesting is the recent finding that UDP-5SGalNAc is utilized for transfer to a glucosamide derivative by $\beta(1-4)$GalT [64]. There are also reports which document the stereoselective galactosylation of acceptors having only a distant structural relationship to the natural N-acetylglucosamine acceptor (Fig. 4). Compounds A, B, C [65] and the C-glycoside D [66] are β-galactosylated at the expected positions in 20–40% yield according to the standard protocol [67], whereas compound E and related ones experience frame-shifted galactosylations [47]. An interesting new application was shown by Wang and co-workers [68]. Upon incubation of $\beta(1-4)$GalT with racemic (±)-conduritol B and UDP-Gal, exclusive galactosylation of the (–)-isomer was observed which has been used to resolve the enantiomers. Although most of the non-natural substrates cited above have pretty low transfer rates, preparatively useful amounts of the desired galactosides have been produced.

A variety of non-mammalian $\beta(1-4)$GalTs are widely distributed in Nature. For example, a cell free extract from *S. hygroscopicus* has been used to galactosylate macrolide antibiotics [69]. Notably, a number of solid-phase galactosylations of immobilized substrates have also been probed successfully [70–72].

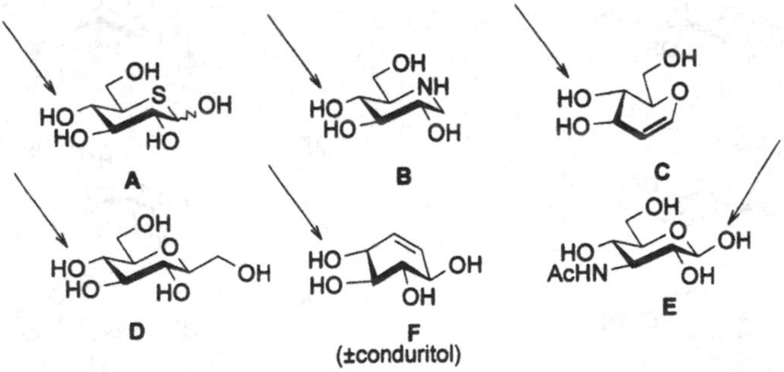

Fig. 4. Non-natural galactose acceptors (*arrows* indicate the site of the galactosyltransfer)

3.1.2
$\beta(1-3)$Galactosyltransferases

The core I $\beta(1-3)$GalT has been isolated from various animal tissues. By transferring a galactose unit in a β-mode onto the 3-OH-group of a terminal, α-linked N-GalNAc (Fig. 5) the enzyme forms the T-antigen (core I). Overproduction of

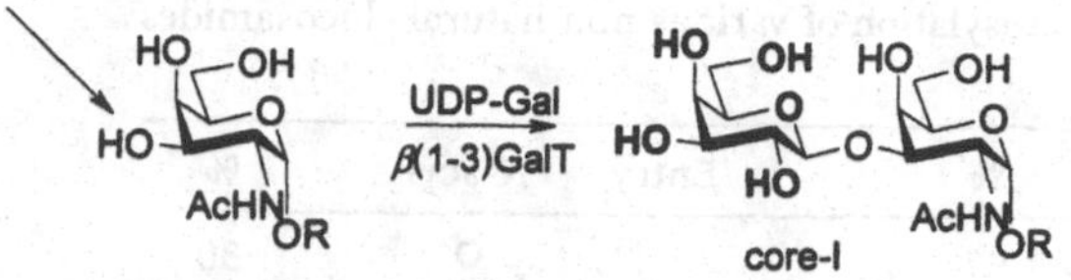

Fig. 5. Enzymatic formation of core-I disaccharide

the latter is characteristic of cancer cells. The rat liver enzyme has been shown to tolerate minor modifications on the acceptor C-6 carbon, and a number of peptides are tolerated at the aglycon position apart from the natural O-linked serine or threonine residues [73].

A rat $\beta(1\text{-}3)$GalT (E.C.2.4.1.62) of different specificity has been cloned and overexpressed in mouse melanoma B16 cells [74]. This enzyme transfers a galactose unit from UDP-Gal onto the 3-OH-group of a GalNAc acceptor in a β-mode to form glycolipids of the ganglioside family, such as $G_{D1b}/G_{M1}/G_{A1}$. The enzyme can be expected to be useful for the synthesis of a variety of non-natural gangliosides and glycolipids but for that purpose has not yet been probed on a preparative scale.

A further $\beta(1-3)$GalT, distinct from the enzyme in melanoma cells, from human brain tissue has been cloned and overexpressed successfully in insect cells [75]. This enzyme transfers a galactose unit from UDP-Gal onto the 3-OH-group of a terminal N-acetylglucosamine in a β-mode to form the type-I (or Lewis^c) disaccharide (Fig. 6).

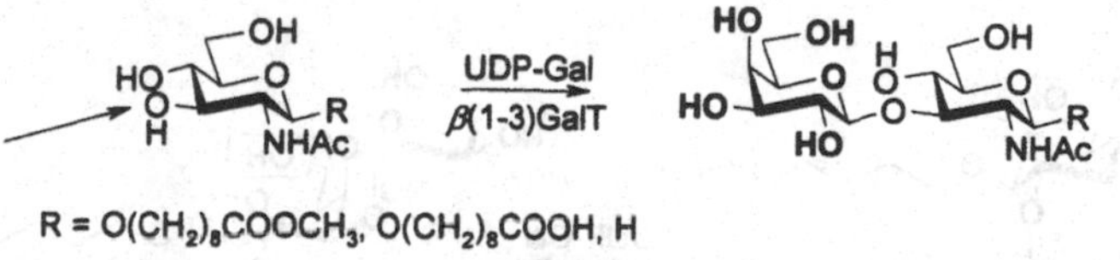

$R = O(CH_2)_8COOCH_3, O(CH_2)_8COOH, H$

Fig. 6. Enzymatic formation of Lewis^c-type disaccharides

A limited study conducted with this enzyme shows its high synthetic potential. Various non-natural N-acyl residues (Table 4) are tolerated by this transferase, despite their close proximity to the site of enzymatic galactosylation [76]. A more thorough investigation may unveil an even broader synthetic scope of this biocatalyst.

3.1.3
$\alpha(1\text{-}3)$Galactosyltransferases

Only two α-GalTs have been more closely explored up to now [28]. The human blood group B transferase (E.C. 2.4.1.37), which is responsible for the synthesis of the blood group B trisaccharide, transfers a galactose unit from UDP-Gal in

Table 4. $\beta(1-3)$Galactosylation of various non-natural glucosamides according Fig. 6

Entry	N-acyl	%	Entry	N-acyl	%
1	 (acetyl)	97	4	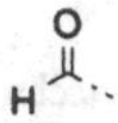 (formyl)	30
2	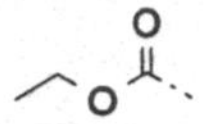 (ethoxycarbonyl)	34	5	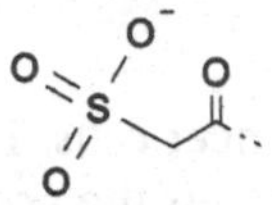 (sulfonyl)	51
3	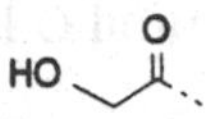 (HO-acetyl)	61	6	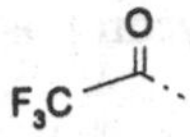 F_3C-acetyl	59

an α-mode onto the 3-OH-group of a terminal galactose acceptor carrying an α-linked fucose at its 2-OH position (Fig. 7).

When studying the minimum structural requirements of an enzyme preparation from human blood, Hindsgaul and coworkers [77, 78] found that the enzyme tolerates only minor changes at the 6-position of the acceptor galactose. Although the relative transfer rates for the non-natural acceptors are far lower than for the parent compound, the enzyme may be applicable for the preparative synthesis of non-natural blood group B derivatives (cf. Fig. 7 and Table 5). Modifications at the fucose moiety have not been tested.

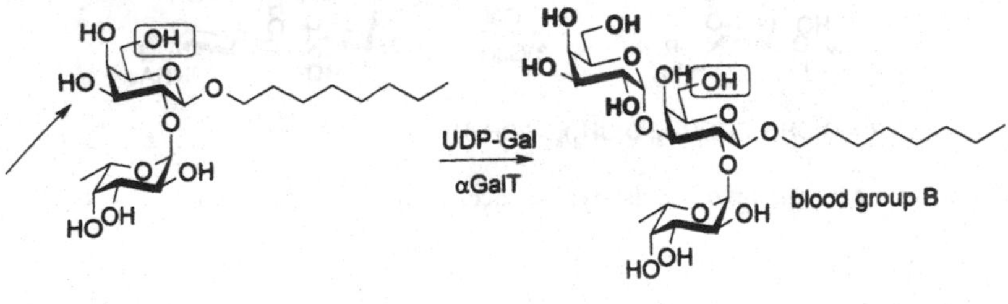

Fig. 7. Enzymatic formation of blood group B trisaccharide (compare Table 5 for structural modifications of acceptor)

Table 5. $\alpha(1-3)$Galactosylations of various non-natural acceptors; modified groups as indicated in Fig. 7

Residue	Rel. rate	Residue	Rel. rate
6-H	35%	6-OMe	3.2%
6-F	43%	6-NH$_2$	2%

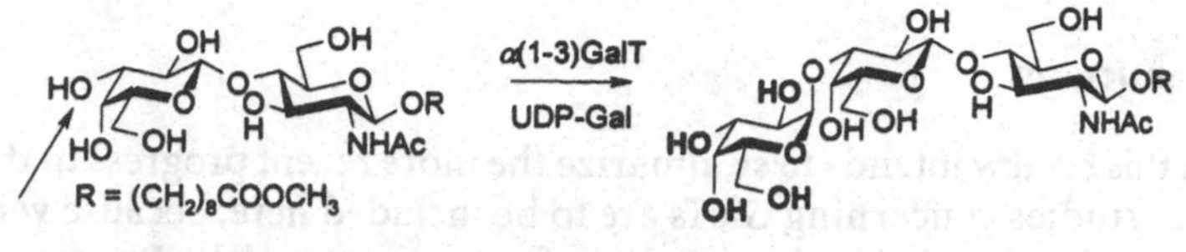

Fig. 8. Enzymatic formation of linear-B trisaccharide

A second mammalian α-GalT (E.C.2.4.1.141) attracted the attention of the glycobiologists. This enzyme catalyzes the α-specific transfer of a galactose unit onto the 3-OH group of a terminal galactose to form the so called linear-B-trisaccharide (Fig. 8).

The trisaccharide has been found to be responsible for hyperacute rejections in the xenotransplantation area [10, 79]. The pig enzyme has been cloned and overexpressed in insect cells, and is currently being evaluated for preparative use [80]. It tolerates a wide range of non-natural acyl replacements of the N-acetyl group of the penultimate GlcNAc moiety (Table 6) [81].

Table 6. α(1-3)Galactosylations with various non-natural N-acyl-glucosamides according Fig. 8: alloc = allyloxycarbonyl; Z = benzyloxycarbonyl

Entry	Acyl	Yield %	(mg)	Entry	Acyl	Yield %	(mg)
1	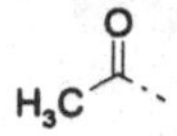 H₃C	38	12.4	7	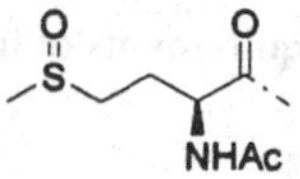 (S, NHAc)	84	23.9
2	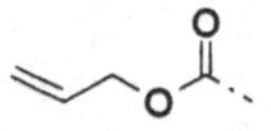 (allyl-O)	44	13.1	8	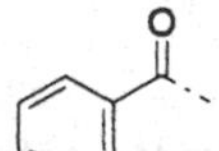 (phenyl)	64	7.2
3	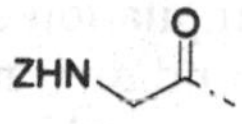 ZHN	42	21.8	9	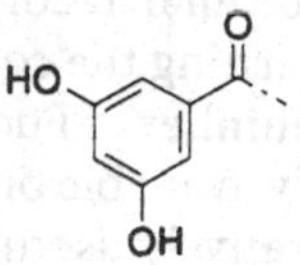 HO, OH	69	17.1
4	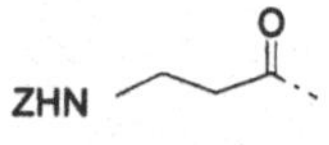 ZHN	49	8.5	10	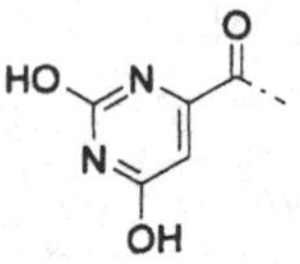 HO, N, N, OH	27	6.6
5	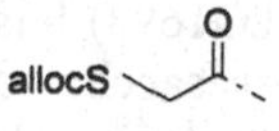 allocS	61	21.7	11	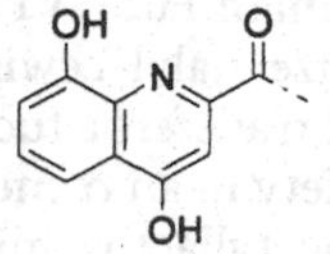 OH, N, OH	64	10.4
6	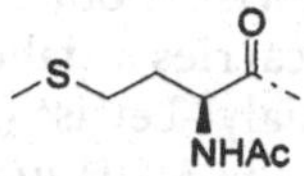 S, NHAc	75	13.8				

3.2
Glucosyltransferases

Although this review intends to summarize the more recent progress in the area, some older studies concerning GlcTs are to be included here, because very little attention has been paid to these types of enzymes recently. Elegant studies exploring modifications of the UDP-Glc donor (Fig. 9) have been performed early on [82]. The investigated GlcTs accept UDP-Glc donors with modifications at the 2-OH-group of the glucose moiety, whereas only minor changes are allowed at the 4- and 6-OH-groups. The 3-OH-group of the donor has been found to be a prerequisite feature for their recognition by the transferases. Unfortunately, these results have mostly been neglected in more recent summaries, although they were seminal for the preparative exploitation of glucosyltransferases.

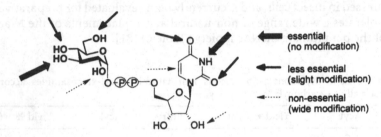

Fig. 9. Functional requirements of investigated glucosyltransferases [82]

3.3
Fucosyltransferases

Fucosides and *fuco*syltransferases – FucT – play key roles in the regulatory pathways of intercellular recognition events [11]. A recent compilation should be consulted concerning the confusing nomenclature of FucTs and their natural acceptors [83]. A number of FucTs have been cloned and overexpressed but none is yet commercially available on a scale for preparative usage. Some rather unexpected and preparatively useful findings from studies with human enzymes are included here [28, 48].

3.3.1
Fucosyltransferase VI

Recombinant human FucT VI [84, 85] (EMBL accession no. LO1698) has been used to synthesize sialyl-Lewis^x and a number of related tetrasaccharides. In vivo, the enzyme transfers a fucose unit from GDP-Fuc [42] onto the 3-OH group of a GlcNAc moiety in an α-mode (Fig. 10).

When an N-acetyllactosamine acceptor is fucosylated Lewis^x is obtained. If the terminal galactose of the N-acetyllactosamine acceptor carries a sialic acid, the fucosylation of the N-acetylglucosamine unit leads to sialyl-Lewis^x (as depicted in Fig. 10) [86, 87]. In consequence, this means that the 3-OH group of

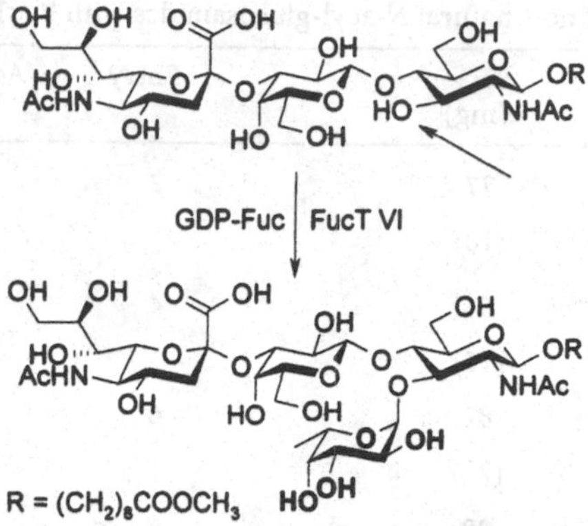

Fig. 10. Enzymatic $\alpha(1$-$3)$fucosylation with FucT VI

the galactose unit will be open to wide modifications. We have explored the usefulness of recombinant FucT VI for the preparation of sialyl-Lewis^x derivatives that have the natural N-acetyl group of the glucosamine acceptor replaced. Although this group is located in the immediate proximity of the OH-group to be fucosylated, an unexpectedly broad range of non-natural N-acyl residues is tolerated by the enzyme (Table 7) [61, 88].

The carbonyl group can be replaced by a thiocarbonyl group (entries 5, 6) or by a sulfonamide (entry 24). The enzyme also accepts the presence of negative (entry 10) or positive charges (entry 11, 13) on the glucosamine N-acyl chain. The natural N-acetyl group can as well be replaced by bulky aromatic or heteroaromatic amides (entries 12, 15–18). Even an exchange of the N-acetyl functionality by highly polar and bulky glycuronamides (entries 19–23) is tolerated. Irrespective of the glycuronamide structures, exclusive α-fucosylation of the glucosamide 3-OH-group is observed, under standard incubation conditions [67].

It is an unexpected finding that the N-acetylglucosamine moiety can be replaced completely (Fig. 11). However, the examples **C** and **D** show the selective fuco-

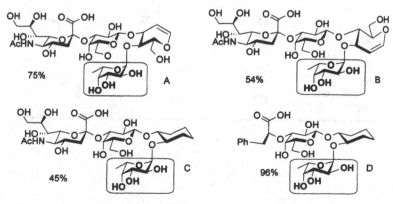

Fig. 11. Enzymatic fucosylations of glycomimetics

Table 7. Fucosylations of non-natural N-acyl-glucosamides with FucT VI (Fig. 10) [61, 88]

Entry	Acyl	% (mg)	Entry	Acyl	% (mg)
1	(formyl structure)	77 (18)	2	(isobutyryl structure)	78 (15)
3	(glycolyl structure)	64 (15)	4	(trifluoroacetyl structure)	96 (26)
5	(thioacetyl structure)	82 (21)	6	(N-ethyl thioamide structure)	50 (7)
7	(allyl carbonate structure)	88 (60)	8	(S-ethyl thioester structure)	50 (13)
9	(N-ethyl amide structure)	83 (8)	10	(SO₃Na structure)	45 (11)
11	(amino structure)	60 (14)	12	(benzyl ester structure)	89 (15)
13	(guanidino structure)	69 (16)	14	(S-allyloxycarbonyl structure)	78 (14)
15	(dihydroxyphenyl ketone structure)	73 (25)	16	(hydroxyquinoline carbonyl structure)	78 (13)
17	(hydroxyquinoline structure)	89 (10)	18	(dihydroxypyrimidine carbonyl structure)	64 (12)
19	(sugar OMe structure)	60 (11)	20	(sugar OMe structure)	32 (6)
21	(sugar OMe structure)	25 (5)	22	(sugar OMe structure)	35 (5)
23	(sugar OMe structure)	73 (8)	24	(methylsulfonyl structure)	91 (23)

sylation at the only OH-group of the glycosylated cyclohexanediol as a N-acetyl-glucosamine mimetic with natural α-selectivity. As example **D** illustrates, the sialic acid moiety can even be substituted by a phenyl-lactic acid residue [56]. This compound only very distantly resembles a natural acceptor, but nonetheless becomes fucosylated in the desired fashion in high yield.

A number of non-natural fucosyl donors have also been probed with this enzyme [89]. As can be seen from Fig. 12, the C-6-atom of fucose is open for modifications. L-Gal and D-Ara are good substrates, whereas replacement of the 2-OH group of the fucose donor apparently is not tolerated by FucT VI. Indeed, Hindsgaul and coworkers succeeded in attaching the blood group A trisaccharide to the 6-position of the fucose donor and proved the biocatalytic transfer of this strange "sugar" [90]. The combination of non-natural donors with non-natural acceptors by enzymatic fucosyltransfer has also been probed which proved to be instrumental in the assembly of a library of sialyl-Lewis^x tetrasaccharides (Fig. 12) [91].

Selected combinations are listed in Table 8. The examples show the synthetic efficiency by which FucT VI attaches non-natural D-Ara or L-Gal donors (non-natural with respect to the transferase) selectively to the 3-OH-group of a series of glucosamide acceptors carrying non-natural aromatic amides.

Table 8. Transfer of non-natural sugars onto non-natural N-acyl glucosamide acceptors (compare Fig. 12)

Entry	Acyl	Sugar	% (mg)	Entry	Acyl	Sugar	% (mg)
1		L-Gal	66 (18.0)	7		D-Ara	81 (10.0)
2		D-Ara	66 (8.0)	8		L-Gal	68 (8.0)
3		L-Gal	71 (13.0)	14		L-Gal	60 (5.0)
4		D-Ara	54 (6.0)	10		D-Ara	87 (11.0)
5		L-Gal	82 (14.0)	11		L-Gal	70 (9.0)
6		D-Ara	46 (8.0)	12		L-Gal	53 (7.0)
				13		D-Ara	62 (8.0)
9		L-Gal	77 (6.0)				

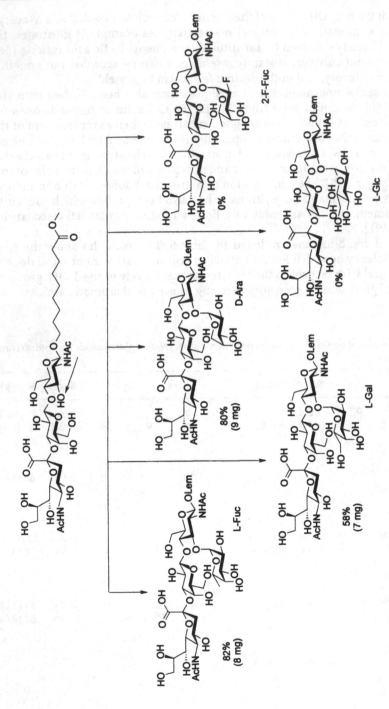

Fig. 12. Fucosylations with FucT VI and non-natural donors

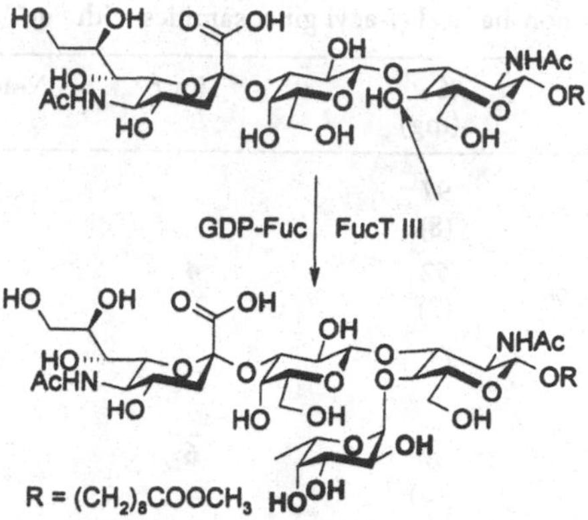

Fig. 13. Enzymatic fucosylations with FucT III to give sialyl-Lewis[a]

3.3.2
Fucosyltransferase III

The FucT III transfers an L-fucose unit from GDP-fucose onto the 4-OH-group of a galactosylated N-acetylglucosamine in an α-mode to give the Lewis[a] trisaccharide or the sialyl-Lewis[a] tetrasaccharide, respectively (Fig. 13).

Using an enzyme isolated from human milk, key polar groups on the acceptor and donor substrates have been decoded [48,92]. The availability of larger quantities of the recombinant form of fucosyltransferase III (Lewis-enzyme, EMBL accession no. X 53578) [93] allowed evaluation of the enzyme for preparative use with non-natural substrates [56,89,94,95].

The replacement of the N-acetyl group of the N-acetylglucosamine unit is widely tolerated (Table 9). Especially striking is again the selective fucosylation of the glycuronamide derivatives (entries 11–15). Heterocyclic substituents on the acceptor apparently do not affect the enzyme either (entries 8–10). Non-natural fucose donors are also recognized by FucT III and transferred in the expected way to form Lewis[a] trisaccharides or sialyl-Lewis[a] tetrasaccharides, respectively (see Fig. 14).

It goes without saying that the combined use of non-natural acceptors and non-natural donors has subsequently been explored (see Table 10) [96].

In contrast to FucT VI, the FucT III transfers L-glucose (entry 4), the charged 2-aminofucose [42] (entry 15) and 2-fluorofucose (entry 2). This is particularly noteworthy with respect to results obtained by Wong and coworkers who have observed that GDP-2-fluorofucose is a potent inhibitor for the functionally closely related FucT V [97].

These examples confirm a greater substrate flexibility of FucT III as compared to FucT VI in their recognition of non-natural substrates. Thus, a large library of sialyl-Lewis[a] derivatives could be easily assembled [96].

Table 9. Fucosylations of non-natural N-acyl glucosamides with FucT III (see Fig. 13)

Entry	N-acyl residue	% (mg)	Entry	N-acyl residue	% (mg)
1	(acetyl structure, CH₃)	97 (8)	2	(allyl carbonate structure)	78 (9)
3	(structure with OH, OH)	52 (7)	4	(structure with OH, F)	100 (6)
5	(structure with OH, OH)	87 (15)	6	(structure with OH, OH)	56 (7)
7	(structure with OMe, OH, OMe)	70 (12)	8	(structure with N, OH, N, OH)	80 (13)
9	(quinoline structure, HO, N)	52 (9)	10	(quinoline structure, HO, N, OH)	76 (20)
11	(sugar structure, HO, HO, OMe, OH)	82 (19)	12	(sugar structure, HO, HO, OMe, OH)	63 (19)
13	(sugar structure, HO, HO, HO, OMe)	79 (20)	14	(sugar structure, OMe)	56 (12)
15	(sugar structure, OMe, OH, OH)	75 (9)			

3.3.3
Purine-Diphosphate-Fucoses

The search for potent and selective fucosyltransferase inhibitors stimulated more profound investigations with non-natural fucose-donors. Purine base-modified fucose donors were synthesized and incubated with FucT III or FucT VI and their respective type-I or type-II acceptors [98]. Surprisingly, both enzymes tolerate an exchange of the guanine by other purine bases having different hydrogen bonding capabilities. These substrate analogs are handled by the transferases like the natural GDP-fucose as is evident by the good isolated product yields (Fig. 15 and Table 11). However, neither of the transferases accepts GTP-fucose.

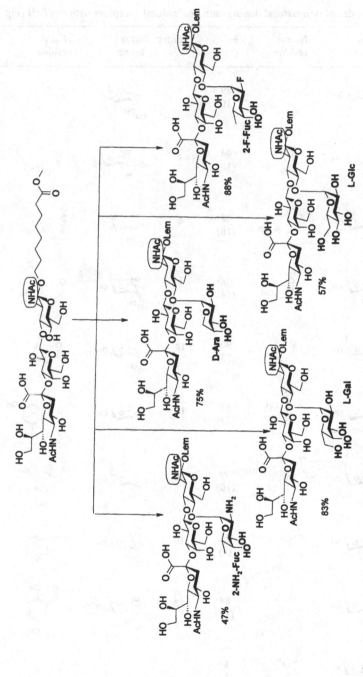

Fig. 14. Enzymatic fucosylations with FucT III and non-natural donors

Table 10. Transfer of non-natural donors onto non-natural acceptors with FucT III (Fig. 14)

Entry	Sugar donor	N-acyl residue	% (mg)	Entry	Sugar donor	N-acyl residue	% (mg)
1			89 (10)	2			73 (11)
3			94 (10)	4			48 (5)
5			82 (18)	6			77 (27)
7			100 (14)	8		SO_3Na	81 (12)
9		NH_2	96 (13)	10			56 (14)
11		$-CH_3$	67 (12)	12			64 (9)
13		OMe	60 (10)	14			72 (10)
15		NH_2	32 (8)	16			87 (18)
17		CH_3	84 (14)	18		CH_3	84 (14)
19			84 (11)				

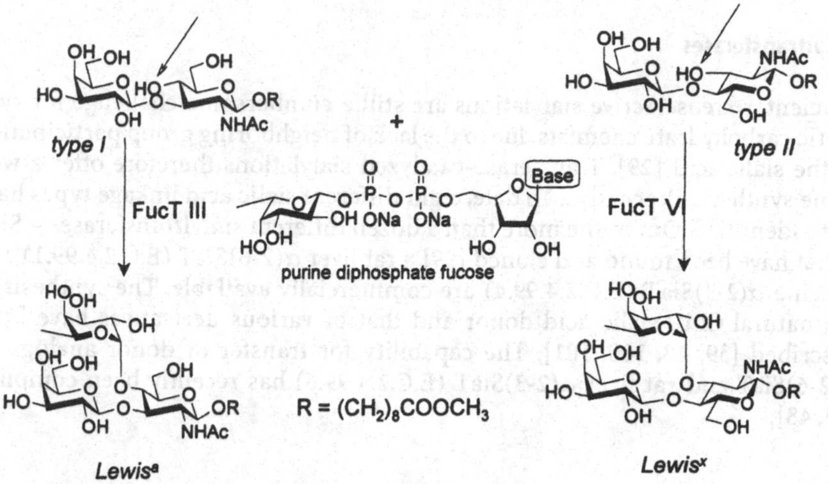

Fig. 15. Probing purine diphosphate fucoses (see also Table 11)

Table 11. Transfer from differently nucleotide activated fucoses with FucT III and FucT VI to form Lewis^a and Lewis^x (according to Fig. 15)

Entry	Nucleotide	Base	Yield of Lewis^a [%]	Yield of Lewis^x [%]
1	GDP	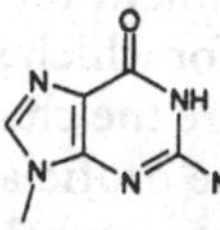	96	83
2	ADP	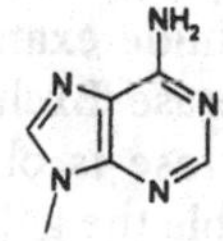	76	60
3	XDP	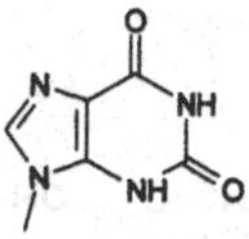	73	62
4	IDP	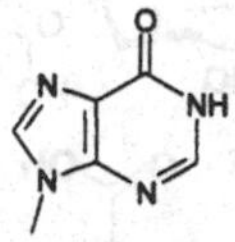	68	72
5	GTP	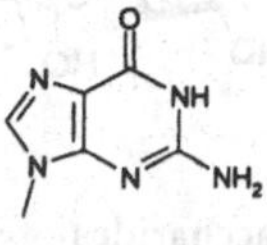	0	0

3.4
Sialyltransferases

Efficient, stereoselective sialylations are still a cumbersome challenge for synthetic carbohydrate chemists due to the lack of neighboring group participation of the sialic acid [29]. Transferase-catalyzed sialylations therefore offer a welcome synthetic alternative. To date, eight different sialic acid linkage types have been identified. Out of the more than a dozen different *sialyltransferases* – SiaT – that have been found and cloned [99] a rat liver α(2-6)SiaT (E.C.2.4.99.1) and porcine α(2-3)SiaT (E.C.2.4.99.4) are commercially available. The synthesis of the natural CMP-sialic acid donor and that of various derivatives have been described [39, 48, 100, 101]. The capability for transfer of donor analogs by α(2-6)SiaT and rat liver α(2-3)SiaT (E.C.2.4.99.6) has recently been compiled [28, 48].

3.4.1
α(2-3)Sialyltransferases

The rat liver α(2-3)SiaT (EMBL accession no. M97754) has been cloned and overexpressed in COS cells [102]. The enzyme transfers a sialic acid unit from CMP-sialic acid onto the 3-OH-group of a terminal β-linked D-galactose in an α-mode. The galactose acceptor itself can be attached either to the 3-OH-group (type I) or the 4-OH-group (type II) of a N-acetylglucosamine subunit (Fig. 16).

It has been found that the N-acetyl group of the GlcNAc moiety on both the type I and type II acceptors is not a key requirement for the rat enzyme [48]. Extensive variations at this position are allowed for which selected examples are listed in Table 12 [61, 95, 103, 104]. Noteworthy are the charged (entries 4, 6, 13) and the sulfonamide (entry 9) replacements. The orotic acid amide (entry 15) does not show inhibitory effects despite its close structural similarity to the cytidine part of the donor (Fig. 2). The glycuronamide examples (entries 16–20) underline the synthetic versatility of the transferase. Exclusive transfer of sialic acid to the 3-OH-group of the terminal galactose is observed, although the galacturonamides (entries 17, 18) closely resemble the galactose acceptor.

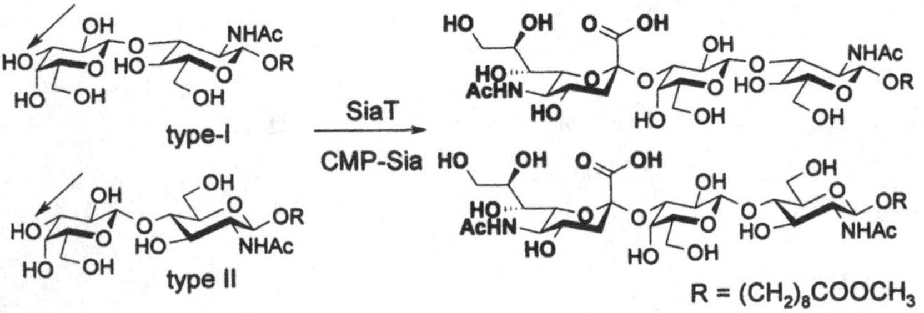

Fig. 16. Enzymatic sialylation of type I and type II disaccharides

Table 12. Enzymatic α(2-3)sialylations of various N-acyl-type I disaccharides with aglycon-O-$(CH_2)_8COOMe$; compare Fig. 16

Entry	Acyl residue	Yield % (mg)	Entry	Acyl residue	Yield % (mg)
1	- C(O)CH₃	75 (81)	2	- C(O)CF₃	89 (39)
3	- C(O)H	55 (10)	4	- C(O)CH₂NH₂	59 (12)
5	- C(O)CH₂OH	79 (14)	6	- C(O)CH₂SO₃Na	53 (13)
7	- C(O)OCH₂CH=CH₂	71 (36)	8	- C(O)Ph	72 (48)
9	- S(O)₂CH₃	94 (28)	10		46 (30)
11		15 (12)	12		60 (31)
13		82 (39)	14		88 (48)
15		73 (56)	16		70 (25)
17		90 (35)	18		93 (30)
19		63 (25)	20		74 (10)

The same holds true for the type II case where the listing is completed by guanine and pteridine examples (Table 13, entries 12–14). These results are by no means trivial since the porcine α(2–3)SiaT had previously been shown to display a considerable substrate specificity. For example, the enzyme discriminates the penultimate N-acetylglucosamine from an N-acetylgalactosamine [105].

The kinetics of porcine α(2-3)SiaT with non-natural donors have been studied. Thus, replacements of the natural N-acetyl group at C-5 on the sialic acid moiety are allowed (Fig. 2) [106]. A wider synthetic potential may be disclosed from the combined incubations of α(2-3)SiaTs in the presence of non-natural acceptors and non-natural donors.

Table 13. Enzymatic $\alpha(2-3)$sialylations of various N-acyl-type II disaccharides with aglycon-$O(CH_2)_8COOMe$; compare Fig. 16

Entry	Acyl residue	Yield % (mg)	Entry	Acyl residue	Yield % (mg)
1	- C(O)CH₃	71 (81)	2	- C(O)H	77 (31)
3	- C(O)CH(CH₃)₂	77 (15)	4	- C(O)OCH₂CH=CH₂	72 (186)
5	- C(O)Ph	81 (16)	6	- C(O)CH₂OH	89 (36)
7	- C(O)CH₂NH₂	49 (16)	8	- C(O)CH₂SO₃Na	92 (23)
9	- C(S)CH₃	77 (24)	10	- C(S)OCH₂CH₃	65 (15)
11	- S(O)₂CH₃	92 (34)	12		86 (20)
13		41 (9)	14		39 (7)
15		74 (19)	16		86 (41)
17		95 (17)	18		87 (20)
19		100 (17)	20		62 (11)
21		35 (17)	22		66 (15)

3.4.2
$\alpha(2\text{-}6)$Sialyltransferases

The $\alpha(2\text{-}6)$Sia-T transfers a sialic acid from CMP-sialic acid onto the 6-OH-group of a terminal galactose in an α-mode (Fig. 17).

Two $\alpha(2\text{-}6)$SiaTs have been described as being able to sialylate non-natural "galactose" residues (see groups marked in Fig. 17 and Table 14) [107]. From the kinetic and semi-preparative studies, the indifference of $\alpha(2-6)$SiaT towards the hydroxylation pattern in the very proximity of the sialylation site is evident.

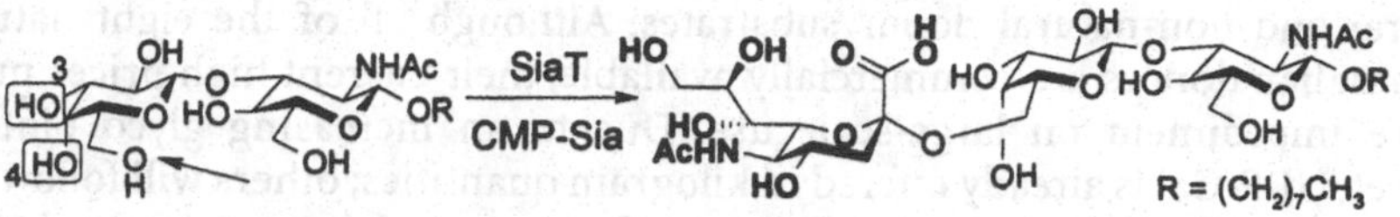

Fig. 17. Enzymatic α(2-6)sialylation (see also Table 14)

Table 14. α(2-6)Sialylation of non-natural type I disaccharides; compare Fig. 17

C-3′	C-4′	[%]	mg	C-3′	C-4′	[%]	mg
OH	OH	79	2.4	OH	F	79	2.5
H	OH	63	2.1	OH	epi-OH	33	1.8
OH	H	63	2.2	OH	OCH₃	13	0.9

3.5
Other Transferases

To date only a limited amount of work has been invested in probing other glycosyltransferases for preparative use [108]. A rather large number of N-acetylglucosaminetransferases have been described. Paulsen and coworkers [109, 110] and Hindsgaul and coworkers [111, 112] have synthesized a number of substrates to decode key polar groups on the UDP-GlcNAc donor and the respective acceptors. These studies may prove useful for future preparative applications [28].

The N-GalNAcT (E.C.2.4.1.165), the blood group A transferase, has also been explored. The Hindsgaul group has studied the kinetics of a variety of acceptor substrates in order to define key hydroxy functions. This work is nicely summarized and should be consulted before using the transferase on a preparative scale [28].

Far less knowledge has been accumulated on the *man*nosyl*t*ransferase family. A yeast β(1–4)-ManT has recently been cloned in *E. coli* [113], and the kinetics of a yeast α(1–2)ManT, also cloned in *E. coli*, with a limited number of non-natural acceptors has been reported [28, 114]. This will be helpful for future preparative applications.

4
Outlook

Transferases will, without doubt, become routine catalysts in the carbohydrate chemist's labs. One prerequisite for a more widespread use of these biocatalysts is their commercial availability. Improved techniques in cloning, overexpression and protein purification should facilitate the future commercialization of glycosyltransferases and successively, their widespread synthetic application. The second requirement for enzymatic glycosylations is an improved access to the natural and non-natural donor substrates. Although all of the eight natural mammalian donors are commercially available, their current high prices put a severe impediment on large-scale use. Due to an increasing glyco biotech market, UDP-Gal is already offered in kilogram quantities; others will follow for sure. Improved protocols for the synthesis of non-natural donors may lead to an ever growing knowledge about the substrate specificities of the various transferases and their preparative applicability. In the hands of glycochemists, this will be a powerful tool to synthesize reliably large numbers of non-natural oligosaccharides for high-throughput screens towards novel carbohydrate-based pharmaceuticals.

The full synthetic potential of these biocatalysts is still beyond imagination. Especially, selective changes in the protein sequences by molecular biology techniques will give catalysts with tailor-made properties. Thus enzymes with enhanced stabilities and with defined high affinities toward unusual substrate patterns, retaining their exclusive regio- and stereoselectivity, may be created.

5
References

1. Kennedy JF (1988) Carbohydrate chemistry. Clarendon Press, Oxford
2. Aspinall GO (1982–1985) The polysaccharides, vols I–III. Academic Press, New York
3. Sandhoff K, Yussuf HKM (1991) TIGG 3(11):152
4. Fukuda M (1994) In: Fukuda M, Hindsgaul O (eds) Molecular glycobiology. IRL-Press, Oxford, p 1
5. Dwek RA (1996) Chem Rev 96:683
6. Miller DJ, Macek MB (1992) Nature 357:589
7. Singhal A, Hakomori S-I (1990) Bioassays 12:223
8. Fukuda M (1992) In: Fukuda M (ed) Cell surface carbohydrates and cell development. CRC Press, Boca Raton, p 127
9. Carbone FR, Gleesen FA (1997) Glycobiology 7:725
10. Sandrin MS, McKenzie IFC (1994) Immunol Rev 141:169
11. Varki A (1993) Glycobiology 3:97
12. Haywood AM (1994) J Virol 68:1
13. Karlsson K-A (1995) Carbohydr Eur 14
14. McCoy JJ, Mann BJ, Petri WA (1994) Infect Immunol 62:3045
15. Muramatsu T (1993) Glycobiology 3:291
16. Witzcak ZI (1995) Curr Med Chem 1:392
17. McAuliffe JC, Hindsgaul O (1997) Chem Ind 3:170
18. Lowe JB (1994) In: Fukuda M, Hindsgaul O (eds) Molecular glycobiology. IRL-Press, Oxford, p 163
19. Giannis A (1994) Angew Chem Int Ed Engl 33:178
20. Weinhold EG, Knowles JR (1992) J Am Chem Soc 114:9270
21. Roy R, Andersson FO, Harms G, Kelm S, Schazer R (1992) Angew Chem Int Ed Engl 31:1478
22. Parker W, Lateef J, Everet ML, Platt JL (1996) Glycobiology 6:499
23. Neethling FA, Joziasse D, Bovin N, Cooper DKC, Oriol R (1996) Transplant Int 9:98
24. Kanie O, Hindsgaul O (1992) Curr Opin Struct Biol 2:674
25. a Khan SH, Hindsgaul O (1994) In: Fukuda M, Hindsgaul O (eds) Molecular glycobiology. IRL-Press, Oxford, p 206
25. b Schachter H (1994) In: Fukuda M, Hindsgaul O (eds) Molecular glycobiology. IRL-Press, Oxford, p 88

26. Wong C-H, Haynie SL, Whitesides GM (1982) J Org Chem 47:5416
27. Augé C, Mathieu C, Mérienne C (1986) Carbohydr Res 151:147
28. Ichikawa J (1997) In: Large DG, Warren CD (eds) Glycopeptides and related Ccompounds. Marcel Dekker, New York, p 79
29. Fukuda M, Hindsgaul O (1994) Molecular glycobiology. IRL-Press, Oxford
30. Wong C-H, Whitesides GM (1994) Enzymes in synthetic organic chemistry. Tetrahedron Organic Chem Series, vol 12. Pergamon Press, Oxford
31. Large DG, Warren CD (1997) Glycopeptides and related compounds. Marcel Dekker, New York
32. Watt GM, Lowden AS, Flitsch SL (1997) Curr Opin Struct Biol 7:652
33. Leloir LF (1971) Science 172:1299
34. a) Unverzagt C, Kunz H, Paulson JC (1990) J Am Chem Soc 112:5893; b) Ichikawa Y, Liu JL-C, Shen G-J, Wong C-H (1991) J Am Chem Soc 113:6300
35. Heidlas JE, Williams KW, Whitesides GM (1992) Acc Chem Res 25:307
36. YAMASA-Corporation 1-23-8, Nihonbashi-Kakigaracho, Chou-ku, Tokyo 103, Japan
37. Chappell MD, Halcomb RL (1997) Tetrahedron 53:11,109
38. Kajihara Y, Ebata T, Koseki K, Kodama H, Matsushita H, Hashimoto H (1995) J Org Chem 60:5732
39. Kittelmann M, Klein T, Kragl U, Wandrey C, Ghisalba O (1995) Appl Microbiol Biotechnol 44:59
40. Adelhorst K, Whitesides GM (1993) Carbohydr Res 242:69
41. Wittmann V, Wong C-H (1997) J Org Chem 62:2144
42. Baisch G, Öhrlein R (1997) Bioorg Med Chem 5:383
43. Pallanca JE, Turner NJ (1993) J Chem Soc Perkin TransI 3017
44. DeLuca C, Lansing M, Creszenzi F, Martini I, Shen G-J, O'Regan M, Wong C-H (1996) Bioorg Med Chem 4:131
45. Ko JH, Shin H-S, Kim YS, Lee DS, Kim C-H (1996) Appl Biochem Biotechnol 60:41
46. Yamazaki T, Warren CD, Herscovic A, Jeanloz RW (1981) Can J Chem 59:2247
47. Gambert U, Thiem J (1997) Top Curr Chem 186:21
48. Palcic MM, Hindsgaul O (1996) TIGG 8:37
49. Guilbert B, Davis NJ, Pearce M, Aplin RT, Flitsch SL (1994) Tetrahedron Asymmetry 5:2163
50. Baisch G, Öhrlein R (1996) Angew Chem Int Ed Engl 35:1812
51. Öhrlein R, Ernst B, Berger EG (1992) Carbohydr Res 236:335
52. Baisch G, Öhrlein R, Ernst B (1996) Bioorg Med Chem Lett 6:749
53. Hayashi M, Tanaka M, Miyauchi H (1996) J Org Chem 61:2938
54. Guilbert B, Khan TH, Flitsch SL (1992) J Chem Soc Chem Commun 1526
55. Goodridge GM, Guilbert B, Flitsch S, Revers L, Webberley MC, Wilson IB (1994) Bioorg Med Chem 2:1243
56. Öhrlein R (1997) Biotrans '97, Abstract C10, La Grande Motte, France
57. Kren V, Augé C, Sedmera P, Havlícek V (1994) J Chem Soc Perkin I 2481
58. Schultz M, Kunz H (1993) Tetrahedron Asymmetry 4:1205
59. Wong C-H, Schuster M, Wang P, Sears P (1993) J Am Chem Soc 115:5893
60. Danieli B, Luisetti M, Schubert-Zsilavecz M, Likussar W, Steurer S, Riva S, Monti D, Reiner J (1997) Helv Chim Acta 80:1153
61. Baisch G, Öhrlein R (1998) Bioorg Med Chem (in print)
62. Kajihara Y, Endo T, Ogasawara H, Kodama H, Hashimoto H (1995) Carbohydr Res 269:273
63. Do K-Y, Cummings RD (1995) J Biol Chem 270:18,447
64. Tsuruta O, Shinohara G, Yuasa H, Hashimoto H (1997) Bioorg Med Chem Lett 7:2523
65. Gautheron-LeNarvor C, Wong C-H (1991) J Chem Soc Chem Commun 1130
66. Panza L, Chiappini Pl, Russo G, Monti D, Riva S (1997) J Chem Soc Perkin I 1255
67. Palcic MM (1994) Methods Enzymol 230:300
68. Yu L, Cabrera R, Ramirez J, Malinoskii VA, Brew K, Wang PG (1995) Tetrahedron Lett 36:2897
69. Sasaki J, Mizoue K, Morimoto S, Omura S (1996) J Antibiotics 49:1110
70. Wiemann T, Taubken N, Zehave U, Thiem J (1994) Carbohydr Res 257:C1–C6

71. Köpper S (1994) Carbohydr Res 265:161
72. Schuster M, Wang P, Paulson JC, Wong C-H (1994) J Am Chem Soc 116:1135
73. Granovsky M, Bielfeldt T, Peters S, Paulsen H, Meldal M, Brockhausen J, Brockhausen I (1994) Eur J Biochem 221:1039
74. Miyazaki H, Fukumoto S, Okada M, Hasegawa T, Furukawa K (1997) J Biol Chem 272:24,794
75. Kolbinger F, Streiff M, Katopodis A (1998) J Biol Chem 273:433
76. Baisch G, Öhrlein R, Streiff M, Kolbinger F (1998) Bioorg Med Chem Lett 8:751
77. Lowary TL, Hindsgaul O (1993) Carbohydr Res 249:163
78. Helland A-C, Hindsgaul O, Palcic MM, Stults CLM, Macher BA (1995) Carbohydr Res 276:91
79. Gustaffson K, Strahan K, Preece A (1994) Immunol Rev 141:59
80. Joziasse DH, Shaper NL, Salyer L, Van den Eijnden DH, Van der Poel AC (1990) Eur J Biochem 191:75
81. Baisch G, Öhrlein R, Kolbinger F, Streiff M (1998) Biorg Med Chem Lett 8:1575
82. Kochetkov NK, Shibaev VN (1974) Izv Akad Nauk SSSR Ser Khim 5:1169
83. Staudacher E (1996) TIGG 8:391
84. Koszdin KL, Bowen BR (1992) Biochem Biophys Res Commun 187:152
85. Ge Z, Chan NWC, Palcic MM, Taylor DE (1997) J Biol Chem 272:21,357
86. Kashem MA, Wlasichuk KB, Gregson JM, Venot AP (1993) Carbohydr Res 250:129
87. Ball GE, O'Neill RA, Shultz JE, Lowe JB, Weston BW, Nagy JO, Brown EG, Hobbs CJ, Bednarsky MD (1992) J Am Chem Soc 114:5449
88. Baisch G, Öhrlein R, Katopodis A, Ernst B (1996) Bioorg Med Chem Lett 6:759
89. Baisch G, Öhrlein R, Streiff M, Katopodis A, Kolbinger F (1997) Bioorg Med Chem Lett 7:2447
90. Hällgren C, Hindsgaul O (1995) J Carbohydr Chem 14:453
91. Baisch G, Öhrlein R, Katopodis A, (1997) Bioorg Med Chem Lett 7:2431
92. Du M, Hindsgaul O (1996) Carbohydr Res 286:87
93. Sasaki K, Kurata K, Funayama K, Nagata M, Watanabe E, Ohta S, Hanai N, Nishi T (1994) J Biol Chem 269:13,730
94. Baisch G, Öhrlein R, Streiff M, Kolbinger F (1998) Bioorg Med Chem Lett 8:161
95. Baisch G, Öhrlein R (1998) Carbohydr Res in print
96. Baisch G, Öhrlein R, Streiff M, Kolbinger F (1998) Bioorg Med Chem Lett 8:755
97. Murray BW, Wittmann V, Burkhart MD, Hung S-C, Wong C-H (1997) Biochemistry 36:823
98. Baisch G, Öhrlein R, Katopodis A (1996) Bioorg Med Chem Lett 6:2953
99. Sasaki K (1996) TIGG 8:195
100. Brossmer R, Gross HJ (1994) Methods Enzymol 247:153
101. Chappell MD, Halcomb RL (1997) Tetrahedron 53:11,109
102. Wen DX, Livingston BD, Medziharadszky MF, Kelm S, Burlingame AL, Paulson JC (1992) J Biol Chem 267:21,011
103. Baisch G, Öhrlein R, Streiff M, Ernst B (1996) Bioorg Med Chem Lett 6:755
104. Baisch G, Öhrlein R, Streiff M (1998) Bioorg Med Chem Lett 8:157
105. Öhrlein R, Hindsgaul O, Palcic MM (1993) Carbohydr Res 244:149
106. Gross HJ, Brossmer R (1995) Glycoconjugate J 12:739
107. Van Dorst JALM, Tikkanen JM, Krezdorn CH, Streiff M, Berger EG, van Kuik JA, Kamerling JP, Vliegenthart JFG (1996) Eur J Biochem 242:674
108. Crawley SC, Palcic MM (1996) Modern Methods Carbohydr Synth 492
109. Paulsen H, Meinjohanns E, Reck F, Brockhausen I (1993) Liebigs Ann 737
110. Reck F, Meinjohanns E, Tan J, Grey AA, Paulsen H, Schachter H (1995) Carbohydr Res 275:221
111. Alton G, Srivastava G, Kaur KJ, Hindsgaul O (1994) Bioorg Med Chem 2:675
112. Ogawa S, Furuya T, Tsunoda H, Hindsgaul O, Stangier K, Palcic MM (1995) Carbohydr Res 271:197
113. Watt GM, Revers L, Webberley MC, Wilson IBH, Flitsch S (1997) Angew Chem Int Ed Engl 36:2354
114. Herrmann GF, Wang P, Shen G-J, Garcia-Junceda E, Khan S, Matta KL, Wong C-H (1994) J Am Chem Soc 59:6356